Lecture Notes in Physics

Edited by H. Araki, Kyoto, J. Ehlers, München, K. Hepp, Zürich
R. Kippenhahn, München, H. A. Weidenmüller, Heidelberg,
J. Wess, Karlsruhe and J. Zittartz, Köln
Managing Editor: W. Beiglböck

306

L. Blitz F. J. Lockman (Eds.)

The Outer Galaxy

Proceedings of a Symposium
Held in Honor of Frank J. Kerr at the University of Maryland,
College Park, May 28–29, 1987

Springer-Verlag
Berlin Heidelberg GmbH

Editors

Leo Blitz
Astronomy Program, University of Maryland
College Park, MD 20742, USA

Felix J. Lockman
National Radio Astronomy Observatory
Edgemont Road, Charlottesville, VA 22903, USA

ISBN 978-3-662-13661-4   ISBN 978-3-540-39285-9 (eBook)
DOI 10.1007/978-3-540-39285-9

2158/3140-543210

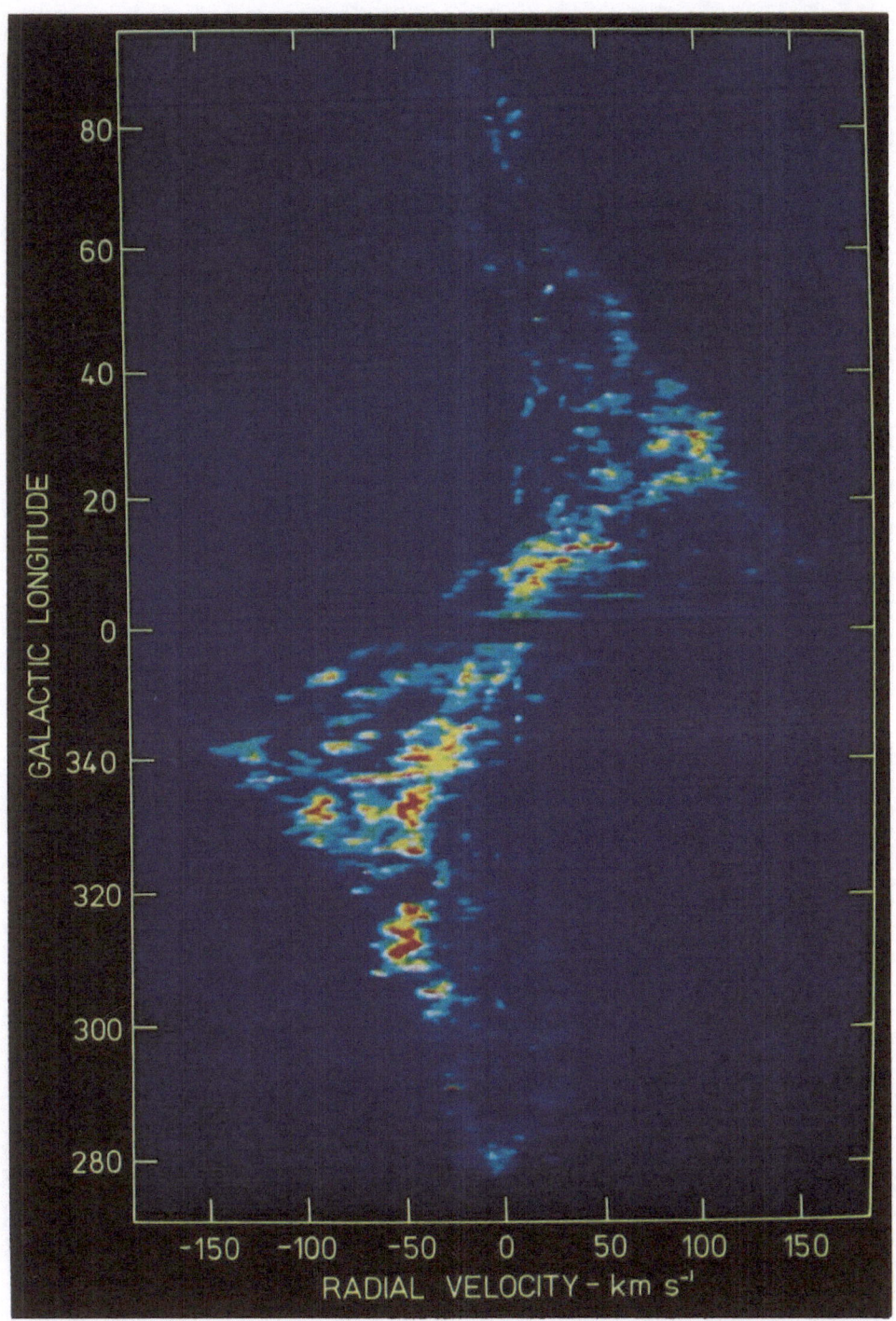

Figure 1 (McCutcheon, Robinson, Manchester, Whiteoak). False-color longitude
-velocity plot of CO emission as observed by CSIRO ($279° < \ell < 358°$, and
$2° < \ell < 13°$) and FCRAO ($358° < \ell < 86°$). The FCRAO results are from Sanders
et al. 1984. The effective spatial and velocity resolutions are 9 arcmin. and
1 kms^{-1}, respectively.

PREFACE

Frank Kerr's distinguished career as both a scientist and an administrator at the University of Maryland has been so inextricably linked with his interest in the structure of our Galaxy that it is hard to mention one without thinking of the other. Therefore, it seemed natural to invite many of the most noted workers in the field of galactic structure to the University for a symposium in Frank's honor on his retirement. We decided to focus on a topic that was of broad current interest, and to which Frank has contributed significantly. This symposium on The Outer Galaxy was the result. A large fraction of active workers in the field of galactic structure came from around the world to share their research in Frank's honor. Most of us, after all, have been friends of Frank for as long as we have been working in this field.

The talks at the meeting covered diverse topics, from the very nearby outer Galaxy (e.g., Magnani; Shuter and Dickman), through processes relevant to the entire Galaxy (e.g. Balbus; Norman and Ikeuchi) to the outer regions of other galaxies (e.g. Thaddeus; Rubin; Verter and Kutner). There was lively debate about issues like the rotational parameters and the mass of the Galaxy (e.g. Schecter et al.; Moffatt). The order of the papers in this volume represents our attempt to organize the different topics and to highlight areas of current debate.

The volume ends with two contributions of a more personal nature, one from Woody Sullivan, who was once Frank's student, and another from Gerard de Vaucouleurs, a longtime friend and colleague. Woody's article was transcribed from a tape recording of his after-dinner talk. It is a fine piece of scholarship, edited only lightly to retain its entertaining flavor.

We are lucky to have had the help of the entire staff of the University of Maryland Astronomy Program, as well as that of all the graduate students during what is normally their summer break, in planning and executing the symposium. We specifically thank Kathleen Briedecker Christian, who served as the receptionist at the meeting, and Dr. John Trasco, who was involved in every detail of the meeting and banquet and who deserves much of the credit for the meeting's success. Maureen Kerr was a valuable conspirator and smuggled out pictures of the young Frank Kerr. Pete Jackson and Ed Grayzeck helped with much of the early scientific organization. Special thanks go to George Kessler and Pat Smiley of the NRAO graphics division for help in preparing this book, and to Elizabeth L.T. Brown for postponing the wedding until after the meeting. Finally, we are grateful for the financial support provided by the Astronomy Program and by the dean, J. Robert Dorfman.

<table>
<tr><td>Leo Blitz</td><td>Felix J. Lockman</td></tr>
<tr><td>Princeton</td><td>Charlottesville</td></tr>
</table>

CONTENTS

I. FUNDAMENTAL PARAMETERS: SCALE, ROTATION AND MASS

THE GALACTIC ROTATION CURVE AT $R > R_0$ FROM OBSERVATIONS OF THE
21 CM LINE OF ATOMIC HYDROGEN

G. R. Knapp
Department of Astrophysical Sciences
Princeton University, Princeton, NJ 08544, U.S.A.

It is a great honour to be invited to talk on the subject of the structure of
the Galaxy on the occasion of Frank Kerr's retirement, for this is a subject to
which he has made so many fundamental contributions. It is also a great pleasure.
I love doing astronomy, and that I get to do it at all is in large part thanks to
his having been willing to put up with me as a graduate student. The following
paper is offered with respect, affection and gratitude.

I INTRODUCTION

The measurement of the galactic rotation curve at distances greater than R_0, the
distance between the Sun and the Galactic center, has received a lot of attention
lately for the usual combination of theoretical and observational reasons. As
equipment sensitivity has improved and new observational tools have become
available, it has become possible to trace the rotation curve of the galaxy we can
study in most detail, our own. These developments have led to increasingly detailed
studies of the distribution and nature of the unseen mass which appears to be
present in all galaxies, while the study of the outer regions of the Galaxy via the
warping and flaring of the HI disk, of metallicity gradients etc., is of great
importance for the understanding of galaxy formation and evolution. Only in the
Galaxy, for example, can the amount of material in molecular clouds, the size
spectrum of molecular clouds and the star formation rate be studied as a function of
galactocentric radius, and the study of the distribution of interstellar matter in
the outer Galaxy requires a well-determined rotation curve.

For $R < R_0$, of course, the rotation curve may be directly determined by
observations of the tangent-point velocities of interstellar HI or CO. It has
recently become possible to measure the rotation curve between $R/R_0 \sim 0.8$ and 2 by

3

observations of the radial velocities and distances of HII regions (Jackson et al. 1979, Blitz, Fich and Stark 1982, and Brand 1986) and of carbon stars (Schechter, this conference). The mass distribution of the Galaxy at large distance $(R/R_0 \sim 5-10)$ can be measured by observations of the distant globular clusters, the Magellanic Stream, the Magellanic Clouds and the dwarf spheroidal galaxies (Lynden-Bell 1983 and this conference; Olsiewski et al. 1986; Little and Tremaine 1987) and the total mass of the Galaxy inferred from the radial velocity of the Galaxy relative to M31 (Kahn and Woltjer 1959, Gunn 1974). Rough information about the mass distribution/rotation curve in the intermediate region $(R/R_0 \sim 2-5)$ can be obtained from the kinematics of the globular cluster system (Lynden-Bell 1982) and of the halo stars (Freeman, this conference) and from observations of HI at large distances, as discussed herein.

As yet, no one has found a method of directly measuring the distance to a given HI feature (although variants on the line width-pressure-mass relationships may work in some cases, e.g. Verschuur, 1987). This would seem to preclude the use of HI to give any information about the rotation curve at $R > R_0$. However, some limited information can be obtained if it is assumed that the galactic HI is in a disk with a characteristic scale length λ. The value of λ can be directly obtained from measurements at $R < R_0$, and can thus be used to give an HI distance scale for $R > R_0$ if λ remains the same. This method of characterising the distance distribution of the galactic HI was used by Knapp et al. (1978) and its application to more recent data will be discussed below. This assumption of constant λ is in principle no different from any other indirect method of distance measurement, where observations of objects of assumed identical properties are compared, though one does have the feeling of being on even shakier ground than usual.

II THE ROTATION CURVE AND HI DISTRIBUTION AT $R < R_0$

Before discussing the observations of the Galaxy at $R > R_0$, it is necessary to summarize some results for $R < R_0$. If the Galactic disk is in pure circular rotation everywhere, and the circular velocity as a function of R is $\Theta(R)$, then the radial velocity V_r along the line of sight latitude b = 0, longitude = ℓ, is

$$V_r = \left[\Theta(R) \, \frac{R_o}{R} - \Theta_o \right] \sin \ell \qquad (1)$$

where Θ_o is the circular velocity at R_o. The radial velocity has a maximum value

$$V_m = \Theta(R_T) - \Theta_o \sin \ell \qquad (2)$$

for an object at the tangent point distance, $R_T/R_o = \sin \ell$. Thus

$$V_r = V_m(R) \, \frac{R_o}{R} \sin \ell \quad . \qquad (3)$$

The heliocentric distance is

$$r = R_o \cos \ell + (R^2 - R_o^2 \, \sin^2 \ell)^{1/2} \qquad (4)$$

Hence the values of r and R found from galactic kinematics scale with R_o, and Θ_o drops out, so that its actual value is irrelevant to the determination of kinematic distances.

Observations of cold gas (HI and CO) in the inner Galaxy ($R \leq R_o$, $0 \leq \ell \leq 90°$, $270 \leq \ell \leq 360°$) then give values for the slope of the rotation curve near the Sun:

$$AR_o = -\frac{1}{2} \left\{ \frac{dV_m}{d(\sin \ell)} \right\}_{R_o} \quad . \qquad (5)$$

Although A is defined locally, it has recently been recognized that on the large scale V_m is linear with $\sin \ell$ over a large range of $\sin \ell$, indicating that the rotation curve is also linear (cf. eq. 2). Thus, AR_o can be determined using eq. 5 by fitting over a wide range of $\sin \ell$. However, the $V_m - \sin \ell$ curve also has wiggles of amplitude ~10%, as is often observed in the rotation curves of other galaxies (Rubin 1979) and thus the slope of the $V_m - \sin \ell$ curve undergoes large changes. Thus, although this slope can be obtained with a high formal accuracy, it is important to specify the range over which slope is obtained. Some recent results from both HI and CO observations are summarized below:

5

	Longitude Range	$\lvert 2AR_o \rvert$	Line	Reference
(1)	30° – 90°	220±3	HI	Gunn et al. 1979
(2)	6° – 90°	185±9	HI	Shuter 1981
(3)	300° – 340°	181±10	CO	McCutcheon et al. 1981
(4)	270° – 345°	214±3	CO	Alvarez et al. 1987
(5)	30° – 90°	217±8	CO	Data from Knapp et al. 1985
(6)	270° – 330°	205±12	HI	Data from Kerr et al. 1986

The data for the results quoted as (5) and (6) above are shown in Figures 1 and 2. In each case, the function fitted is

$$\lvert V_m \rvert \; - \; \Delta V \; = \; \lvert 2AR_o \rvert \; (1 \; - \; \lvert \sin \ell \rvert) \tag{6}$$

where the value of ΔV, assumed constant with longitude, is included to correct for broadening of the line by line-of-sight turbulent motion in the gas (cf. Gunn et al.

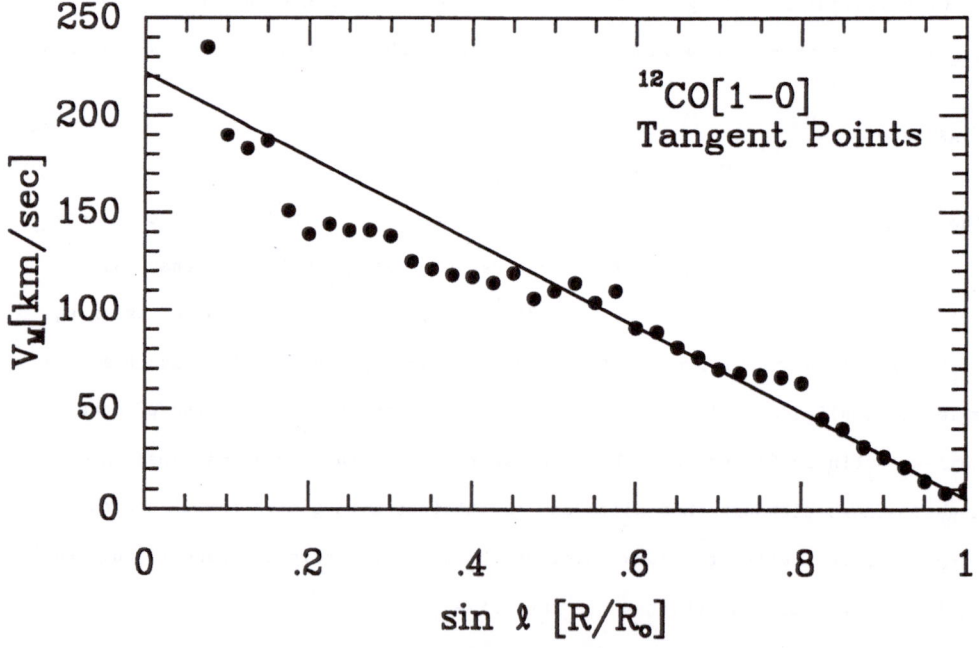

Figure 1: Tangent-point velocities V_m of CO(1-0) emission in the first galactic quadrant, ℓ = 4°-90° (Knapp et al. 1985). The best-fit straight line corresponds to $V_m - 5 = 217 (1 - \sin \ell)$, and is fit between $\sin \ell$ = 0.5 and 1.

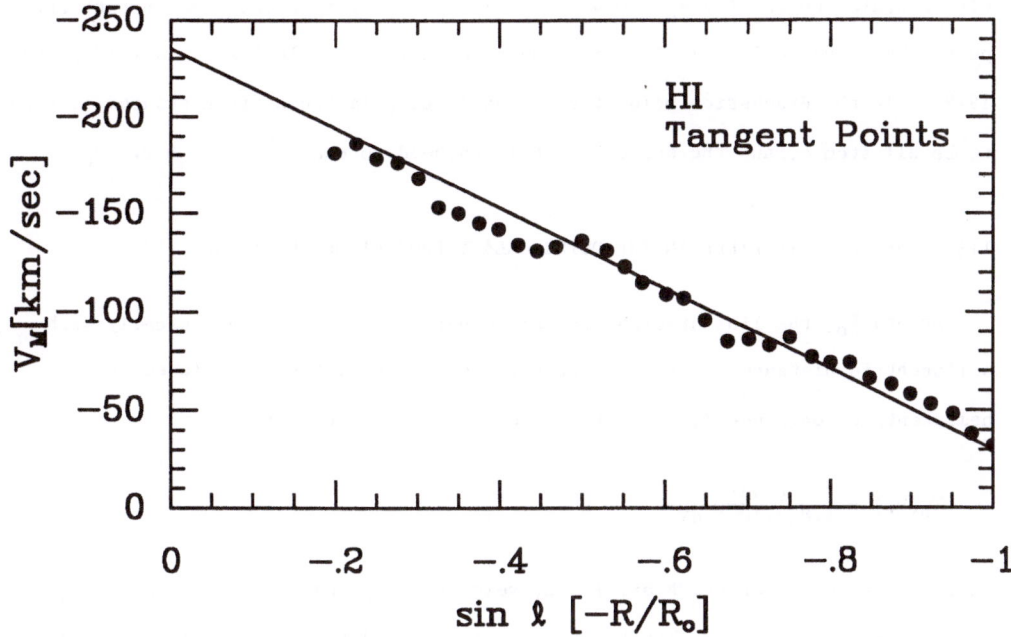

Figure 2: Tangent point velocities V_m of the HI emission in the fourth galactic quadrant, $\ell = 270° - 348°.5$ (from the data of Kerr et al. 1986). The best-fit straight line corresponds to $|V_m| - 30 = 205 (1 - |\sin \ell|)$, and is fit between $\sin \ell = -0.5$ to -1.

1979). For both the first and fourth galactic quadrants, the HI and CO kinematics are observed to agree very well. It is clear from examination of Figures 1 and 2 that the relatively small-amplitude fluctuations in the V_m − $\sin \ell$ curve lead to large fluctuations in the slope of the curve, i.e. in $2AR_0$. For example, the difference between the values of $2AR_0$ found by Gunn et al. (1979) and Shuter (1981) is due almost entirely to the fit having been made over different longitude ranges. Thus caution in the interpretation of these curves needs to be exercised. Tentatively, values of $|2AR_0| = 200$ -220 km/sec seem reasonable, and, if the slope of the rotation curve does not change abruptly at R_0, this should also be the value to some distance beyond the solar circle.

A second result from the inner Galaxy is the azimuthally-smoothed surface density distribution. The HI and CO observations of Burton and Gordon (1978) show that the total hydrogen surface density $\Sigma_H(R)$ ($H = HI + 2H_2$) is roughly exponential with a scale length of $\lambda/R_0 \sim 0.4$. The HI scale height is roughly constant at ~ 120 pc (Jackson and Kellman 1974) while the molecular gas scale height is ~ 50 pc (Stark 1979). In the discussion below the gas at large galactocentric distances is assumed to be all atomic, and the scale length is assumed to continue as $\lambda = 0.4\ R_0$.

III. THE VELOCITY FIELD IN THE OUTER GALAXY FROM HI OBSERVATIONS

Beyond R_0, the line-of-sight radial velocity V_r changes monotonically with heliocentric distance r. For gas at galactocentric distance R, viewed at heliocentric longitude ℓ, the brightness temperature at V_r is

$$T_b(V_r) = \frac{n(R)dr}{1.823\times10^{18}\ dV_r} \qquad K \qquad (7)$$

where distances r, dr and R are in cm, velocities V_r and dV_r are in km/sec, and HI density n(R) is in cm^{-3}. The HI emission is assumed to be optically thin. Equation (7) may be rewritten

$$T_b(V_r) \sim \frac{\Sigma(R)}{(2\pi)^{1/2}\ h(R)} / \left(\frac{dV_r}{dr}\right) \qquad (8)$$

where h(R) is the (gaussian) scale height of the HI disk. The observations of Burton and Gordon (1978) give

$$\Sigma(R) = \Sigma_0\ \exp\left((1 - \frac{R}{R_0})/\lambda\right) \qquad cm^{-2} \qquad (9)$$

with $\lambda = 0.4\ R_0$ and Σ_0, the surface density at R_0, $=\ 6.6\times10^{20}\ cm^{-2}$.

The observational situation at large radii is complicated by the warping and flaring of the HI disk (Kerr 1957). The former can be allowed for by following the maximum intensity in latitude (hence HI surveys with wide latitude coverage, such as those of Weaver and Williams (1974a,b) and Kerr et al. 1986 are essential). The latitude flaring may be roughly described by

$$\frac{h}{R_0} = a\left(\frac{10(kpc)}{R_0}\right) + b\left(\frac{R}{R_0} - 1\right) \qquad (R/R_0 > 1) \qquad (10)$$

where a = 0.012. The value of b can be found for a given velocity-distance model from the latitude extent of the HI at each velocity. (The effects of stray radiation on the measurement of HI emission in the outer Galaxy have been discussed by Henderson et al. (1982).) For example, for a flat rotation curve with $\Theta(R) = \Theta_0 = 250$ km/sec, b = 0.05.

The rotation curve can then be modelled and the prediction of equation (8) compared with observations. Power law rotation curves were discussed by Gunn et al. (1979). Here, linear curves are considered, i.e. curves with a combination of flat and solid-body components. Such curves may be characterized by

$$\frac{\Theta(R) - \Theta_0}{R - R_0} = \alpha \tag{11}$$

Then $|2AR_0| = (\alpha R_0 - \Theta_0)$ (12)

and $V_r = (\alpha R_0 - \Theta_0) (1 - \frac{R}{R_0}) \sin \ell$ (13)

The brightness temperature T_b depends on dV_r/dr (equation 8).
Now $dV_r/dr = (dVr/dR)(dR/dr)$, and

$$\frac{dV_r}{dR} = \frac{R_0}{R^2} (\alpha R_0 - \Theta_0) \sin \ell$$

$$\frac{dR}{dr} = \frac{[R^2 - R_0^2 \sin^2\ell]^{1/2}}{R} \tag{14}$$

It can thus be seen that T_b is monotonic with V_r (an example is shown in Figure 3). Also, the curve of V_r versus ℓ for a given brightness temperature T_b has its maximum amplitude at $\ell = 90°$ and $270°$ and is symmetric in all four quadrants. Thus the observational plot of V_r vs ℓ at a given brightness temperature T_b can be compared with model plots as a diagnostic of the shape of the rotation curve in the outer Galaxy (cf. equation 13). Since the HI layer is turbulent, the model HI line profile at each longitude is smoothed by a gaussian of dispersion 12 km/sec (which corresponds to the observed intrinsic line width of 30 km/sec) (cf. Figure 3).

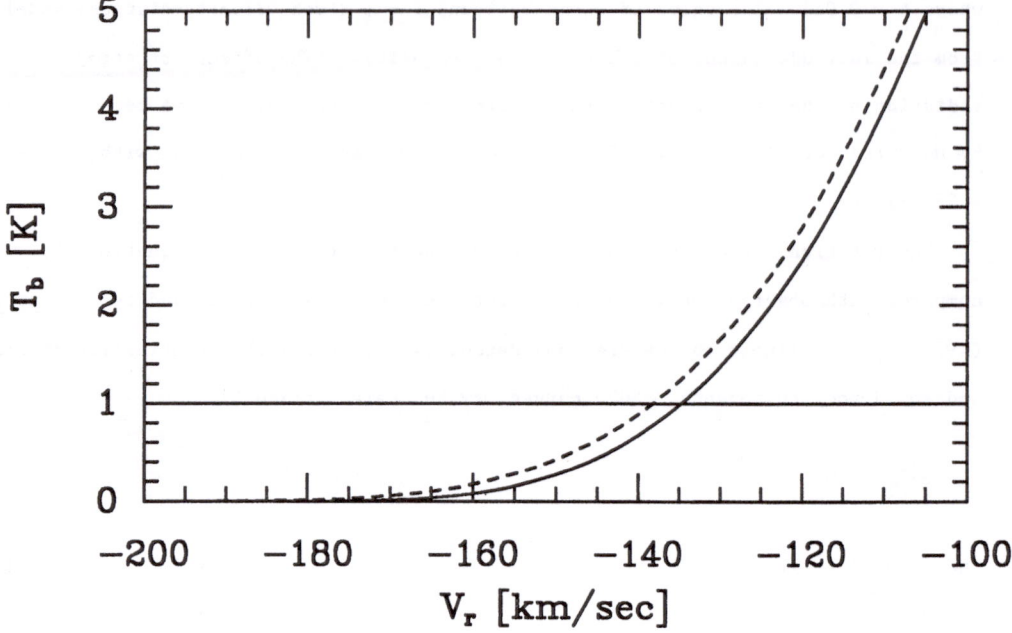

Figure 3:   (Solid line) HI brightness temperature T_b (K) versus velocity with
respect to the local standard of rest V_r (km/sec), from the model in
equations 8-11, with R_o = 8.5 kpc, θ_o = 220 km/sec, α = 0 km/sec/kpc,
ℓ = 90°. The dashed curve is the model smoothed by a gaussian with a
one-dimensional velocity dispersion of 12 km/sec (see text).

The HI emission from the outer Galaxy shows structure due to density and

velocity fluctuations (Burton 1972, Henderson et al. 1982, and Kulkarni et al.

1982). To minimize the effect of these, gas at large distances only is considered.

As illustrated in Figure 4, the velocity increases very slowly with distance at

large distances, and so the effects of velocity and density irregularities

(including spiral arms!) are smoothed out. The brightness temperature contour

chosen for the comparison is the T_b = 1 K contour (Figures 3 and 4) which can be

measured from the velocity - latitude HI maps of Weaver and Williams (1974a,b) and

Kerr et al. (1986).

Examples of the models are shown in Figure 5, where the velocity at 1 K

brightness temperature for both the smoothed and unsmoothed profiles are shown. The

data are shown in Figure 6 (ℓ = 0° to 180°) and Figure 7 (ℓ = 180° to 360°). The

data for ℓ = 90° to 180° (Figure 6) are uncertain because of the presence of high-

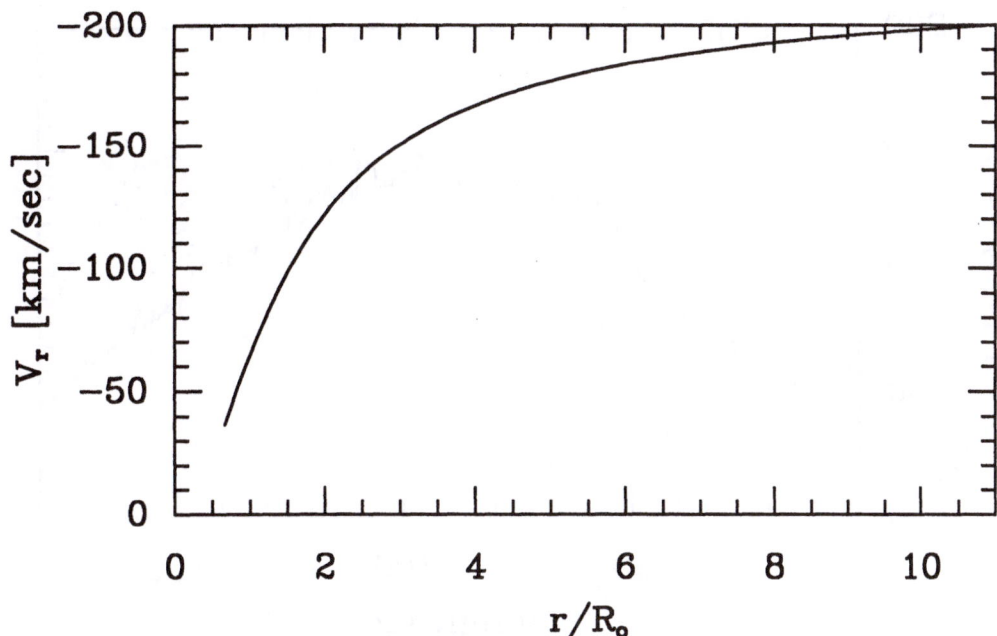

Figure 4: Velocity relative to the LSR, V_r (km/sec) versus heliocentric distance r/R_o, at $\ell = 90°$, for a rotation curve described by $R_o = 8.5$ kpc, $\Theta_o = 220$ km/sec, $\alpha = 0$ (flat rotation curve).

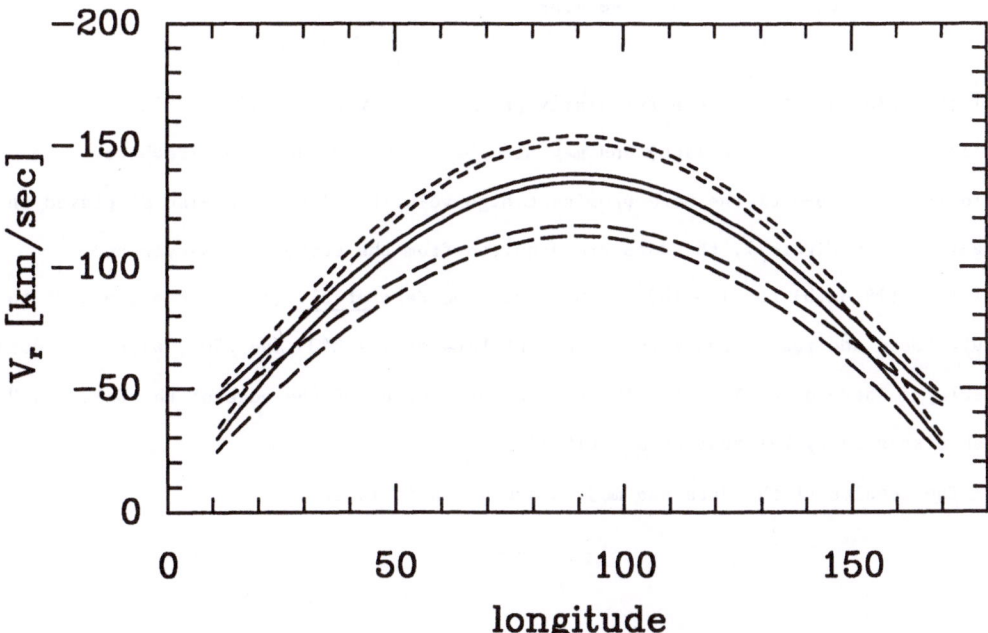

Figure 5: Velocity at which the HI brightness temperature T_b falls to 1K for gas at large galactocentric radii versus galactic longitude. Solid line: $\Theta_o=220$, $R_o=8.5$, $\alpha=0$. Short-dashed line $\Theta_o=250$, $R_o=10$, $\alpha=0$. Long-dashed line $\Theta_o=250$, $R_o=10$, $\alpha=7$. In each case, the two curves represent the T_b point on the unsmoothed (lower curve) and smoothed (upper curve) line profiles.

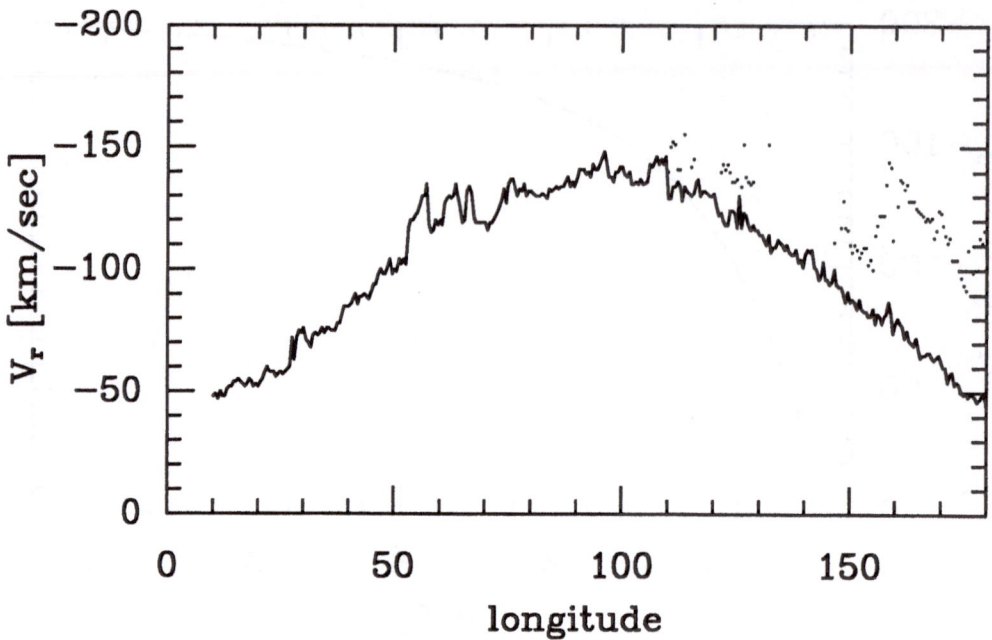

Figure 6: Velocity at which T_b falls to 1 K for $\ell = 10° - 180°$ from the data of Weaver and Williams (1974a,b). The dots represent high-velocity clouds seen on these maps.

velocity clouds, which are particularly prominent between $\ell = 90°$ and $180°$ (cf. Muller et al. 1966; Oort 1966) and may also be confusing the data elsewhere. The velocities of some of the more prominent high-velocity clouds are also displayed in Figure 6. In Figure 7, the data are acquired from the surveys of Weaver and Williams (1974a,b) for $\ell = 180° -250°$, and from Kerr et al. (1986) from $\ell = 240°$ to $350°$. Note the really beautiful agreement between $\ell = 240°$ and $250°$, where the data overlap. These data also clearly show the scalloping of the edge of the Galaxy's HI disk discussed by Kulkarni et al. (1982).

The results of the data and models are given in Table 2.

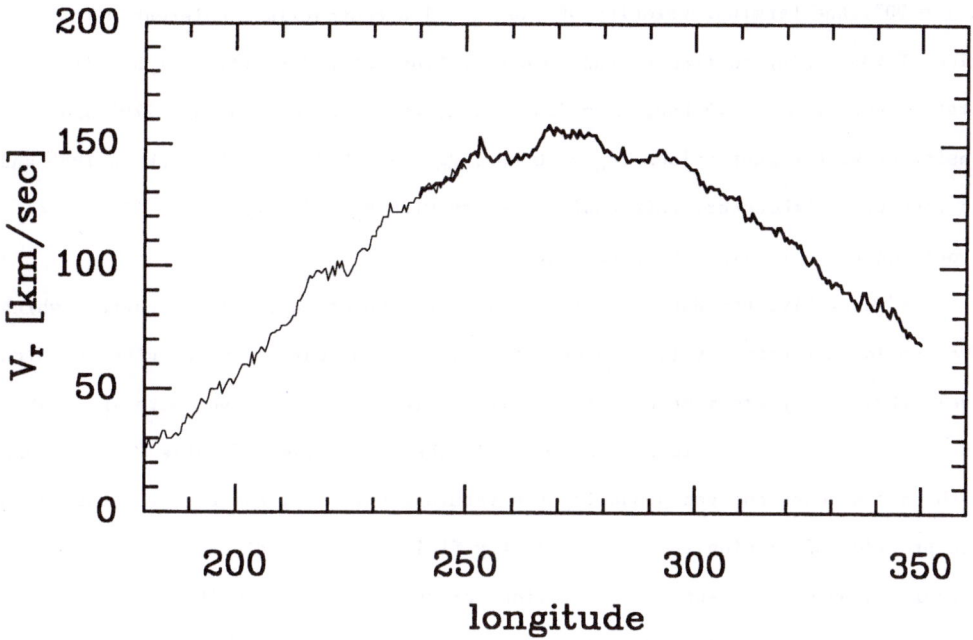

Figure 7: Velocity at which T_b falls to 1 K for $\ell=180°$ to $350°$. The data are from Weaver and Williams (1974a,b) for $\ell=180°$ to $250°$ (light line) and from Kerr et al. (1986) for $\ell=240°$ to $350°$ (heavy line).

Table 2: Model and Observed HI Terminal Velocities (T_b = 1 K) at ℓ = 90° and ℓ = 270°

Model

Θ_0	R_0	α	$(\alpha R_0 - \Theta_0)$	$|V|$	R(kpc)
220	8.5	0	−220	138.3	23
250	10	0	−250	154.0	26
220	8.5	−3	−245.5	152.0	22
250	10	−2	−270.0	164.0	25
250	10	+7	−180	117.0	28

Data: ℓ = 90° (Weaver and Williams 136
            1974)

      ℓ = 270° (Kerr et al. 1986) 152

At $\ell = 90°$, the terminal velocity of the $T_b = 1$ K contour is consistent with the value of $2AR_o$ inferred from HI and CO observations at $R < R_o$ (i.e. with a flat rotation curve, $R_o = 8.5$ kpc, $\Theta_o = 220$ km/sec) while the data at $\ell = 270°$ are consistent with higher values ($R_o = 10$ kpc, $\Theta_o = 250$ km/sec — cf. Jackson 1985 and the results of Schechter, this conference) or with $R_o = 8.5$ kpc, $\Theta_o = 220$ km/sec and a rotation curve falling at large distances.

More generally, the data appear to rule out both strongly rising curves (which give too low a terminal velocity) and strongly falling curves (which give too high a velocity). They are most consistent with a flat rotation curve, with $\Theta_o = 2AR_o \sim$ 210 – 250 km/sec. The values of R listed in the last column of Table 2 correspond to the distance of the gas producing the terminal emission, and confirm that the HI data provide information about the velocity field at larger distances than can yet be explored by measurements of HII regions or stars. It is possible of course to make more sensitive HI observations. Such observations at $\ell = 90°$ and $\ell = 225°$ (Knapp et al. 1978) suggest that the rotation curve is probably approximately flat out to much larger radii ($\sim$ 30–40 kpc).

As can be seen from Equation 13, V_r at a given brightness temperature T_b is almost linear in $\sin\ell$. The data and models in Figures 5–7 plotted this way are shown in Figure 8. For the models, only the smoothed values are plotted. It can be seen that in the first and third quadrants ($\ell = 10° - 90°$, $\ell = 180° - 270°$) the data fit the circular rotation models quite well, and suggest $\Theta \sim 210 - 230$ km/sec. However, in the second and forth quadrants ($\ell = 90° - 180°$, $\ell = 270° - 350°$) there are large-scale systematic deviations – towards $\ell = 180°$, negative velocity gas is seen and towards $\ell = 360°$ positive-velocity gas is seen. Between $\ell = 180°$ and $270°$, the terminal velocity (which is positive) is consistent with circular rotation, but between $\ell = \sim 120°$ and $180°$, the terminal velocity (negative) is systematically higher than values given by circular rotation. Some of the deviations in Figure 8 could be due to changes in the scale length, i.e. in the distribution of HI gas (cf. Henderson et al. 1982) but deviations towards $\ell = 180°$ must mean the existence of non-circular motions in the gas. It is likely that such non-circular motions are indeed present and not that an adjustment is needed in the LSR – examination of

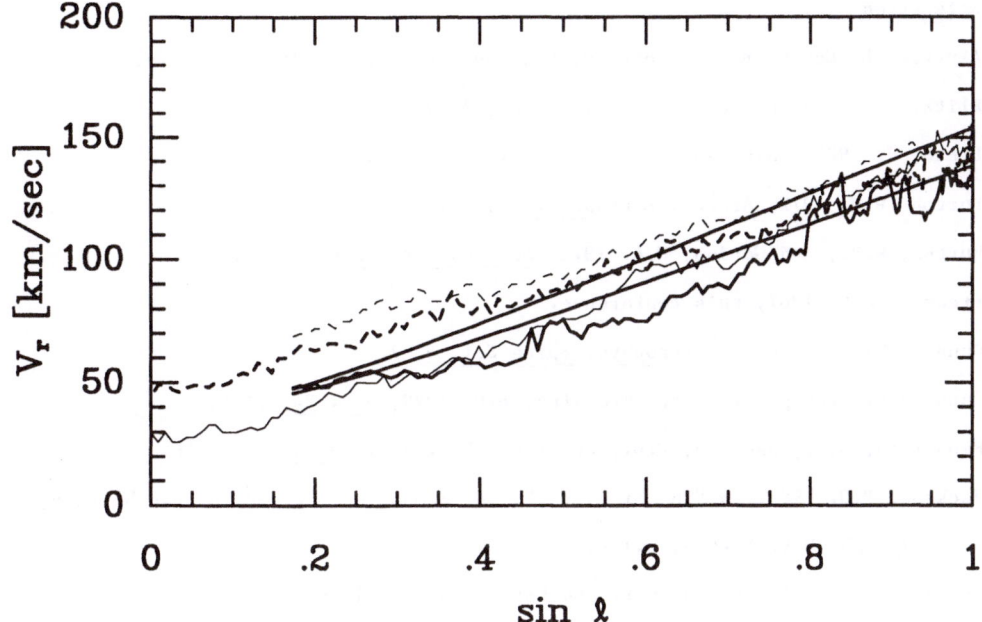

Figure 8: Terminal velocity of the outer Galactic HI, at the $T_b=1$ K contour, versus $|\sin \ell|$. The two model curves shown are flat curves ($\alpha=0$) with (1) $\Theta_0=220$ km/sec, $R_0=8.5$ kpc (lower curve) and (2) $\Theta_0=250$ km/sec, $R_0=10$ kpc (upper curve). The data are: (1) $\ell=10°-90°$ (heavy solid line) (2) $\ell=90°-180°$ (heavy dashed line) (3) $\ell=180°-270°$ (light solid line) (4) $\ell=270°-350°$ (light dashed line), and are taken from the surveys of Weaver and Williams (1974a,b) and Kerr et al. (1986).

latitude-velocity HI maps in this region (e.g. the map at $\ell = 179°.5$ of Weaver and Williams 1974b) shows that, while the bulk of the HI emission is indeed centered at 0 km/sec, there is a second component at ~ -20 km/sec. The large latitude extent of this feature is often cited as evidence that it is nearby: however, since the HI scale height increases with distance (cf. Equation 10) we have no information about the distance of this gas, and it could well be in the outer Galaxy. The possibility of non-circular motions and the nature of high velocity gas remain to be clarified.

In summary, though, observations of the velocity of weak HI emission in the outer Galaxy suggests that the rotation curve stays roughly flat at $\sim 210 - 230$ km/sec out to distances of $\sim 25 - 30$ kpc and probably well beyond.

This research is supported by the National Science Foundation via grant AST86-02698 to Princeton University.

References

Alvarez, H. Cohen, R.S., Bronfman, C., and Thaddeus, P. 1987, preprint.

Blitz, L., Fich, M., and Stark A.A. 1982, Ap.J. Suppl 49, 183.

Brand, J. 1986, Ph.D. Thesis, University of Leiden.

Burton, W.B. 1972, Astron. Astrophys. 19, 51.

Burton, W.B., and Gordon, M.A. 1978, Astron. Astrophys. 63, 7.

Freeman, K.C. 1987, this conference.

Gunn, J.E. 1974, Comm. Astrophys. Space Sci. 6, 7.

Gunn, J.E., Knapp, G.R., and Tremaine, S.D. 1979, A.J. 84, 1181.

Henderson, A.P., Jackson, P.D., and Kerr, F.J. 1982, Ap.J. 263, 116.

Jackson, P.D. 1985, in "The Milky Way", IAU Symposium 106, ed. H. van Woerden, R.J.
    Allen and W.B. Burton, p.179.

Jackson, P.D., Fitzgerald, M.P. and Moffat, A.F.J. 1979, in "Large Scale
    Characteristics of the Galaxy", IAU Symposium 84, ed. W.B. Burton, D. Reidel
    Co., p221.

Jackson, P.D. and Kellman, S.A. 1974, Ap.J. 190, 53.

Kahn, F.D., and Woltjer, L. 1959, Ap.J. 130, 705.

Kerr, F.J. 1957, A.J. 62, 93.

Kerr, F.J., Bowers, P.F., Jackson, P.D., and Kerr, M. 1986, Astron. Astrophys.
    Suppl. 66, 373.

Knapp, G.R., Tremaine, S.D., and Gunn, J.E. 1978, A.J. 83, 1585.

Knapp, G.R., Stark, A.A., and Wilson, R.W. 1985, A.J. 90, 254.

Kulkarni, S., Blitz, L., and Heiles, C. 1982, Ap.J. (Letters) 259, L63.

Little, B., and Tremaine, S.D. 1987, preprint.

Lynden-Bell, D. 1983, in "Kinematics, Dynamics and Structure of the Milky Way," ed.
    W.L.H. Shuter, D. Reidel Co., p. 349.

Lynden-Bell, D. 1987, this conference.

McCutcheon, W.W., Robinson, B.J., Whiteoak, J.B., and Manchester, R.N. 1983, in
    "Kinematics, Dynamics and Structure of the Milky Way," ed. W.L.H. Shuter, D.
    Reidel Co., 165.

Muller, C.A., Raimond, E., Schwarz, U.J., and Tolbert, C.R. 1966, <u>B.A.N. Suppl.</u> 1, 213.

Olszewski, E.W., Peterson, R.C., and Aaronson, M. 1986, <u>Ap.J. (Letters)</u> 302, L45.

Oort, J.H. 1966, <u>B.A.N.</u> 18, 421.

Rubin, V.C. 1979, in "The Large-Scale Characterisitics of the Galaxy" I.A.U. Symposium 84, ed. W. B. Burton, p211.

Schechter, P.L. 1987, in preparation.

Shuter, W.L.H. 1981, <u>M.N.R.A.S.</u> 194, 851.

Stark, A.A. 1979, Ph.D. Thesis, Princeton University.

Verschuur, G.L. 1987, in preparation.

Weaver, H.F., and Williams, D.R.W. 1974a, <u>Astron. Astrophys. Suppl. Ser.</u> 17, 1.

Weaver, H.F., and Williams, D.R.W. 1974b, <u>Astron. Astrophys. Suppl. Ser.</u> 17, 251.

GROPING FOR TRUTH FROM THE GALAXY'S OUTERMOST SATELLITES

D. Lynden-Bell
Institute of Astronomy, The Observatories,
Cambridge CB3 0HA

Three explanations of Kerr's bending of the Galaxy's outer disk are briefly considered.

The motions of distant globular clusters and satellites of the Galaxy give a best estimate of its mass of 2.5×10^{11} $M_\odot$ out to 120 kpc. This estimate depends on the isotropy of the motions of distant objects which will be untrue if the distant dwarf spheroidal galaxies are fragile to tidal disruption. The large velocity dispersions recently found in them imply that they are sufficiently rugged to resist the tides. However, critical discussion of the stellar velocity observations shows that the observers of Carina have underestimated their errors so that its true velocity dispersion is much smaller. This implies that Carina's mass-to-light ratio is stellar and that it _is_ tidally fragile.

When measurements close to the observational limits are required, two or three independent astronomers observing the _same_ stars with independent equipment and reduction procedures are needed. Three independent sets of measurements of the same stars allow external errors to be found for each set.

The high velocity at the tip of the Magellanic Stream is hard to reproduce in low mass Galaxy models unless non-gravitational forces act on the gas.

Direct evidence favouring a large mass for the Milky Way is not yet forthcoming but accurate velocities for Eridanus, Leo I and Palomar 14 might well produce it.

It is shown that a mass of 10^{12} $M_\odot$ for the Galaxy is not in serious conflict with available data and would allow concordance with the Mass of the Local Group determined from the timing argument.

----

1. Introduction

I have come in tribute to Frank Kerr whose work has been seminal in generating interest in the dynamics of the outer Galaxy, the Magellanic Stream and the Local Group. I hope to show how all these and the dwarf spheroidal Galaxies play vital roles in our assessment of the Galaxy's mass and the nature of dark matter. Vital parts of this discussion rest on work done with P.J. Godwin and A. Kulessa.

The second section gives the timing argument of Kahn & Woltjer and like them, considers the origin of the Galaxy's warp. The third section reviews work aimed at determining the mass of the Galaxy from the radial velocities and distances of its distant satellites and shows that a small mass is preferred if the motions are isotropic. A preference for more circular orbits would occur if satellites on diving orbits were tidally torn apart. However, dwarf spheroidal galaxies with larger velocity dispersions due to dark matter would be heavier and harder to disrupt. These considerations lead to section four where we discuss the observations of internal velocity dispersions of the dwarf spheroidals. It is shown that the velocity dispersion of 6 km/sec for Carina is an overestimate. The best from current data being only 1.1 km/sec. This gives Carina a stellar mass-to-light ratio and casts doubt on the high values claimed for those other dwarf spheroidals which rest on similar low dispersions. Implications for the nature of dark matter are briefly considered.

Finally, in section 5 we assess the current evidence on the mass of the Galaxy, emphasise its uncertainty and the critical data still needed.

2. The Timing Argument and the Galaxy's Warp

Kerr[1,2,3] discovered the Galaxy's warp in 1956. Stimulated in part by this, Kahn & Woltjer[4] wrote their classic paper which first gave the timing argument on the mass of the Local Group. If it is assumed that the Galaxy and M31 emerged from the Big Bang and that their gravity has been responsible for holding them back and reversing their initial expansion, then their separation r, their relative velocity $\dot{r}=v$ and the time t since the Big Bang obey[5]

$$\left(\frac{2GM}{r^3}\right)^{\frac{1}{2}} t + 1.2 \left(\frac{vt}{r}\right) = \pi/2 \tag{1}$$

where M is the sum of the masses of M31 and the Galaxy.

This approximation is good to 5% for $v \leq 0$. Equation (1) may be rewritten as an equation for M

$$M = \frac{r^3}{2G} \left(\frac{\pi}{2t} - 1.2 \frac{v}{r}\right)^2 \tag{2}$$

Inserting v = -123 km/sec and r = 0.7 Mpc, we find that for

$t = 2.10^{10}$ this yields $M = 3 \times 10^{12} M_\odot$
while $t = 10^{10}$ gives $M = 5 \times 10^{12} M_\odot$.

Both these values are so large that over 90% of the mass must be of some dark form if the estimates are correct. However, it is the conjecture that the approaching motion of the galaxies is caused by their gravity, that has led to this large estimate. Either the Galaxy or M31 could have been given a significant ejection velocity from some other group and the motion of approach could be due to chance.

In the same paper Kahn and Woltjer suggested that the warp of the Galaxy's disk might be caused by an intergalactic wind blowing on the inclined galaxy. If the wind blows directly onto the galactic disk both edges will be blown back so Khan and Woltjer considered a gaseous spherical halo around the galaxy whose pressure distribution is modified by the wind. For an idealised spherical halo this can give distortions of the right sense. However, the direction of the line of nodes of galactic bending is toward a galactocentric longitude of $75°$ which is $45°$ away from Andromeda towards which the Galaxy is assumed to fall.

Kerr[1,2] had earlier proposed a direct tidal interaction from the Magellanic Clouds but the masses seemed too small. It is perhaps worth reconsidering this simple explanation now that heavy haloes are in vogue because these enhance tides both directly by giving more mass and indirectly by disturbing the halo that distorts the disk[6]. However, the precessions induced in the disk are likely to destroy the simple tidal distortion.

I think I first heard Kerr when he addressed the Royal Astronomical Society. I was at once stimulated to think about the warp and considered the problem of the free precession of a galaxy. To illustrate this I used to throw an old British penny in the air to demonstrate the wobble as it rotated. The new pennies are too small but here is one from 1918 made at the Old Mint on Tower Hill at about the time that Frank was born not far away in St. Albans, England. It has seen the same vicisitudes of war and peace as you have Frank. It still demonstrates precessional wobble well, so I hope you will keep it as a reminder of this pleasant occasion. A galaxy is not a rigid body so the effect of the precessional wobble is to bend it[7]. Although Hunter & Toomre[8] later showed that galaxies with gently tapered edges would not precess regularly, nevertheless galaxies with sharper edges do so. Such motions do predict that the line of nodes of the bending is independent of radius and this is a pronounced property of the warps observed[7].

3. The Mass of the Galaxy measured from distant satellites

Figure 1 is taken from my contribution to the Vatican study week in 1981. The points denote globular clusters and the graph plots $\log(3 V_\ell{}^2)$ where V_ℓ is the line of sight velocity in the Galaxy's system of rest against $\log R$, the galactocentric distance. One may prove that for any bounded orbit in any potential the time averages $\langle \underline{v}^2 \rangle$ and $\langle V_c{}^2 \rangle$ are equal[9]. Here $V_c{}^2 = -\underline{r}.\partial\psi/\partial\underline{r}$ and ψ is the Galaxy's gravitational potential. For isotropic velocities $\langle \underline{v}^2 \rangle = \langle 3 v_\ell^2 \rangle$. The heavy points are averages taken over the columns indicated top and bottom. In my naivety I believed this diagram showed that the Galaxy had a V_c = constant velocity curve out to about 100 kpc. However, I realised that the measurements of the outermost systems mainly found from photographic spectra by Hartwick & Sargent[10] were very difficult to make. Not only are the stars very distant but the metal lines of these very deficient

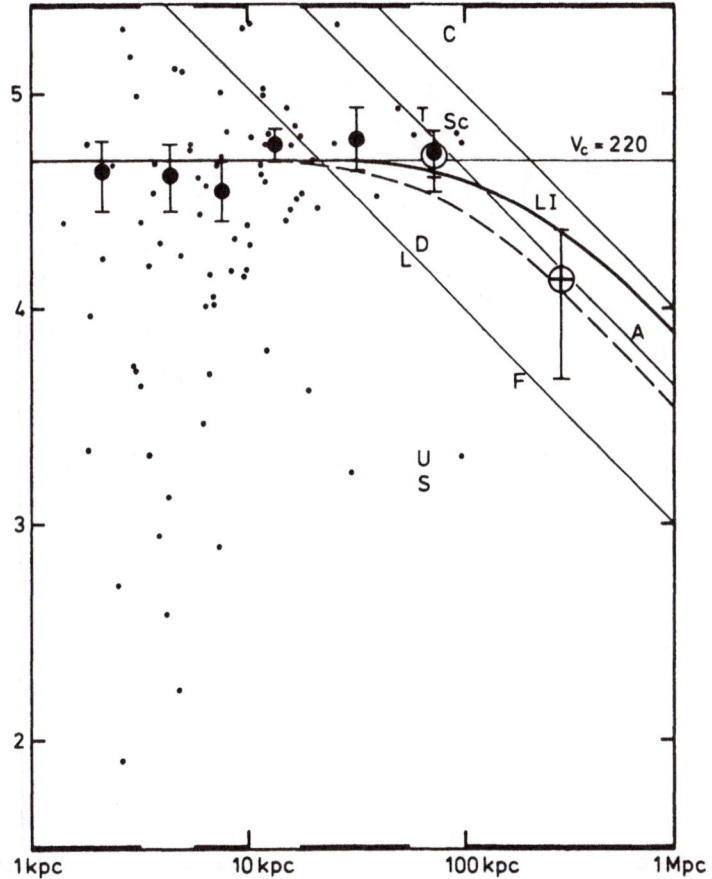

Fig. 1. The 1981 data with $\log(3v^2)$ versus $\log r$ for the line of sight velocities of globular clusters and satellites of the Galaxy. The hard dots denote r.m.s. averages of globular clusters in bins denoted in the margin.
C = Carina, SC = Sculptor, T = Tip of Magellanic Stream, D = Draco, L = LMC, U = Ursa Minor, S = SMC, LI = Leo I, F = Fornax, A = M31.

systems are weak. Later I realised that an apparently flat rotation could result from velocity errors that increased with distance coupled with a falling rotation law. Noticing that Carina by itself needed a very massive Galaxy, I rang up Russell Cannon who explained that the velocity was a rough one taken under difficult circumstances and that it should be remeasured if it was of any theoretical importance. The remeasurement[11] resulted in Carina falling from its position as the fastest moving satellite of the Galaxy to being one of the slowest moving. Unfortunately our velocity of 240 km/sec had a small correction applied twice and Godwin's reworking of the reductions now gives 229±5 km/sec heliocentric. Meanwhile Mark Aaronson[12] among others[13] was busy improving the velocities of the distant

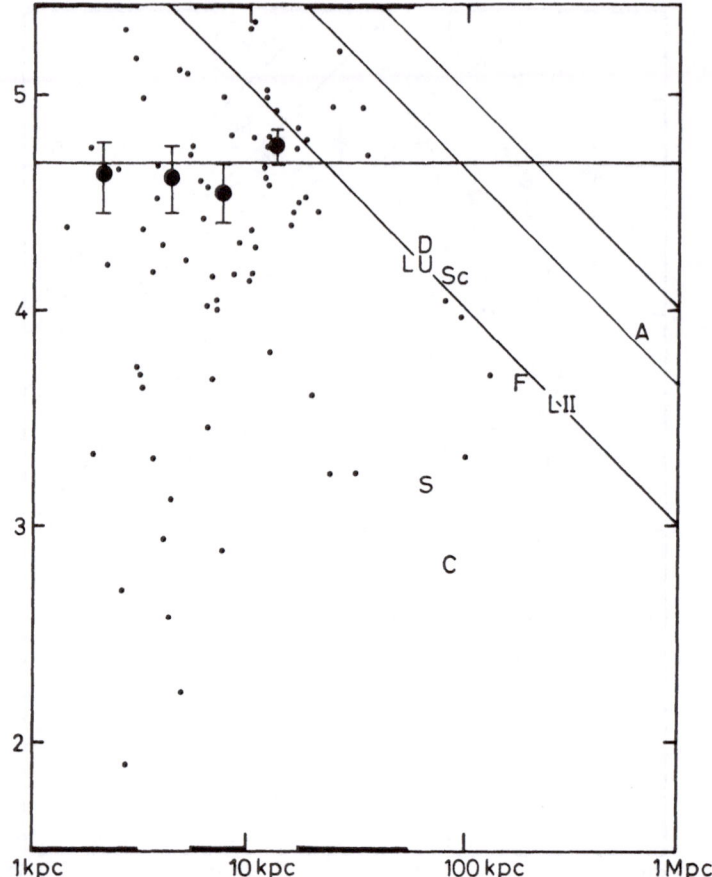

Fig. 2. As Figure 1 but with 1986 data.

satellites so we were able to give a discussion of the mass of the Galaxy determined
from distant satellites[11]. On the assumption of isotropic velocities these gave both
a low circular velocity and a low total mass.   However, the dwarf spheroidal
satellites are fragile objects so we argued that those on deeply diving orbits would
have been disrupted. This might lead to the survivors having preferentially more
circular orbits which we showed could easily increase the mass estimate by a factor
4.   More recently Little & Tremaine[14] have confirmed our results using both distant
globulars and dwarf spheroidals and a more thorough statistical method, see Table 1.

TABLE 1

Estimates of circular velocity and Galactic mass from objects at beyond 50 kpc

L–B, C & G	Little & Tremaine
$\langle 3V_\ell^2 \rangle = \langle V_c^2 \rangle^{\frac{1}{2}} = 106$ km/s	107 km/s
$M = 2\langle(V_r^2 - \varepsilon^2)r\rangle/\langle e^2\rangle$	Distribution fitting
$M = 2.6 \pm 0.8 \times 10^{11} M_\odot$	$2.4\,{}^{+1.3}_{-0.7} \times 10^{11} M_\odot$
95% Upper limit	$M < 5.2 \times 10^{11} M_\odot$

The above numbers assume isotropic velocities $\langle e^2 \rangle = \frac{1}{2}$

Olszewski, Peterson & Aaronson[15] provided data improvement which forms the basis for Table 2 and Figure 2 which repeats Figure 1 with modern data. In their deduction of a somewhat higher mass they used the less certain data for systems for which they give alternative possibilities for the velocity. We have amended their Table in two respects. Firstly, Armandroff & Da Costa's[16] new result on Sculptor gives it a heliocentric velocity of 107 ± 2 in place of the 198 km/sec of Hartwick & Sargent[10] and the 20 km/sec of Richer & Westerlund[17]. Why the velocity from the 3 Carbon stars of Richer & Westerlund should be so different I do not know. Secondly, a new velocity for Leo II taken with the Isaac Newton Telescope gives a heliocentric velocity of 90 km/sec which agrees well with that given earlier, so the note saying this velocity is uncertain has been removed. Looking down Table 2 one notices a most remarkable correlation in the last columns. Whenever $v^2 r$ is in double figures there is a note saying that the data is uncertain. With this in mind it seems wise to take the view that whenever there are alternative velocities given in Table 2 then the authors are prepared to admit that either value may be WRONG. If that is the case it is not clear that either velocity is RIGHT. The only certain thing is that the velocity needs redetermination and on that basis we have refused to use either alternative. The mean of the remaining values of $v^2 r$ is $2.7\pm0.5 \times 10^5$ which leads to a mass of the Galaxy of $M = 2.5 \pm .5 \times 10^{11} M_\odot$ if we use an isotropic velocity distribution. This mass is derived from objects at between 50 and 220 kpc from the Galactic Centre. A. Kulessa has adapted the method of Little & Tremaine[14] to give the mass out to 220 kpc on the assumption of isotropic orbits in a potential that behaves as $r^{-\beta}$. For $\beta=1$ this reduces to the Newtonian case but more generally it gives halo or total matter densities proportional to $r^{-(\beta+2)}$. Figure 3 shows that the most probable value of β is close to the Newtonian $\beta=1$ but that a lower β coupled with a larger mass is not ruled out so that Little & Tremaine's conclusions that with

TABLE 2

From Table 2 of Olszewski   Peterson & Aaronson 1986 (Amended)

	v_{GC}	r_{GC}	$v^2_{GC} r_{GC}/10^5 (km/s)^2 kpc$	Notes
AM1	-42	117	1.8	
Eridanus	{-137, -217	81	16, 39	1
N2419	-25	96	0.6	
Pal 3	-59	94	3.2	
Pal 4	+54	108	3.1	
Pal 14	{+167, +94	75	20, 5.1	1
Pal 15	{+32, +150	95	0.9, 20	1,2
LMC	76	50	2.9	
SMC	28	61	0.5	
Draco	-95	75	6.8	
Ursa Minor	-88	65	5.0	
Sculptor	+74*	79	4.3*	
Carina	14	92	0.2	
Fornax	-34	140	1.6	
Leo I	{+77, +150:	220	12, 48:*	1,2
Leo II	41[†]	220	2.3[†]	
		Average omitting uncertain values	2.7±0.5	

Notes: 1. Uncertain Velocity   2. Uncertain Distance
* Amended   [†]Confirmed by new measurement

90% confidence $M < 5.10^{11}$ $M_\odot$ can be relaxed somewhat if other potentials are used. They also argued that the high velocity dispersions of the dwarf spheroidals show them to be rugged rather than tidally fragile so that our earlier argument against isotropic velocities did not hold. In the next section we show that the velocity dispersions of dwarf spheroidals are subject to error, so that those systems are likely to be tidally fragile after all.

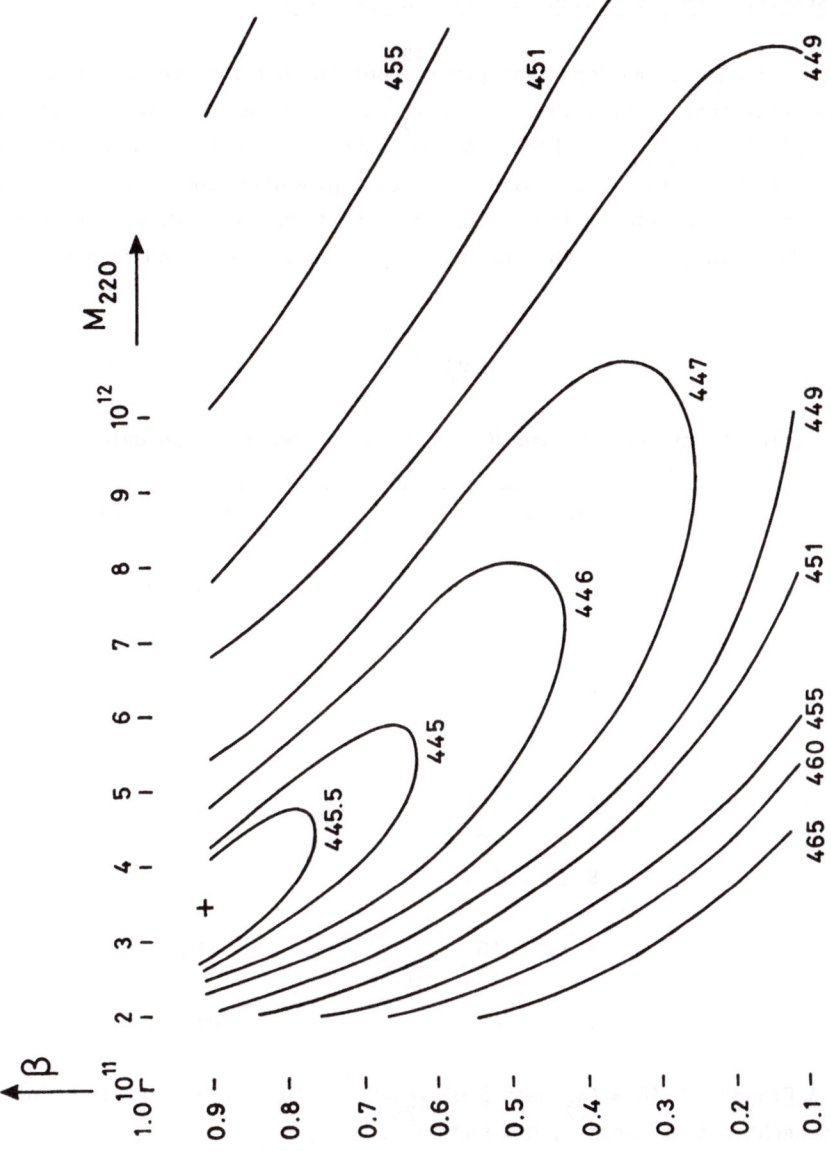

Fig. 3. A. Kulessa's contours of equal probability for the Mass out to 220 kpc plotted against β where the potential of the Galaxy varies as $r^{-\beta}$.

4. Overestimates of Dwarf spheroidal velocity dispersions

Tremaine & Gunn in an important paper[18] pointed out that there was a limit on how light an elementary particle could be if it was to be the prime constituent of dark matter. Lin & Faber[19] and Faber & Lin[20] showed the dwarf spheroidal might be critically important for this test while Aaronson[12] provided the first hint of a measured velocity dispersion. Table 3 gives current data according to da Costa. I have added a determination of the dispersion of Carina suggested from our data by Godwin.

TABLE 3

Velocity dispersions and M/L$_v$ ratios for dwarf spheroidals

	σ/km/s		M/L$_v$ (solar units)
Fornax	6 ± 2	SF	1 ± 0.5
	8 ± 3	AO	
Sculptor	6 ± 3	SF	6 ± 2
	6.2 ± 1.2	AD	
Carina	6 ± 3	CSA	10 ± 6
	6 ± 2	SF	
	8 ± 4	G	
Draco	9 ± 2	AO	37 ± 11
Ursa Minor	11 ± 3	AO	100 ± 40

SF = Seitzer & Frogel[21], AO = Aaronson & Olszewski[13], AD = Armandroff & Da Costa[16], CSA = Cook, Schechter & Aaronson[22], G = Godwin[23].

I shall concentrate on the Carina data as we have all the individual star velocities and they have been determined by three different sets of observers using different telescopes and reduction techniques. Paul Schechter kindly communicated to us the individual stars' velocities determined from his spectra by Cook & Aaronson. These are preliminary in the sense that it was hoped to determine them again using another reduction package. Table 4 gives the differences of the velocities of the five stars from their mean according to the 3 sets of observers Seitzer & Frogel (F), Aaronson,

Cook & Schechter(S), Lynden-Bell, Cannon & Godwin(G). Any variability in the stars will be treated as a random error in the velocities as it does not contribute to the velocity dispersion of Carina as a whole that we wish to measure. By considering the differences of the velocities as measured by different observers we can determine the external error of measurement + variability. In fact

$$\frac{1}{N-1} \Sigma(v_G - v_S)^2 = \epsilon_G^2 + \epsilon_S^2 = 108.0 \ (25) \ (km/s)^2$$

$$\frac{1}{N-1} \Sigma(v_F - v_G)^2 = \epsilon_F^2 + \epsilon_G^2 = 81.8 \ (20) \ (km/s)^2 \qquad (1)$$

$$\frac{1}{N-1} \Sigma(v_S - v_F)^2 = \epsilon_S^2 + \epsilon_F^2 = 54.7 \ (13) \ (km/s)^2$$

These values can be compared with the observers' estimates of their accuracies from the internal consistency of repeat observations which are $\epsilon_G = 4$, $\epsilon_S = 3$, $\epsilon_F = 2$ and lead to the numbers in brackets above. It is clear that the numbers in brackets are very discrepant with those beside them and that the errors can not be blamed on a single observer. Solving for the ϵ one obtains $\epsilon_G = 8.2$, $\epsilon_S = 6.4$ and $\epsilon_F = 3.8$ but from so few stars these will have significant errors. Evidently the old adage that the error in km/s is twice as big as the observer thinks is rather well born out by these data. We may now use our estimates of the ϵ to estimate Carina's velocity dispersion from the more accurate observations.

$$\frac{1}{N-1} \Sigma v_S^2 = \epsilon_S^2 + \sigma_c^2 = 33.7 \Rightarrow \sigma_c^2 = -7.3 \qquad (2)$$

$$\frac{1}{N-1} \Sigma v_F^2 = \epsilon_F^2 + \sigma_c^2 = 23.4 \Rightarrow \sigma_c^2 = +9.0 \qquad (3)$$

Evidently these results are not of high accuracy. A serious contribution to their error stems from the errors in our determinations of the errors ϵ. The worst of these come with the large value of ϵ_G which is hardly surprising since our measurements were taken to determine the mean velocity of Carina rather than its internal velocity. It is possible to find σ_c^2 by ignoring our (G) results completely. From the last two equations

$$\epsilon_S^2 + \epsilon_F^2 + 2\sigma_c^2 = 57.1 \qquad (4)$$

Subtracting the value of $\epsilon_S^2 + \epsilon_F^2$ from equation (1) we deduce

$$\sigma_c^2 = 1.2 \ (km/s)^2 \qquad \sigma_c = 1.1 \ km/s$$

This value should be more accurate as it uses only the more accurate observers' results. Using the less accurate results above also we get $\sigma_c = 1.1 \, ^{+1.7}_{-1.1} \, km/s$.

Since the old value of σ_c was 6, it is evident that σ_c^2, which occurs in mass determinations, should be reduced by a factor ~30. Such a reduction would put Carina's mass-to-light ratio at less than 1 and eliminate the case for dark matter within it.

Returning to Table 3 we have now found that not only the 'large' dwarf spheroidal Fornax but also one of the smallest Carina have stellar mass-to-light ratios. It is somewhat odd if the others have the very large ones so often quoted. It would seem more likely that the observers have underestimated their velocity errors by factors of 2 or more. The observed velocity dispersions are similar to the 6 km/s wrongly interpreted in Carina. Some of the data are from red giants rather than Carbon stars but the acid test of star by star comparison with different observations has not yet been done.

5. The Galaxy's uncertain mass and observations needed

In a Keplerian orbit the maximum radial velocity as seen from the point mass is GM e/h and the distance of closest approach is $r_{min} = \dfrac{h^2}{GM(1+e)}$. Hence

$$GM = (v_r)^2_{max} \; r_{min} \left(\frac{1+e}{e^2}\right)$$

where e is the eccentricity. At the tip of the Magellanic Stream the radial velocity in the Galaxy's system of rest is 200 km/s. This gives

$$M = 8 \times 10^{11} \left(\frac{1+e}{2e^2}\right) M_o \left(\frac{r_{min}}{45 \text{ kpc}}\right)$$

For e=1 the bracket is 1 while it is 3 for e=$\frac{1}{2}$. Thus unless the tip of the Magellanic Stream is on an orbit that dives deeply into the galaxy with $r_{min} \ll 45$ kpc, we are forced to a high mass Galaxy. However, the tip of the Magellanic Stream is sparse gas that could have been affected by the Galaxy's halo. It could be much closer than 45 kpc and e could be near 1.

In summarising what we know of the Galaxy's mass, the work of Henderson, Jackson & Kerr[24] among others shows that the Galaxy extends to at least 15 kpc from the centre with a roughly constant circular velocity. Taking this to be 220 km/s the mass can not be significantly less than 2×10^{11} $M_\odot$. The data from the distant satellites out to 220 kpc give best values about 2.5×10^{11} $M_\odot$ assuming isotropic velocities but to do not rule out heavy haloes with $\rho_H \propto r^{-2.6}$ and masses of nearly 10^{12} $M_\odot$ out to 200 kpc. Such models are favoured by the Local Group's timing argument and the high velocity at the tip of the Magellanic Stream. However, the arguments for a heavy halo are indirect. It is most important to find an object with well determined distance and well determined velocity which clearly requires a heavy halo. Likely candidates are listed in Table 4. Refined velocities and distances for these objects are urgently needed.

Frank Kerr's work has been large scale thorough and definitive. I apologise that the best I can offer you is a question mark over the Galaxy's mass. Although Carina is but one dwarf spheroidal, the serious doubts over their velocity dispersions and the existence of substantial quantities of dark matter in them must be resolved by more persistent and more accurate observations.

TABLE 4

Distant clusters in need of good velocities[25]

	r_{GC}/kpc	l	b	
Eridanus	90.2	218.11	−41.33	
NGC 1841	38.3	297.02	−30.15	
Arp Madore 2	61.5	248.13	−05.88	
ESO 093-SC08	56.6	293.51	−04.04	} Obscured
BH 176	78.3	328.42	04.34	
Pal 14	69.9	028.76	42.18	
Pal 15	62.3	018.87	24.29	
Ter 8	40.4	005.76	−24.56	

REFERENCES

1. Kerr, F.J. 1956, Proc. Radio Astr. Symposium Sydney, cf. AJ 1957 62, 93.
2. Kerr, F.J., Hindman, J.V. & Carpenter, M.S. 1957, Nature 180, 677.
3. Kerr, F.J. 1962, M.N.R.A.S. 123, 327.
4. Kahn, F.D. & Woltjer, L. 1959, Ap. J. 130, 705.
5. Lynden-Bell, D. 1981, Observatory 101, 111.
6. Lynden-Bell, D. 1985, IAU Symposium 10 The Milky Way Galaxy, eds. H. van Woerden et al., p.467.
7. Lynden-Bell, D. 1965, M.N.R.A.S. 129, 299.
8. Hunter, C. & Toomre, A. 1969, Ap. J. 155, 747.
9. Lynden-Bell, D. 1981, Observatory 101, 200.
10. Hartwick, F.D.A. & Sargent, W.L.W. 1978, Ap. J. 221, 512.
11. Lynden-Bell, D., Cannon, R.D. & Godwin, P.J. 1983, M.N.R.A.S. 204, 87p.
12. Aaronson, M. 1983, Ap. J. 266, L11.
13. Aaronson, M. & Olszewski, E.W. 1987, IAU Symposium 117, Princeton: Dark Matter in the Universe, eds. J. Kormendy & G.R. Knapp.
14. Little, B. & Tremaine, S. 1987, Ap. J. in press.
15. Olszewski, E.W., Peterson, R. & Aaronson, M. 1986, Ap. J. 302, L45.

16. Armandroff, T.E. & Da Costa, G. 1986, A.J. $\underline{92}$, 777.

17. Richer, A.B. & Westerlund, B.E. 1983, Ap. J. $\underline{264}$, 114.

18. Tremaine, S. & Gunn, J.E. 1978, Phys. Rev. Lett. $\underline{42}$, 407.

19. Lin, D.N.C. & Faber, S.M. 1983, Ap. J. $\underline{266}$, L21.

20. Faber, S.M. & Lin, D.N.C. 1983, Ap. J. $\underline{266}$, L17.

21. Seitzer, P. & Frogel, J.A. 1985, A.J. $\underline{90}$, 1796.

22. Cook, K., Schechter, P. & Aaronson, M. 1983, Bull. Amer. Astron. Soc. $\underline{15}$, 907.

23. Godwin, P.J. & Lynden-Bell, D. 1987

24. Henderson, A.P., Jackson, P.D. & Kerr, F.J. 1982, Ap. J. $\underline{263}$, 116.

25. Webbink, R. 1985, in Dynamics of Star Clusters, IAU Symposium 113, eds.
    J. Goodman & P. Hut, (Dordrecht, Reidel), p. 541.

CARBON STARS AT $2R_0$ AND THE ROTATION OF THE MILKY WAY

Paul L. Schechter,[1,2] Marc Aaronson,[2,3,4]
Kem H. Cook,[1,2,3] and Victor M. Blanco[5]

1. Introduction and General Approach

The present study, using carbon stars to measure differential rotation in the outer Milky Way, grew out of a somewhat improbable convergence of interests and experience on the part of its authors. Two of us, Aaronson and Schechter, had collaborated in a study of the systematic deceleration of galaxies by the Local Supercluster, and as part of that study, had searched for rotation about its center (Aaronson et al. 1982). Blanco and collaborators (Blanco et al. 1980) had shown the carbon stars in the Large Magellanic Cloud exhibit a relatively small spread in absolute magnitude. Aaronson (1983) had been using carbon stars to study the internal dynamics of dwarf spheroidals, and with Cook and Schechter had begun a program to obtain velocities in the Carina dwarf (Cook et al. 1983). As a secondary program, when the weather, moonlight, seeing or sidereal time would not permit work on Carina, velocities were obtained for carbon stars in the outer Milky Way.

Carbon stars exhibit a range of properties which make them particularly well suited for such a study. They are abundant, as evidenced by Westerlund's (1971) discovery of approximately one per square degree near the galactic plane. They are readily identified on objective prism plates. They are intrinsically bright, making it possible to work through several magnitudes of galactic absorption. They are old enough to have lost memory of any systematic velocities with which they might have been born, but still relatively young, with smaller random velocities than older tracers. Their spectra are exceedingly rich, making it easy to obtain accurate velocities with relatively low signal to noise.

_____

[1]Mount Wilson and Las Campanas Observatories
[2]Visting Astronomer, CTIO
[3]Steward Observatory
[4]deceased 30 April 1987
[5]Cerro Tololo Inter-American Observatory

Over the course of the last four years, objective prism plates were obtained with the Curtis Schmidt telescope for six fields in the southern Milky Way, chosen because it was suspected, for a variety of reasons, that they might be unusually transparent. JHK photometry was then obtained at CTIO for the carbon stars identified in these fields, and velocities were obtained for a subset of these at Las Campanas. At present we have both photometry and velocities for 166 stars, and photometry alone for another 240 stars. We are now obtaining complementary data for a set of six northern fields. A comparable number of carbon stars have been identified (most of them for the first time), and are being followed up with photometry and spectroscopy.

The 166 stars for which we have both velocities and photometry were drawn from fields which span a deliberately narrow range of galactic longitude, with $235 < \ell < 315°$, and thus are not adequate for a determination of the solar motion with respect to the system of carbon stars. We have therefore included in our analysis a sample of 102 bright, nearby carbon stars, with a more uniform spread in longitude, which permits simultaneous solution for a carbon star standard of rest and the more interesting parameters. The JHK photometry for these comes from the work of Walker (1980), Cohen et al. (1981), and Noguchi et al. (1981). The velocities for the nearby stars are drawn largely from the work of Sanford (1944), Dean (1976) and Walker (1979).

The photometry provides relative heliocentric distances to the carbon stars. The velocities give the differential rotation at some galactocentric radius. One must use the distance to the galactic center to obtain a galactocentric radius from a heliocentric radius. We must therefore either put our relative distances onto an absolute scale, and specify the distance to the galactic center on a consistent distance scale, or we must determine the distance to the galactic center on our relative scale. We have chosen the latter course. By observing carbon stars in the fourth quadrant which lie beyond the tangent point and near the solar circle (as judged from the fact that they have velocities which are nearly zero) we can measure the apparent magnitude of a carbon star at a geometric multiple of R_0 which depends only upon the longitude of the star. This approach has the advantage that it guarantees the consistency of the distances to the carbon stars and the distance to the galactic center.

2. Some Details

In practice we fit a model to the set of observed values of m, v and ℓ (where m is the apparent K magnitude), assuming a constant rotation curve with amplitude v_0 and taking as a free parameter the apparent magnitude m_0 that a carbon star would have at a distance R_0. The assumption of a constant rotation curve is not as

restrictive as it might at first seem, since any linear rotation curve has an
associated constant rotation curve which would produce the same observed quantities.
The parameter v_0 is then equal to the product $2AR_0$, where A is the familiar Oort
constant.   In its simplest form our model consists of three equations,

$$v_{lsr} = v_0(1/r - 1)\sin\ell,$$
$$r = \sqrt{(1 - 2x\cos\ell + x^2)}, \text{ and}$$
$$x = 10^{[(m - m_0)/5]},$$

where r is galactocentric distance in units of R_0. The parameters v_0 and m_0 are
varied so as to minimize differences between the observed and predicted velocities.
It is a straightforward generalization of the model to add a term which allows for a
linear dependence of absolute magnitude on J-K color. The color coefficient, c_{jk},
can be treated as a free parameter. One can also make explicit the conversion from
heliocentric velocity to a velocity with respect to a local standard of rest,
leaving the option of treating the components of the solar motion as additional free
parameters. Caldwell and Coulson (1987) give a detailed treatment of these points.

If one restricts attention to the subset of brighter stars for which the
velocity residuals are dominated by the random velocities of the stars, one finds an
rms line of sight velocity dispersion averaged over longitude of 19 km/s. Turning
attention to the more distant subset, and assuming a velocity dispersion of 19 km/s
at the solar circle which drops to half that value at $2R_0$, there is an excess
velocity residual which can be accounted for if the carbon stars exhibit an rms
dispersion in absolute magnitude, σ, of 0.6 mag. This is consistent with the spread
in absolute magnitude observed by Cohen et al. (1981) in the Magellanic Clouds. At
large galactocentric distances, the random velocities and the spread in absolute
magnitude contribute equally to the scatter about the assumed model.

There is however, a complication which makes the spread in absolute magnitudes
a far more serious problem than the random velocities: Malmquist's effect. The
mean absolute magnitude at a given apparent magnitude will in general be different
from the mean absolute magnitude in a volume limited sample by $-\sigma^2 d\ell n A(m)/dm$, where
A(m) gives the differential counts as a function of apparent magnitude along the
appropriate line of sight. At brighter apparent magnitudes A(m) is governed by the
volume effect, increasing as $10^{0.6m}$, making the mean absolute magnitudes brighter by
roughly 0.5 mag. At fainter apparent magnitudes, A(m) reaches a maximum and begins
to decrease as the volume effect is dominated by the scale length and scale height
of the galactic disk and by the increasing effects of galactic absorption. The
Malmquist correction therefore changes sign at faint apparent magnitudes.

33

Since the scale length of the galactic disk and the scale height for carbon stars (which may be a function of galactocentric radius) are at best poorly known, and since absorption is quite patchy, it is difficult to construct a smooth approximation to A(m) for each direction in which we observe a carbon star. Our current best effort has been to assume that the volume density of carbon stars varies as $x^2\exp(-x/s)$, where x is again heliocentric distance. We adopt a value of 0.33 for the scale parameter s, which produces a smooth approximation to A(m) which looks something like the distribution actually observed when summed over all fields. Since our adopted smooth function is specified in terms of x, and since Malmquist's correction is applied to m_0, x must now be solved for iteratively. Convergence is quite rapid.

3. Some Results

We have applied the approach outlined above to the combined near and far samples of carbon stars, 268 in all, treating m_0, v_0, $v_{\Theta x}$, $v_{\Theta y}$, and c_{jk} as free parameters. To minimize the covariance between m_0 and c_{jk}, we let m_0 refer to a star with J-K = 1.95, a typical color. The results are summarized in Table 1. We include a covariance matrix to give some idea of the degree to which the changing result for one parameter influences the result for the others.

Table 1: Parameters, Uncertainties, and Covariances

parameter	value	uncertainty	covariances				
$2AR_0$	248	16	1.00	0.75	-0.44	0.20	-0.20
m_0	6.69	0.16		1.00	-0.28	0.59	0.01
$v_{\Theta x}$	12.5	2.5			1.00	-0.15	-0.01
$v_{\Theta y}$	15.1	2.4				1.00	-0.09
c_{jk}	-1.09	0.29					1.00

The value for m_0 in Table 1 may be used to compute galactocentric positions for the 400 stars in the 6 program fields. These are shown in Figure 1, where the filled symbols indicate the 166 stars for which velocities have been measured and which were used in obtaining the results given above. While more than a dozen stars would appear to lie at distances greater than $2R_0$ from the galactic center, a few of these, probably those which appear to be most distant, will turn out not to be carbon stars when spectra are available for the complete sample.

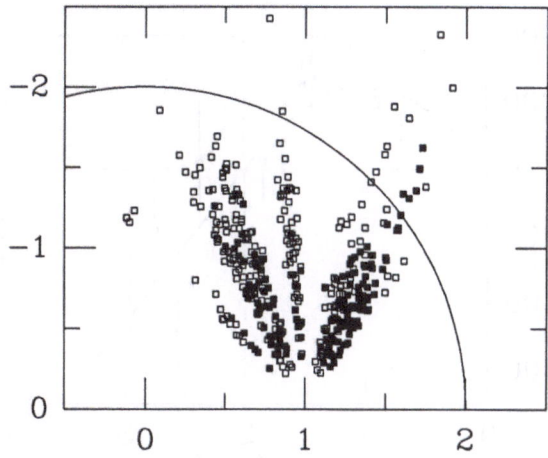

Fig. 1.--The distribution stars for which JHK photometry has been obtained as part of the present program, projected onto the galactic plane. The galactic center is at (0,0), the Sun at (1,0), and the ordinate is negative toward $\ell = 270°$. The filled symbols are the stars for which velocities have been measured.

The vast majority of stars in the fourth quadrant lie in two fields, centered at $\ell = 285°$ and $\ell = 295°$. Even at $\ell = 295°$, the extreme velocity expected at the tangent point is only 26 km/s. Since the random velocities in the line of sight are only slightly smaller than this, many stars are needed for an accurate determination of the apparent magnitude at which the rotation curve crosses zero beyond the tangent point. Some confidence in our result can be gained by taking all those stars with $285 < \ell < 315°$ and plotting $v_{lsr}/\{\sin\ell[1 - 1/\sqrt{(1 - \cos^2\ell)]\}}$ against $x/\cos\ell$. The expected locii for different values of ℓ lie very close to each other in such a plot, and begin to differ significantly only when the abscissa is greater than 3. Figure 2 shows such a plot, where, to reduce the scatter, the stars have been averaged in groups of four, with one point plotted per group. Also plotted in the figure is the expected locus for $\ell = 295°$.

If the adopted value of m_0 is incorrect, then the abscissas for the plotted points are all in error by some constant factor. A rough idea of the uncertainty in m_0 may be had by considering how large a factor might be applied to those abscissas without producing a poor fit to the smooth curve.

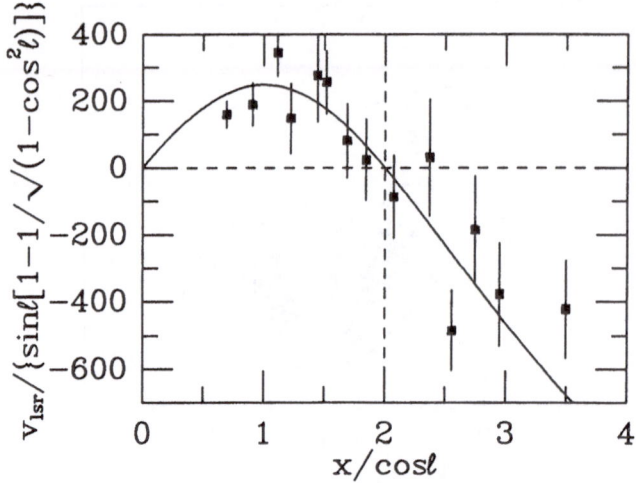

Fig. 2.--Radial velocity and heliocentric distance scaled so that observations at $|\ell| < 90°$ lie roughly along the same locus for values of the abscissa less than 3. The smooth curve is drawn for a linear rotation curve with $2AR_0$ = 248 km/s observed along ℓ = 295°. Each filled symbol is an average of data for four stars, lying in the range $285 < \ell < 315°$.

Some measure of the deviation of the rotation curve from the assumed linear form can be had by plotting $-v_{lsr}/\sin\ell$ against $(1 - 1/r)$. Figure 3 shows such a plot for the 166 stars in the fainter sample, with each point an average of the data for 12 stars. The straight line gives the best fitting model; its slope is v_0. With the exception of the outermost point, the deviations from linearity are exceedingly small.

At face value, the deviation of the outermost point from the best fit would indicate that the rotation curve rises suddenly at $2R_0$. An alternative explanation is that our Malmquist correction, which involves both an assumed A(m) and an adopted σ, is incorrect at the faintest apparent magnitudes. Another possibility is that carbon stars at $2R_0$ are fainter than those at R_0, due either to a different mean metallicity or a different mean age. A small effect of the sort that is seen is expected because the distribution of absolute magnitudes is not symmetric, as is implicitly assumed in applying Malmquist's correction, but is skewed toward the faint end.

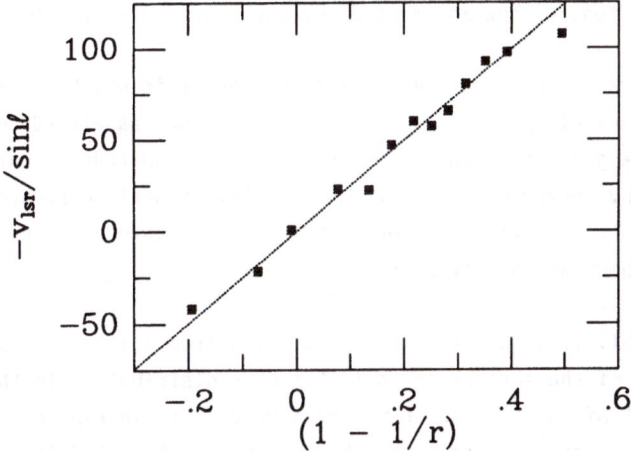

Fig. 3.--Radial velocity and galactocentric distance scaled so that a linear rotation curve would produce a straight line. Each filled symbol is an average of data for 12 stars. Galactocentric distances of R_0 and $2R_0$ are located at abscissas of 0 and 0.5, respectively.

The reader is cautioned against using the points plotted in Figure 3 to estimate the degree to which v_0 is constrained by the data. If a larger value of m_0 were adopted, smaller values of x and r would be computed for those stars beyond the solar circle and the slope v_0 would be larger. The correlation of the uncertainties in v_0 and m_0 is reflected in the relatively large value of their mutual correlation coefficient given in Table 1.

4. Discussion

The principal result of our efforts thus far is that the rotation curve appears to be quite flat over the range $0.8 < r < 2.0$, with $2AR_0 = 248 +/- 16$ km/s. This value for $2AR_0$ is larger than values derived interior to the solar circle using H I and CO (cf. Gunn et al. 1978, Knapp 1987). In the absence of other data, this difference between the results inside and outside the solar circle might be interpreted as a negative second derivative for the Milky Way rotation curve. It should be noted, however, that Caldwell and Coulson derive a smaller value for $2AR_0$ over the range $0.7 < r < 1.5$, which is closer to, but still larger than, the H I and CO results.

The large value of $2AR_0$ is unsettling from another point of view as well: if it is interpreted as the amplitude of a flat rotation curve, it is large compared to the velocity dispersions measured in various extreme population II objects (Norris et al. 1985, Bahcall and Casertano 1986, Sandage and Fouts 1987). While the relation between the circular velocity and the observed velocity dispersions depends upon number density gradients in the tracer and possible anisotropies in the velocity dispersion tensor, the apparent discrepancy warrants attention.

There are several assumptions implicit in our analysis which, if relaxed, might yield different values of m_0 and v_0. The first of these, as already mentioned, is the possibility of a galactic gradient in the absolute magnitude of carbon stars. This can be tested by observing stars at the same heliocentric distance but at different galactocentric distances. Toward this end, we have obtained spectra for most of the stars shown as open symbols in Figure 1.

A second possibility is a large scale deviation from axisymmetry of the galactic potential, of the sort expected if the mass distribution in the halo were triaxial. Since most of the stars with measured velocities in Figure 1 lie in a fairly narrow range of galactic azimuth, they are not well suited to looking for such deviations. We have therefore begun an extension of our work to quadrants I and II in the northern Milky Way.

We suspect that the ultimate limitation of our approach is the spread in absolute magnitude of our carbon stars. While it is possible to make an approximate correction for this spread, the uncertainties inherent in such a correction will come to dominate those associated with the size of the sample as the sample size increases. We will continue to look for some additional observable quantity which might improve our ability to determine distances to these otherwise nearly ideal tracers of galactic rotation.

The support of the U.S. National Science Foundation via grant AST83-18504 is gratefully acknowledged. Cerro Tololo Inter-American Observatory, National Optical Astronomy Observatories, is operated by the Association of Universities for Research in Astronomy, Inc., under contract with the National Science Foundation.

REFERENCES

Aaronson, M. 1983 Ap.J.(Letters), 266,L11.

Aaronson, M., Huchra, J., Mould, J., Schechter, P.L, and Tully, R.B. 1982, Ap.J., 258, 64.

Bahcall, J.N. and Casertano, S. 1986 Ap.J., 308,347.

Blanco, V.M., McCarthy, M.F. and Blanco, B.M. Ap.J., 242,938.

Caldwell, J.A.R., and Coulson, I.M. 1987 A.J., 93,1090.

Cohen, J.G., Frogel, J.A., Persson, S.E. and Elias, J.H. 1981 Ap.J., 249,481.

Cook, K., Schechter, P. and Aaronson, M. 1983 B.A.A.S., 15,907.

Dean, C.A. 1976 A.J., 81,364.

Gunn, J.E., Knapp, G.R., and Tremaine, S.D. 1979 A.J., 84,1181.

Knapp, G.R., 1987 present proceedings.

Noguchi, K., Kawara, K., Kobayashi, Y., Okuda, H., Sato, S. and Oishi, M. 1981
    P.A.S.J., 33,373.

Norris, J., Bessel, M., and Pickles, A.J. 1985 Ap.J.Suppl., 58,463.

Sandage, A. and Fouts, G. 1987, A.J., 93,74.

Sanford, R.F. 1944 Ap.J., 99,145.

Walker, A.R. 1980, M.N.R.A.S., 190,543.

Walker, A.R. 1979, South African Ast. Obs. Circ., Vol.1, No.4., p.112.

Westerlund, B. 1971, Astr.Ap.Suppl., 4,51.

THE VELOCITY FIELD OF THE OUTER GALAXY

Jan Brand[1,+], Leo Blitz[2] and Jan Wouterloot[3]
1. Sterrewacht, Leiden
2. Astronomy Program, University of Maryland
3. Max-Planck-Institut für Radioastronomie, Bonn
+. Now at MPIfR, Bonn

ABSTRACT

Results are presented of a study of the outer Galaxy velocity field. We compiled a catalogue of HII regions and reflection nebulae with $230^{o} \leqslant l \leqslant 305^{o}$ and $|b| \leqslant 15^{o}$. For these objects heliocentric distances and radial velocities have been determined. By combining these data with similar data from the literature, for $l < 230^{o}$, we construct the velocity field of the outer Galaxy in the second and third quadrants. This field includes non-circular motions and is therefore very useful for the determination of kinematic distances.

A curve is fitted to the data, and it is shown that:
1. The rotational velocity is slightly rising as a function of galactocentric radius R, out to the last observed point (R≈20 kpc);
2. The rotation curves derived from objects with $0^{o} \leqslant l < 180^{o}$ and $180^{o} \leqslant l < 360^{o}$ respectively are identical, which implies the velocity field is axi-symmetric.

Non-circular motions are revealed by subtracting the velocity field, determined by the rotation curve, from the observed velocity field. The distribution of the velocity residuals, projected on the galactic plane, is consistent with a streaming pattern as predicted by the density wave theory.

1. INTRODUCTION

The rotation curve of the Galaxy is the azimuthally smoothed average of the velocity field, and presupposes rotation in circular orbits.

The method that is used to determine the rotation curve depends on the galactic quadrant that is investigated. Kwee et al. (1954) and Kerr (1962) were the first to measure the rotation curve in the inner Galaxy, using the HI gas at the tangent points. Schmidt (1965) used the results of Kwee et al. in addition to other (stellar) data to construct a mass model of the Galaxy. Outside the solar circle, this model results in a gradually decreasing rotational velocity. The Schmidt rotation curve has been the canonical rotation curve of our Galaxy for a long time.

The HI velocity fields of external galaxies, measured by Bosma (1978), indicate that the great majority of the rotation curves are flat or rising out to large distances from the centers of the galaxies. But also for our own Galaxy there was mounting evidence that the rotation curve outside the solar circle is not falling: from O-B5 stars (Rubin et al., 1962 and Rubin and Burley, 1964); from Cepheids (Kraft and Schmidt, 1963); from HII regions (Georgelin and Georgelin, 1976); from HI (Knapp et al., 1978); from globular clusters (Hartwick and Sargent, 1978).

The past decade, a number of studies using HII regions have shown that the rotation curve of the outer Galaxy rises to at least $2R_o$ (Jackson et al., 1979; Blitz, 1979; Chini and Wink, 1984; Fich et al., 1987). Other studies have confirmed this trend (e.g. Schneider and Terzian, 1983). Presently, the flat or slightly rising rotation curve has taken the place of the Schmidt curve. An important consequence of this is that the mass of the Galaxy is much larger than what was previously thought.

Most of the studies mentioned above are in general limited to nebulae with $l \lesssim 240^o$, and it is assumed that the rotation curve is the same at all longitudes. Verification of this assumption requires an independent measurement of the southern hemisphere velocity field in the outer Galaxy. The present paper summarizes the results of such an independent measurement and discusses the overall velocity field of the outer Galaxy, which is derived by combining data from all four galactic quadrants.

2. METHOD

A rotation curve is derived by independently determining the distances and velocities of a suitably chosen set of objects. The most useful objects for the outer Galaxy are HII regions and their associated molecular clouds (see Blitz, 1979). Reflection nebulae can also be used. The nebula guides one to the stars associated with it, the distances of which can be determined (spectro-) photometrically; for HII regions the stars are young, and intrinsically bright, which allows them to be seen over large distances. The velocity is that of the molecular material, that is associated with almost all HII regions and reflection nebulae. The molecular clouds are massive ($\gtrsim 10^4$ M_o), and as a consequence are little affected by small gravitational perturbations; they are easily and accurately detected in CO.

Existing catalogues of HII regions are incomplete in listing small nebulae (diameter$\lesssim$5-10 arcmin), which are potentially the most distant. Therefore, a catalogue of galactic nebulae has been compiled by searching ESO/SRC J-prints, preferentially for small nebulosities. The catalogue contains 400 regions with $230^o \lesssim l \lesssim 305^o$ and with $|b| \lesssim 10^o$ to 15° (Brand et al., 1986).

3. OBSERVATIONS

A. Distances to the exciting stars.

The basic material from which the distances of the nebulae are derived are the colours and magnitudes of the associated stars. We used the VBLUW photometer on the

41

Dutch 91-cm telescope at ESO (Chile) to observe about 1400 stars in and around 223 nebulae (Brand, 1986). Some additional observations were carried out in the UBV-system on the ESO 1-m telescope.

The distance determination procedure involves transforming the observed colours into T_{eff} and log g using model stellar atmospheres, and was tested on well-studied star clusters; the uncertainty in the distance modulus of a single star was found to be about 25% (Brand and Wouterloot, 1987). The largest heliocentric distance derived is about 10 kpc.

B. Velocity of the associated molecular material.

Of the 400 objects in the catalogue, 308 (77%) were observed in CO by Brand et al. (1987), using the 4-m millimeter wave telescope at CSIRO in Sydney. Of these, 76% were detected. In 194 cases (63% of those observed) the CO emission could be associated with the nebula.

The potentially very distant clouds (V_r(CO)$\geqslant$50 km/s; 5% of all observed sources) were mapped to confirm association with the optical nebulae; in other cases, positions off the nebula were observed, to verify that the CO emission decreases in intensity away from the nebula.

3. RESULTS AND DISCUSSION

A. The sample.

A study of the large-scale characteristics of the Galaxy (such as the velocity field) requires as large a galactocentric azimuth coverage as possible. Therefore, our data set is extended by including nebulae from Blitz et al. (1982) and Chini and Wink (1984). For the inner Galaxy we used HI data at the tangent points (Burton, pers. comm.); the innermost 2 kpc are disregarded, because of the extreme non-circu-

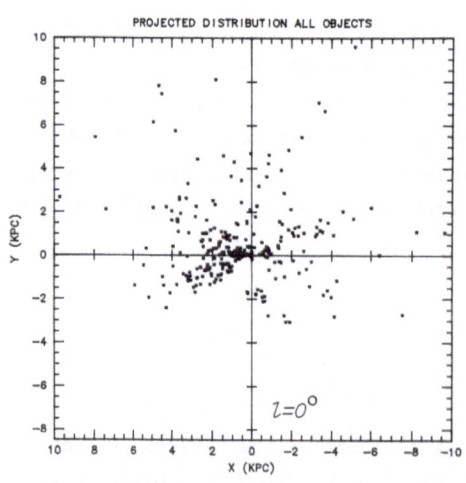

$l=90°$

$l=0°$

Figure 1.

Projected distribution of all objects in the combined data set for which a (spectro-) photometric distance is a-vailable. The Sun is at (X,Y)=(0,0).

larity of the velocities of the stars and gas close to the center. We have grouped all nebulae in this sample that are close in space and velocity into kinematically distinct complexes. The distribution of all nebulae for which we have a distance is shown in Figure 1. Apart from a small section in the fourth quadrant, the coverage of the galactic plane is quite extensive.

B. Observed velocity field.

Figure 2 shows the two-dimensional velocity field (radial velocity V_r). All objects have been redistributed in 1x1 kpc^2 bins, to give smoother contours. Because the observed velocities are shown, Figure 2 *includes non-circular motions*, and can be used to determine kinematic distances to objects for which l and V_r are known.

C. The rotation curve.

Assuming circular orbits for the gas and stars, the measured radial velocity V_r of an object at galactic coordinates l and b is transformed into the circular rotational velocity Θ; the galactocentric distance R is calculated from the heliocentric

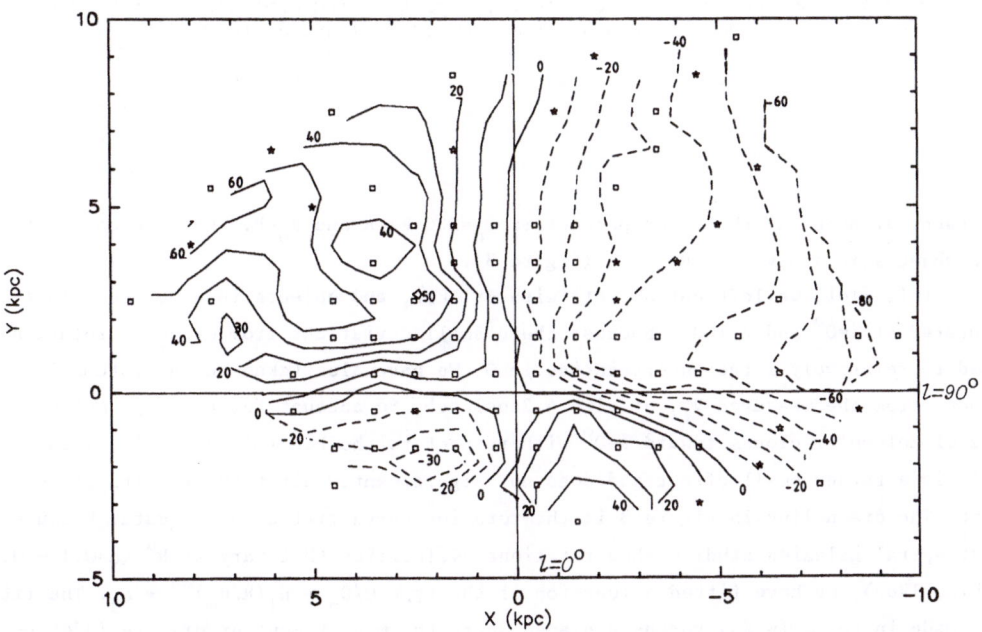

Figure 2. *Observed radial velocities, projected on the galactic plane. The Sun is at (X,Y)=(0,0). Contour values range from −80 km/s to +70 km/s in steps of 10 km/s. To obtain a smooth velocity field, data were averaged in 1x1 kpc^2 bins (bin centers are shown as □). For some locations in the galactic plane data are lacking or sparse; here some points are added (☆), interpolated in order to produce well-behaved contours in these regions. The velocity field shown here includes non-circular motions, and allows determination of kinematic distances.*

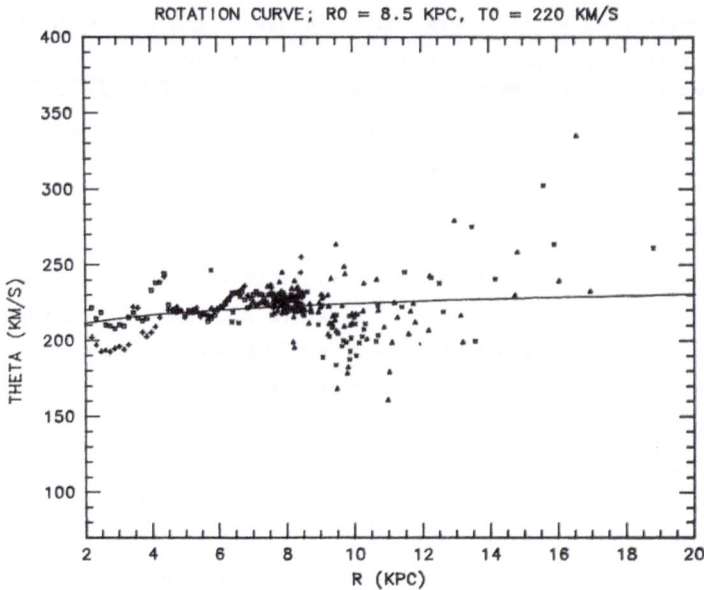

Figure 3. *Circular velocity Θ as a function of galactocentric distance R for all objects in the limited data set (see text), together with Burton's inner Galaxy HI data. Distinction is made between Northern ($0^{\circ} \leq l < 180^{\circ}$) and Southern ($180^{\circ} \leq l < 360^{\circ}$) Hemisphere data: Northern HI (+), Southern HI (□), Northern HII/CO (x), Southern HII/CO (Δ). The drawn line is the power law curve fitted through the data.*

distance d. Adopting the solar parameters Θ_o=220 km/s and R_o=8.5 kpc, we can plot the objects in a Θ versus R graph (Figure 3).

In this plot, we left out objects with d<1 kpc, and objects in a 30° cone in l, centered at $l=0^{\circ}$ and $l=180^{\circ}$, because there small deviations from circular rotation lead to relatively large uncertainties in Θ. We have also taken out a systematic component from the measured V_r, of size $4.2\cos l\cos b$, to account for the apparent streaming of molecular clouds toward $l=0^{\circ}$ with respect to the LSR (Blitz <u>et al.</u>, 1980). This is a rather small effect and does not significantly alter the results presented here. The drawn line in Figure 3 is the rotation curve fitted to the data. Because most spiral galaxies studied show rotational velocities that vary as R^{α} ($\alpha \approx 0.1 - 0.2$; Rubin, 1983), we have fitted a function of the type $\Theta/\Theta_o = a_1(R/R_o)^{a_2} + a_3$. The fit was made in the ω (=Θ/R) versus R plane, where the measurement errors are independent, and the points were weighted inversely with their uncertainties.

The rotational velocity increases slightly with R. Note that the outer Galaxy data points suggest a steeper rise than what is implied by the fitted curve. The distances to most of these regions depend on one or two stars only and as a consequence have a larger uncertainty and a smaller weight. Binning the data in R shows the significance of this rise, however (see Brand, 1986). The mass of the Galaxy, implied by this curve is about 4.5×10^{11} M$_\odot$.

Making a separate fit for objects with $0^\circ \leqslant l < 180^\circ$ and $180^\circ \leqslant l < 360^\circ$ respectively, shows that the northern and southern hemisphere rotation curves are identical. This implies that the asymmetry in the outer Galaxy HI distribution (derived under the assumption of an axi-symmetric velocity field; Henderson _et al._, 1982) is probably spatial (i.e. not a consequence of an asymmetry in the velocity field).

D. Velocity residuals and non-circular motions.

The small-scale fluctuations in the data that are not followed by the fitted curve (Figure 3) are the consequence of non-circular motions.

The residual velocity (ΔV_r) of an object is defined as the difference between its observed radial velocity and that which is expected from the rotation curve (here, as before, one systematic term has been removed in advance). These non-circular motions are a combination of a random (cloud-cloud velocity dispersion; σ_{rand}) and a systematic component (streaming; σ_{str}). Both should average out to zero if the number of objects is large and if they cover a large enough area of the galactic disk. Both conditions are met in our sample, and we find $<\Delta V_r> = -0.3 \pm 0.9$ (m.e.) km/s. The dispersion around this value is $\sigma_{res} = 12.1$ km/s. Following the argument above, we can write $\sigma_{res}^2 = \sigma_{str}^2 + \sigma_{rand}^2$. Adopting $\sigma_{rand} = 4$ km/s (Liszt and Burton, 1983) leads to

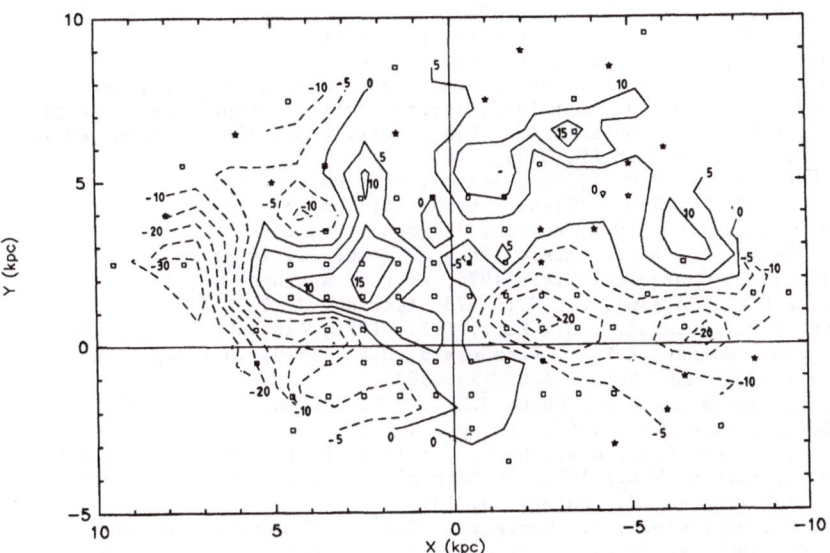

Figure 4. _Radial velocity residuals, projected on the galactic plane. The Sun is at $(X,Y)=(0,0)$. Contour values range from -30 km/s to +15 km/s in steps of 5 km/s. The data were averaged in 1x1 kpc^2 bins. Symbols □ and ★ as in Figure 2._

σ_{str}=11.4 km/s. If the streaming motions are isotropic, we find $\sigma_{str} \approx$20 km/s. This compares well with what is found in external galaxies (e.g. Marcelin et al., 1985).

Binning the residuals in 1x1 kpc^2 bins, and projecting them onto the galactic plane, leads to Figure 4. This iso-velocity contour plot demonstrates that the residuals in the molecular gas are not randomly scattered over the plane, but rather they are organized in a large-scale pattern. The same is found for stars, HI and HII gas, as discussed by Burton and Bania (1974a, b). They showed that the HI residuals can be reproduced in a satisfactory way by assuming the local gas is immersed in a flow pattern as predicted by the linear density wave theory. Their results show a good overall agreement with the data in Figure 4, although the details differ. This is a good indication for organized streaming motions, as predicted by the density wave theory, from these kinematic data. No obvious large-scale structure is present in the distribution of the young stars in the data set, however (see Brand, 1986).

ACKNOWLEDGEMENTS

This work was supported in part by The Netherlands Foundation for Astronomical Research (ASTRON), grant no. 407/83 from the NATO Scientific Affairs Division, grants nos. AST 80-21283 and AST 83-15276 from the U.S. National Science Foundation, by the Kerkhoven-Bosscha Fonds, and the Alfred P. Sloan Foundation (L.B.).

REFERENCES

Blitz, L.: 1979, Astroph. J. Lett. 231, L115
Blitz, L., Fich, M., Stark, A.A.: 1980, in I.A.U. Symp. 87, p. 213
Blitz, L., Fich, M., Stark, A.A.: 1982, Astroph. J. Suppl. Ser. 49, 183
Bosma, A.: 1978, Ph.D. Thesis, University Groningen
Brand, J.: 1986, Ph.D. Thesis, University Leiden
Brand, J., Blitz, L., Wouterloot, J.G.A.: 1986, Astron. Astroph. Suppl. Ser. 65, 537
Brand, J., Wouterloot, J.G.A.: 1987, Astron. Astroph. Suppl. Ser., in press
Brand, J., Blitz, L., Wouterloot, J.G.A., Kerr, F.J.: 1987, Astron. Astroph. Suppl. Ser. 68, 1 (Erratum in 69, 343)
Burton, W.B., Bania, T.M.: 1974a, Astron. Astroph. 33, 425
Burton, W.B., Bania, T.M.: 1974b, Astron. Astroph. 34, 75
Chini, R., Wink, J.E.: 1984, Astron. Astroph. 139, L5
Fich, M., Blitz, L., Stark, A.A.: 1987, Astroph. J., in press
Georgelin, Y.M., Georgelin, Y.P.: 1976, Astron. Astroph. 49, 57
Hartwick, F.D.A., Sargent, W.L.W.: 1978, Astroph. J. 221, 512
Henderson, A.P., Jackson, P.D., Kerr, F.J.: 1982, Astroph. J. 263, 116
Jackson, P.D., Moffat, A.F.J., FitzGerald, M.P.: 1979, in I.A.U. Symp. 84, p. 221
Kerr, F.J.: 1962, Mon. Not. R.A.S. 123, 327
Knapp, G.R., Tremaine, S.D., Gunn, J.E.: 1978, Astron. J. 83, 1585
Kraft, R.P., Schmidt, M.: 1963, Astroph. J. 137, 249
Kwee, K.K., Muller, C.A., Westerhout, G.: 1954, Bull. Astron. Inst. Neth. 12, 211
Liszt, H.S., Burton, W.B.: 1983, in "Kinematics, Dynamics, and Structure of the Milky Way" (ed. W.L.H. Shuter), p. 135
Marcelin, M., Boulesteix, J., Georgelin, Y.P.: 1985, Astron. Astroph. 151, 144
Schmidt, M.: 1965, in "Galactic Structure" (St. Stel. Sys. Vol. V), p.513
Schneider, S.E., Terzian, Y.: 1983, Astroph. J. Lett. 274, L61
Rubin, V.C. et al.: 1962, Astron. J. 67, 491
Rubin, V.C., Burley, J.: 1964, Astron. J. 69, 80
Rubin, V.C.: 1983, in I.A.U. Symp. 100, p. 3

THE IMPORTANCE OF STELLAR DISTANCES IN DETERMINING THE
ROTATION CURVE OF THE OUTER GALAXY

Anthony F.J. Moffat

Département de physique, Université de Montréal;

and Observatoire Astronomique du mont Mégantic

C.P. 6128, Succ. A, Montréal, PQ H3C 3J7, Canada

Abstract

A comparison is made of the effect of random errors in radial velocity and in distance of HII regions on the determination of the Galactic rotation curve beyond the solar circle. Even if more precise CO velocities are not available for the associated nebulosity, <u>stellar</u> radial velocities <u>can</u> be used with no significant disadvantage for galactic longitudes $\ell \lesssim 135°$ and $\ell \gtrsim 225°$, due to the unavoidably large errors in the photometric stellar distances. To improve the distance estimates, a plea is made for high quality stellar photometry using CCD detectors, especially of the main sequences of young <u>groups</u> of luminous stars. A key object is the 10 kpc distant HII region S289 at $\ell = 219°$.

1. Effects of Random Errors on $\Theta(R)$

The determination of the galactic rotation curve $\Theta(R)$ well beyond the solar circle ($R \gg R_0$) requires young, luminous disk objects with reliably measured local standard of rest (LSR) radial velocities V_r <u>and</u> distances from the sun r. Clearly, one should select a class of objects for which random errors in V_r and r (σ_{V_r} and σ_r) are as small as possible. Experience shows that HII regions are probably the best candidates, V_r's being derived from nebular emission lines (Hα or especially CO) and r from photometry and spectroscopy of their exciting OB stars. Note that most HII regions in the outer Galaxy have probably already been identified from previous deep, optical Schmidt-telescope surveys (cf. Fich 1987). The best HII regions are those that contain <u>several</u> stars, preferably forming a cluster-like main sequence in a colour-magnitude diagram (cf. Moffat, FitzGerald, and Jackson 1979: MFJ). HII regions are known to reach out to R $\approx$ 18 kpc (assuming R $\approx$ 8.5 kpc), which is normally taken to be the outer limit of the optical disk.

Assuming these objects to follow circular rotation, one has the usual relation in the LSR system:

$$V_r = R_0(\Theta/R - \Theta_0/R_0) \sin \ell, \quad \text{with } R^2 = R_0^2 + r^2 - 2r\,R_0 \cos \ell.$$

This can be rewritten as

$$\Theta(R) = (V_r/\sin \ell + \Theta_0) R/R_0.$$

For a given direction ℓ, errors in V_r and r, assumed to be mutually independent, will project to errors in Θ of σ_1 and σ_2, respectively, where

$$\sigma_\Theta^2 = \sigma_1^2 + \sigma_2^2, \text{ and}$$

$$\sigma_1 \equiv \sigma_{V_r} \frac{\partial \Theta}{\partial V_r} = \sigma_{V_r} R/(R_0 \sin \ell),$$

$$\sigma_2 \equiv \sigma_r \frac{\partial \Theta}{\partial r} = \sigma_r (r - R_0 \cos \ell) \Theta/R^2.$$

Figure 1 shows σ_1 and σ_2 versus R for:

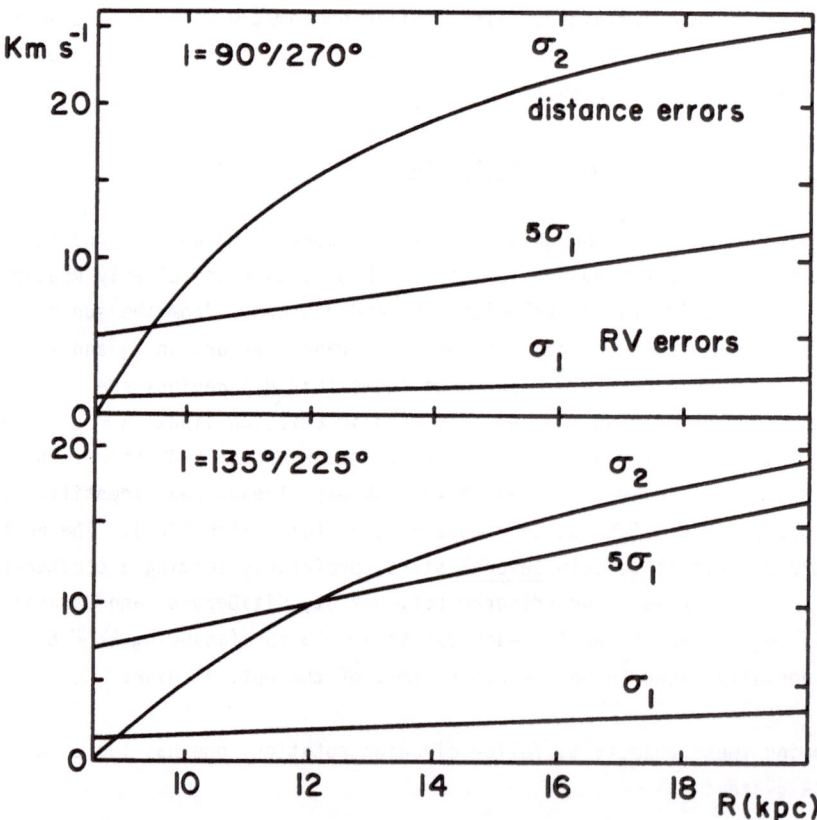

Figure 1: Error components of $\Theta(R)$ due to random errors in radial velocities (σ_1) and distances (σ_2) for two pairs of values of Galactic longitude, for $\sigma_{V_r} = 1$ km s^{-1}, and for $\sigma_{DM} = 0.3$ mag.

$R_0 = 8.5$ kpc,

$\Theta(R) = \Theta_0 = 220$ km s^{-1} (flat rotation curve),

$\sigma_{V_r} = 1$ km s^{-1} (typical for CO V_r's) and

$\quad\quad = 5$ km s^{-1} (typical for stellar V_r's based on ~ 4 stars),

$\sigma_r/r = \sigma_{DM}/(5 \log e)$, with error in distance modulus $\sigma_{DM} = 0.3$ mag (the smallest value possible?) for a small star cluster, and

$\ell = 90°/270°$ and $135°/225°$ (representative longitudes).

We note that:

(a) for $\ell = 90°/270°$, the error in $\Theta(R)$ is completely dominated by uncertainties in the stellar distances (σ_2) beyond $R \approx 11$ kpc,

(b) for $\ell = 135°/225°$, both sources of error are of similar magnitude when V_r and r are based entirely on stars; when CO V_r's are used, errors from stellar distances dominate ($\sigma_2 \gg \sigma_1$), and

(c) for $\ell \rightarrow 180°$, $\Theta(R)$ becomes indeterminable, as expected, even with good V_r's.

We conclude that:

(i) at the outer limit of the optical disk ($R \approx 18$ kpc), random errors amount to ~ 10% in $\Theta(R)$ in the best cases, and

(ii) even if more precise CO V_r's are not available (e.g. nebular emission too faint), stellar V_r's can be used with no significant disadvantage for $\ell \lesssim 135°$ and $\ell \gtrsim 225°$ and possibly even closer to $\ell = 180°$. I am not sure that this latter point has always been appreciated in the recent literature.

2. Improving $\Theta(R)$

(a) From radial velocities

Recent derivations of $\Theta(R)$ using CO V_r's (eg. Blitz 1979; Brand et al. 1987) have excluded HII regions that are too faint to be detected in CO. Now it seems that on the basis of the above remarks, there is little reason not to include at least some of these regions using stellar V_r's as long as there are enough stars in each region. Beyond this, it is difficult to predict any vast improvements in σ_{V_r} and hence σ_1 over present techniques.

(b) From distances

To improve distance estimates, it is possible that careful imagery with CCD detectors and crowded field photometry algorithms may lead to significant improvements. This may apply particularly to the more useful groups of stars embedded in HII nebulosity, where previous techniques (photographic and photoelectric diaphragm photometry) are less accurate.

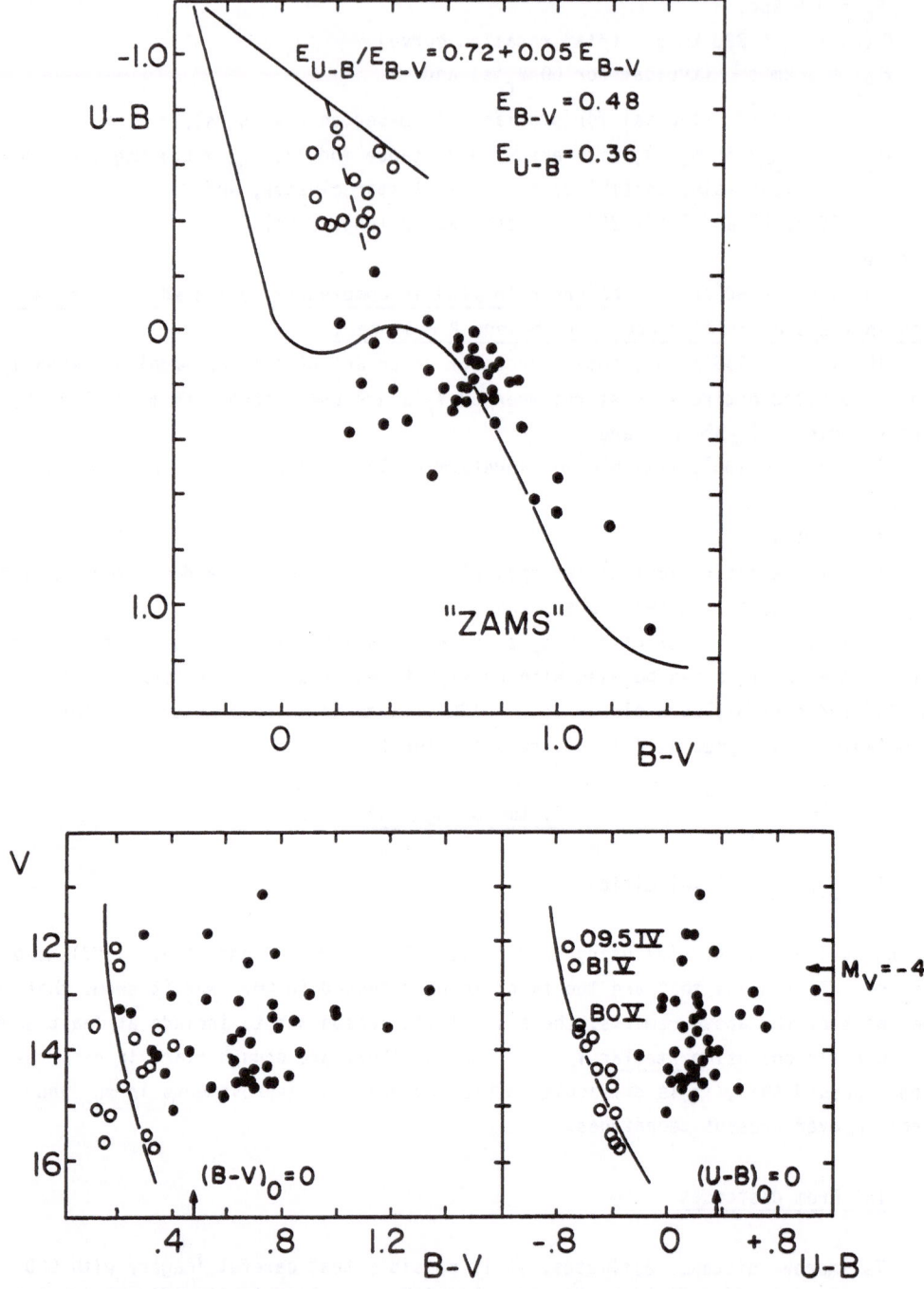

Figure 2: Photographic colour-colour and colour-magnitude diagrams of all stars down to U = 15.5 in a 10' diameter field centred on S 289. Blue cluster members are identified by larger symbols, some with MK spectral types. The zero age main sequence of Schmidt-Kaler (1982) has been fitted with $V - M_V = 16.5$ and $E_{B-V} = 0.48$.

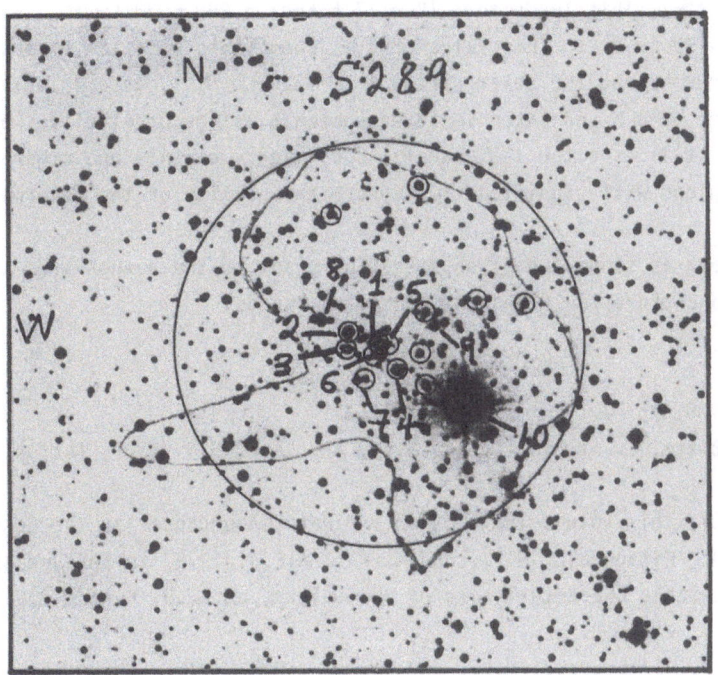

Figure 3: Chart of the faint HII region S 289 (1 = 218.8°, b = -4.6°) from MFJ showing the 10' diameter circle in which all stars to U = 15.5 have now been measured.
The 14 cluster members are circled. The bright star (no. 10) is a foreground A
star. The irregular region of nebular emission is outlined.

(c) Test case: S 289

Of all the HII regions beyond the solar circle, S 289 stands out for its large
distance and the homogeneous group of moderately hot stars it contains. As a sequel
to the pioneering study of MFJ and Jackson, Fitzgerald and Moffat (1979: JFM), and
as a prelude to a future CCD study, I have obtained UBV plates of this region at
the mont Mégantic Observatory. The photographic magnitudes were calibrated by the
photoelectric data of MFJ. The resulting colour-colour and colour-magnitude diagrams are shown in Fig. 2 and a chart of S 289 in Fig. 3.

From the clearly defined cluster main sequence of 14 member stars in Fig. 2, I
deduce $V_0 - M_v$ = 15.0 ± 0.3, hence r = 10.0 ± 1.4 kpc and R = 17.5 ± 1.3 kpc. This
places S 289 among the few, very distant known HII regions near the outer limit of
the optical disk. Because its distance is well determined, S 289 is a key object.
CCD photometry of S 289 to even fainter magnitudes could probably improve the distance estimate.

Since there are no published CO or H α V_r's for this faint HII region, we use
the mean stellar LSR velocity of JFM based on repeated spectra of several stars: V_r

= 47 ± 7 km s^{-1}. This leads to Θ (R = 17.5 kpc) = 298 ± 33 km s^{-1}, with σ_1 = 23 km s^{-1} and σ_2 = 23 km s^{-1}. This value of Θ is compatible with the rising Galactic rotation curve beyond the solar circle (Θ_0 = 220 km s^{-1}), as originally found out to this distance by JFM and later improved somewhat using CO velocities by Blitz and coworkers. It is expected that careful CCD imagery of this and other key cluster-like HII regions will significantly improve the quality of the determination of Θ (R).

This work is supported by an operating grant to the author from the Natural Sciences and Engineering Research Council of Canada.

References

Blitz, L. 1979, Ap. J. (Lett.), 231, L115.

Brand, J., Blitz, L., Wouterloot, J.G.A., and Kerr, F.J. 1987, Astron. Ap., Sup., 68, 1.

Fich, M. 1987, this volume (Extinction and Metal Abundances in the Outer Galaxy).

Jackson, P.D., Fitzgerald, M.P., and Moffat, A.F.J. 1979, in IAU Symposium No. 84, The Large-Scale Characteristics of the Galaxy, ed. W.B. Burton (Dordrecht: Reidel), p. 221.

Moffat, A.F.J., Fitzgerald, M.P., and Jackson, P.D. 1979, Astron. Ap. Sup., 38, 197.

Schmidt-Kaler, Th. 1982, Landolt-Bornstein, ed. K.H. Hellwege (Berlin: Springer), New Ser., Group 6, Vol. 2b, p. 1.

METAL WEAK STARS AND THE GALACTIC CIRCULAR VELOCITY

K.C. Freeman

Mount Stromlo and Siding Spring Observatories

The Australian National University

Abstract

The metal weak halo stars are a slowly rotating population. If we assume that their rotation is in the same sense as that of the disk, then their solar motion gives a lower limit to the galactic circular velocity. For a sample of nearby non-kinematically selected metal weak stars, this limit is about 175 ± 12 km s^{-1}. However, about 25 percent of the metal weak stars belong not to the halo but to a rapidly rotating "thick disk" population. Excluding these stars from the halo sample gives a more useful lower limit of 220 ± 10 km s^{-1} on the galactic circular velocity.

Introduction

The metal weak stars of the galactic halo are part of a slowly rotating population (see Norris 1986 for an overview): for an adopted galactic circular velocity Θ_o of 220 km s^{-1}, the mean rotation of objects with [Fe/H] < -1.2 is only about 40 km s^{-1}. Conversely, the solar motion of these slowly rotating metal weak stars can be used to derive a lower limit on Θ_o, if we accept that the rotation of the halo is in the same sense as the rotation of the galactic disk. This lower limit on Θ_o (of about 180 km s^{-1} from the numbers just presented) is not very useful, because there is little evidence that Θ_o could be so low (see Kerr and Lynden-Bell 1986 for a review). However, Norris et al (1985: hereafter NBP) investigated a sample of spectroscopically selected nearby metal weak stars, and showed that about 25 percent of these stars belong to a rapidly rotating "thick disk" population that had not been

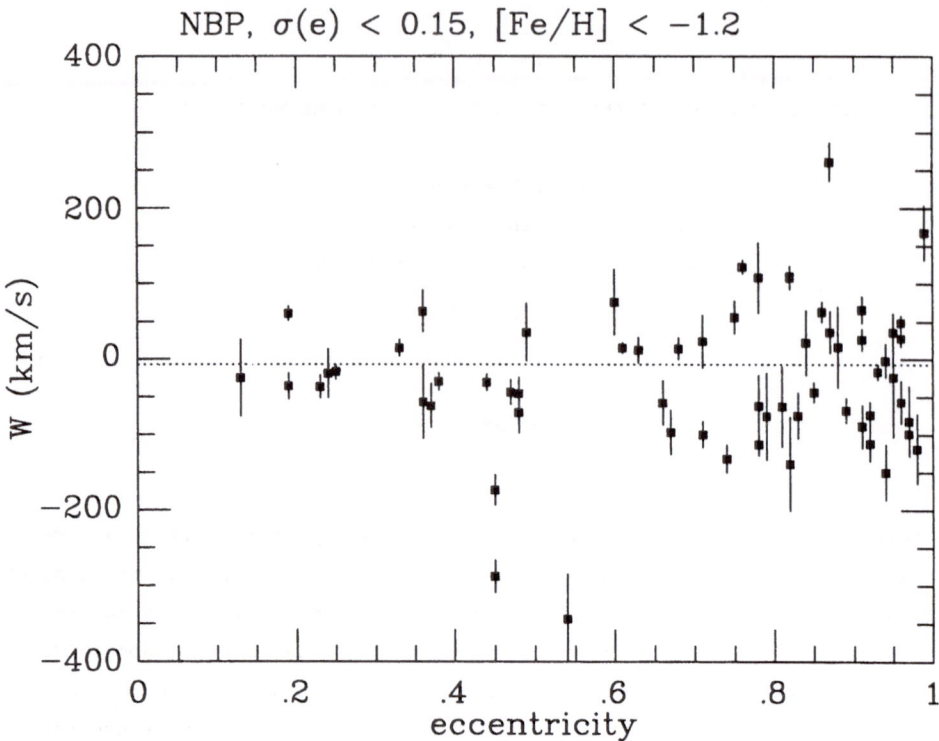

NBP, $\sigma(e) < 0.15$, $[Fe/H] < -1.2$

Figure 1: The W-velocity, eccentricity distribution for the NBP stars with $\sigma(e) < 0.15$ and $[Fe/H] < -1.2$. The dotted line is at $W = -7$ km s^{-1}.

previously identified in the earlier samples of kinematically selected metal weak stars. If we wish to derive the solar motion for the metal weak <u>halo</u> alone, then these "disk" stars should obviously be removed from the sample. The corresponding limit on Θ_o, which is the subject of this paper, becomes more interesting.

In this discussion we will use a subsample of 63 of the NBP stars, with $[Fe/H] < -1.2$ and with accurate heliocentric UVW velocities (defined in the usual sense), such that the orbital eccentricities e (as defined by Eggen et al 1962, hereafter ELS) have standard errors $\sigma(e) < 0.15$. We take the peculiar motion of the sun relative to the LSR to be $(-10, +15, +7)$ km s^{-1}. Figure 1 shows the (W-e) plane for these stars. We see that stars with low e < 0.4 also have a low vertical velocity dispersion σ_W. The table below compares the mean value of V and σ_W for the whole subsample and for stars with e < 0.4.

all: $\langle V \rangle$ = -191 ± 13 km s^{-1} σ_W = 100 ± 9 km s^{-1}

e < 0.4:   - 41 ± 10 km s^{-1} 45 ± 10 km s^{-1}

The mean kinematics for the whole subsample are typical for the galactic halo: low rotation and high σ_W. However, the stars with e < 0.4 are clearly part of a rapidly rotating disk population, as shown by NBP, although they are all metal weak. (We note here that a set of low-e stars need not have rapid mean rotation, but in this sample **all** stars with e < 0.4 have prograde orbits.)

We wish to derive the solar motion for a sample of metal weak stars that belong to the slowly rotating halo; clearly, stars of the metal-weak thick disk (which we will refer to as disk stars hereafter) have no place in such a sample. How do we remove these disk stars from our sample without biasing the derived solar motion for the remaining halo stars ? We present three partly independent procedures which give consistent results for the mean V motion of the slowly rotating halo.

Procedure and Results

Guided by Figure 1, we form a tentative halo sample from stars with e $\geqslant$ 0.5 plus any stars of lower e that have |W| > 100 km s^{-1}. For this halo sample,

$\langle V \rangle$ = -235 ± 10 km s^{-1} σ_W = 113 ± 11 km s^{-1}

Figure 2 shows the (U,V) distribution for this halo sample (filled symbols) and for the remaining disk stars (open symbols). Taking all the stars together, this distribution appears asymmetric. There are more stars of low |U| at the right side of the figure than at the left. This excess comes from the rapidly rotating disk stars which have a small velocity dispersion and, as mentioned above, have no retrograde counterpart.

For a pure well-mixed slowly rotating halo population, we would expect the (U,V) distribution to be approximately symmetrical about U = 10 km s^{-1} and a value of V somewhere between -200 and -250 km s^{-1} (depending on the value of Θ_o and the rotation rate of the slowly rotating halo). We can use this expectation of symmetry as the basis of the <u>first procedure</u> for removing the metal weak disk stars from our halo sample.

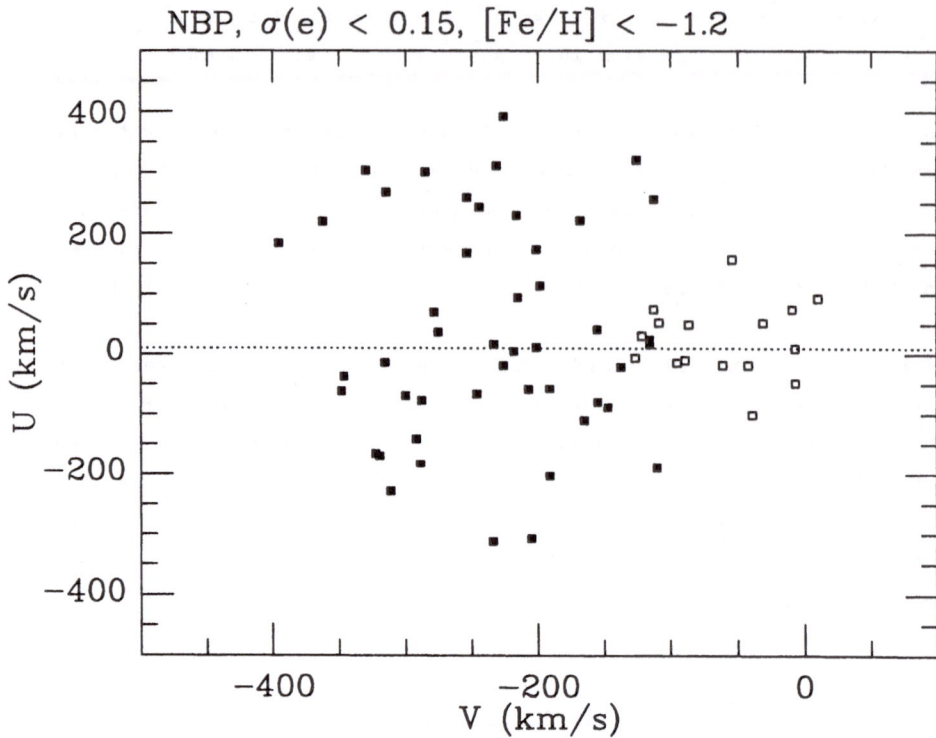

Figure 2: The U-velocity, V-velocity distribution for the NBP stars. The <u>open symbols</u> represent stars of the metal weak disk, defined tentatively by e < 0.5 <u>and</u> $|W| < 100$ km s^{-1}. The <u>filled symbols</u> represent the remaining halo stars. Note that all of the disk stars have prograde orbits. The dotted line is at $U = 10$ km s^{-1}.

We will take disk stars to be those with e < e_{max} <u>and</u> $|W| < 100$ km s^{-1}, and we wish to find the most appropriate value of e_{max}. If e_{max} is taken too small, then the halo sample will still be contaminated by disk stars. If e_{max} is taken too large, then the value of <V> for the remaining halo sample will be biased towards the circular velocity used to calculate the orbital eccentricities (eg the ELS eccentricities are calculated for $\Theta_o = 250$ km s^{-1}). We use the following procedure to estimate the appropriate value of e_{max}. For each value of e_{max}, exclude the disk stars and calculate <V> for the remaining halo sample. Then calculate $(\Sigma U^2)_L$, defined as (ΣU^2) for halo stars with V < <V>, and similarly $(\Sigma U^2)_R$ for stars with V > <V>. For a sample of stars that has a symmetrical (U,V) distribution in the above sense, the ratio $(\Sigma U^2)_L / ((\Sigma U^2)_R$ has an expectation value of unity. Figure 3 shows how this

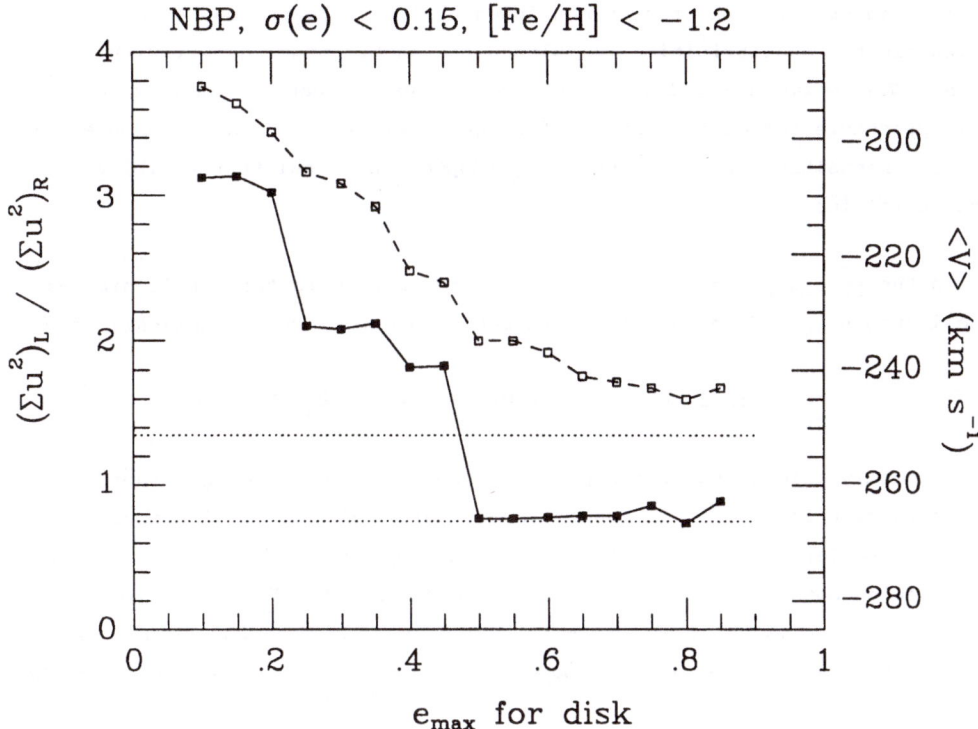

Figure 3: The <u>filled symbols</u> and solid line show how the ratio $(\Sigma u^2)_L/((\Sigma u^2)_R$ for the NBP <u>halo</u> stars varies with the maximum eccentricity adopted for the metal weak disk stars. The dotted lines show the 50% probability values for this ratio for samples drawn from a symmetrical gaussian distribution in (U,V). The halo sample becomes symmetrical in (U,V) if stars with $e < 0.50$ are excluded from the sample. The <u>open symbols</u> and broken line show how $\langle V \rangle$ for the halo sample varies with e_{max}.

ratio changes with the value of e_{max} that is taken to define the disk stars in our sample of stars. For small e_{max}, the ratio is about 3, because the presence of the disk stars seen at the right side of Figure 2 makes the (U,V) distribution asymmetrical. The horizontal dotted lines in Figure 3 show the 50 percent probability values of this ratio for samples of this size drawn from a symmetrical gaussian distribution in (U,V). Figure 3 also shows how $\langle V \rangle$ for the halo sample changes with e_{max}. From Figure 3, $e_{max} = 0.50$ appears to be an appropriate value: for larger values of e_{max}, the remaining halo sample has an approximately symmetrical distribution in the (U,V) plane. The corresponding value of $\langle V \rangle$ is -235 ± 10 km s^{-1} for the halo sample defined in this way.

Because of the risk of bias mentioned earlier, the procedure was repeated using orbital eccentricities calculated for an ELS model galaxy with $\Theta_o = 220$ km s^{-1}. The value of $<V>$ for the halo sample was unchanged. We conclude that this procedure for estimating $<V>$ for the slowly rotating halo component is fairly insensitive to the value of Θ_o adopted to calculate the orbital eccentricities.

In the second procedure, we identify the disk stars from their orbital inclinations i, defined as if the galactic potential were spherical: ie

$$\sin i \quad = \quad |W+7|/\sqrt{\{(W+7)^2 + (V+15+\Theta_o)^2\}}$$

For randomly orientated orbits passing through the solar neighborhood, the expected distribution of i is uniform. Figure 4 shows the distribution of i for our sample, for $\Theta_o = 250$ km s^{-1}. The disk stars lie in a well-defined low inclination mode that extends over the range $0 \leqslant i \leqslant 35°$. For the remaining halo stars, the value of $<V>$ is -239 ± 11 km s^{-1} for $\Theta_o = 250$ km s^{-1}, or -234 ± 10 km s^{-1} for $\Theta_o = 220$ km s^{-1}. Again, the procedure is insensitive to the adopted value of Θ_o.

The third procedure, which is independent of Θ_o, was suggested by Jesper Sommer-Larsen . From Figures 1 and 2, we see that the disk stars lie within an ellipsoidal region of velocity space given approximately by

$$\{(U-10)/150\}^2 + \{(W+7)/90\}^2 + \{(V+50)/80\}^2 = 1;$$

the units are km s^{-1} as usual. We define a corresponding retrograde region

$$\{(U-10)/150\}^2 + \{(W+7)/90\}^2 + \{(V+50+2\Delta)/80\}^2 = 1$$

which we wish to locate symmetrically relative to the mean V for the halo. Any star lying within either of these ellipsoidal regions is removed from the halo sample. We adjust Δ until the mean V velocity of the two ellipsoidal regions (ie $<V> = -\Delta-50$ km s^{-1}) is the same as the mean V velocity for the halo sample. If the slowly rotating halo is symmetrical in V about some mean value $<V>$, then this procedure gives an estimate for $<V>$. The value is $<V> = -227 \pm 10$ km s^{-1}, where the error is derived from simulations of this procedure.

These three procedures were also applied to larger samples of nearby non-kinematically selected metal weak stars, compiled by John Norris and by

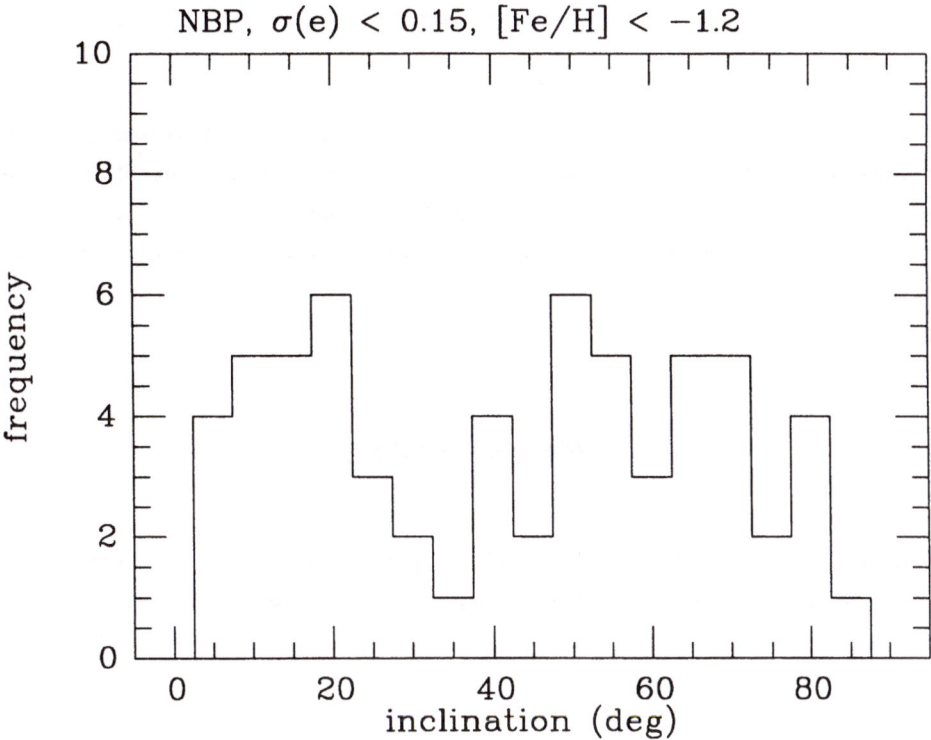

<u>Figure 4</u>: The frequency distribution of orbital inclinations for the NBP stars, calculated as if the galactic potential were spherical. The adopted value of Θ_o is 250 km s^{-1}. The stars of the metal weak disk lie in the low inclination mode, between 0 and 35° inclination.

Jesper Sommer-Larsen. The velocity errors for these samples are somewhat larger than for the restricted NBP sample used above. The results are within the errors of those for the NBP sample: the values of <V> for the halo are slightly more positive, which is in the sense expected if there is some residual contamination by disk stars, resulting from the larger velocity errors.

Conclusion

We adopt $\langle V \rangle = -235 \pm 10$ km s^{-1} as the mean heliocentric V-velocity of the metal weak halo, after removal of the rapidly rotating component of metal weak stars identified by NBP. Then, unless the metal weak halo is in retrograde rotation, the lower limit on the galactic circular velocity near the sun is 220 ± 10 km s^{-1}.

Acknowledgements

I am grateful to John Norris and Jesper Sommer-Larsen for discussions, suggestions and for the use of their compilations of metal weak stars.

References

Eggen, O.J., Lynden-Bell, D., Sandage, A. 1962. Astrophys.J., 136, 748-66.

Kerr, F.J., Lynden-Bell, D. 1986. Mon.Not.R.astr.Soc., 221, 1023-38.

Norris, J.E. 1986. Astrophys.J.Suppl., 61, 667-98.

Norris, J.E., Bessell, M.S., Pickles, A.J. 1985. Astrophys.J.Suppl., 58, 463-92.

A WAVY ROTATION CURVE AND CONSEQUENCES THEREOF

Paris Pişmiş
Instituto de Astronomia - UNAM

ABSTRACT

On the basis of two observationally supported assumptions, namely
that our galaxy has a double logarithmic spiral structure and that the
rotation curve has undulations due to the higher rotational velocity
at the spiral arms, it is shown that the rotation curve will depend on
direction from the galactic center, and Oort's parameter A will be direction-dependent as well.

I. INTRODUCTION

In this paper I shall first present evidences for the existence
of waves in the rotation curves of galaxies in general and of our Galaxy in particular; following that, I shall discuss some effects that
an undulating rotation curve of a spiral will have on the derived local as well as global parameters of the Galaxy.

The premises on which my arguments will be based are essentially
the following:

1)   Spiral structure is coexistent with an otherwise smooth distribution of an older population.

2)   Spiral galaxies, in particular the later types, including our Galaxy, show a wavy rotation curve of which the maxima and minima are
related to the arm and interarm regions, respectively. Premise 1 is
widely accepted and needs no further justification. Premise 2 merits
attention, to which the following section will be devoted.

II. PREMISE 2. DISCUSSION OF ROTATION CURVES.

In the early sixties rotation curves of around 30 spiral galaxies
were given in a series of papers by Burbidge, Burbidge and Prendergast
(for a listing and references see Burbidge and Burbidge 1975). Despite the scatter of the points there was indication of irregularities
in the form of waves along the mean rotation curves of a large percentage of the spirals. I had pointed out at the time that such a circum-
stance was expected (excluding the nuclear region) if spiral galaxies
were composite systems where the different subsystems -spiral features

61

vs. a smooth older population- rotated at different speeds. Thus at
the interarm one would observe essentially the slower rotating subsystem (220 - 230 km s^{-1}) while at the arms one would observe essentially the faster rotating objects ≈250 km s^{-1} (Pişmiş 1965, 1966). In
a later communication I argued, based on the hydrodynamical flow equation along the radial direction, that waves in the rotation curve were
entirely compatible with a steady state galaxy -a direct consequence
of the co-existence of two or more populations. A crucial test was
also suggested allowing to distinguish between my interpretation of
the waves and the alternative explanation in terms of streaming motions along spiral arms predicted by the density wave theory of spiral
structure (Pişmiş 1974). The observed effect may well be a combination of the two.

The work summarized above did not find echo at the time; even
the physical reality of the waves was doubted. However, the existence
of "undulations" in the rotation curves of Sc galaxies was later established with high precision velocity data by Rubin and co-workers (for
a review see Rubin 1983; also Pişmiş 1981). Further evidence for undulations is found in the H I velocity fields of a number of galaxies
(Bosma 1978). The detailed velocity structure of M 81 by Visser (1978)
gives clear indication that the waves are associated with spiral structure. The amplitude of the waves may be up to 20-40 km s^{-1} (≈20 km s^{-1}
in NGC 2998, Rubin et al. 1978 ; ≈40 km s^{-1} in NGC 2903, Simkin 1975).

An undulating rotation law is also detected in the Galaxy. Burton and Gordon (1978) showed that their CO velocities within the first
quadrant agree in general with the H I velocities; both are consistent
with a wavy rotation law. Earlier work within the solar circle by Kerr
(1962) also shows variation in the H I velocity-distance relation in
the form of waves; the undulations may be partly due to an assumed pau-
city of atomic hydrogen between the spiral arms, but a sizable fraction of the dips may also be due to the lower rotation of hydrogen residing there.

The extensive survey of CO cloud velocities by Blitz and co-workers (Blitz, 1979) is consistent with the Sun being at the minimum of a
wave. Georgelins' (1976) velocities of Galactic H II regions are also
consistent with undulating rotation curves. The data from O type stars
catalogued by Cruz-González et al. (1974) support the suggestion that
the Sun is located, again, at the minimum of a wave (Pişmiş 1981).

It is worth mentioning here that the rotation curve seems to be
higher, based on velocities from northern (l°=0° to 180°) hemisphere

objects, as compared with those from the southern one; Kerr (1964) pointed out for the first time this north-south asymmetry. The Georgelins find the like tendency using H II region Fabry-Pérot velocities. A rotation curve from O-star data showed a similar discrepancy (Pişmiş 1981). An interpretation advanced by Kerr (1962) is that the kinematics of the Galaxy have an expansional component of around 7 km s^{-1} along with rotation. This author has also found an expansion of the spiral arms (of 4 km s^{-1}) from the center of the galaxy past the Sun based on motions of the OB associations (Pişmiş 1960).

III. CONSEQUENCES DERIVED FROM A WAVY ROTATION CURVE

1)   The Rotation Curve is not unique; it is direction dependent. To show this we consider a model spiral galaxy with two identical logarithmic spirals, symmetrical with respect to the center (the double spiral can be seen in Fig. 1 given later). The equations of the spirals can be represented as follows:

$$R = a \, e^{b(\phi+n\pi)} \quad \begin{array}{l} \text{spiral 1, for } n = 0, \\[6pt] \text{spiral 2, for } n = 1 \end{array} \tag{1}$$

R, ϕ are polar coordinates, a, b, constants, and ϕ increases in the direction of the opening of the spiral. Take any direction ϕ and envision the run of the rotation velocity with distance from the center. With no loss of generality we can use the angular velocity ω throughout, instead of the linear velocity. Along a direction, ϕ_1=constant, the intersections with the spiral structure will be R_{1p} where p=0, 1, ... Now, along a neighboring direction ϕ_2 with $\phi_2 > \phi_1$, the intersections with the spiral features R_{2p} will be such that

$$R_{2p} > R_{1p}$$

This is clearly seen from expression (1).

The intersections, as we assumed, mark the maxima of the waves (the spiral arms) in the rotation curve; thus, as ϕ increases, the waves will be gradually displaced outwards from the center until they come into coincidence at $\phi_1 + \pi$. It is clear that the rotation curve in a spiral galaxy is angle-dependent.

In an ideal external spiral $\omega(R)$ can be obtained all along the apparent "major axis", and the rotation curve at the two sides of the center will be identical since the galactocentric directions differ by π. If an average is taken over rotation curves with $\Delta\phi < \pi$ degrees, the waves which fall at different distances from the center will be gradually effaced as ϕ increases but at the same time the scatter around the average rotation curve will increase becoming comparable to the amplitude of the waves. However, the situation in our Galaxy is different as will be discussed below.

2) <u>The Rotation Law within the solar circle not completely obtainable through terminal velocities</u>.

From our vantage point we are not able to observe the terminal velocities outside of a central angle of $\approx \pm 60°$ at best on each side of the Sun-center line. Figure 1 illustrates this point.

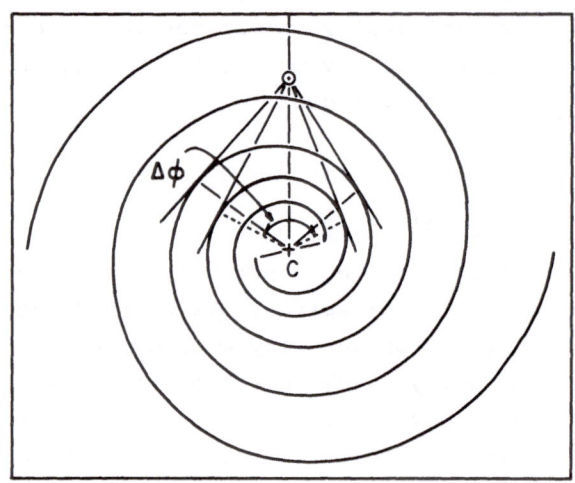

Figure 1. Two symmetrically placed logarithmic spirals with 12° pitch angle to model our Galaxy. The position of the Sun is that of the Georgelins' (1976). The figure also illustrates the argument that it is not possible to observe terminal velocities of spiral arms outside of $\phi \approx \pm 60°$.

Thus in the rotation curve taken as an average within the accessible region, the waves will not be altogether effaced and residual waves will appear though with reduced amplitude. This may account for the shallow waves of the $\omega(R)$ curves of the Galaxy (see for example Fig. 5 in Kraft and Schmidt 1963).

3) <u>Oort's parameter A is direction dependent</u>.

We consider the well-known relation:

$$A = - \tfrac{1}{2} R_0 \left(\frac{d\omega}{dR}\right)_0 .$$

A is thus proportional to the gradient of the angular rotation curve $\omega(R)$ at the Sun. $\frac{d\omega}{dR}$ may be obtained from a free-hand $\omega(R)$ or by fitting a function to represent the curve. For a smooth axis-symmetric velocity field this procedure can be adopted without question. But in a wavy and direction dependent rotation $\frac{d\omega}{dR}$ is rather $\frac{\partial\omega(R,\phi)}{\partial R}$; therefore A will also be direction-dependent. In Figure 2 we present the situation graphically with a surface engendered by the extremities of the vectors $\omega(R)$ placed perpendicular to the underlying plane of the galaxy with the double spiral as sketched in Fig. 1.

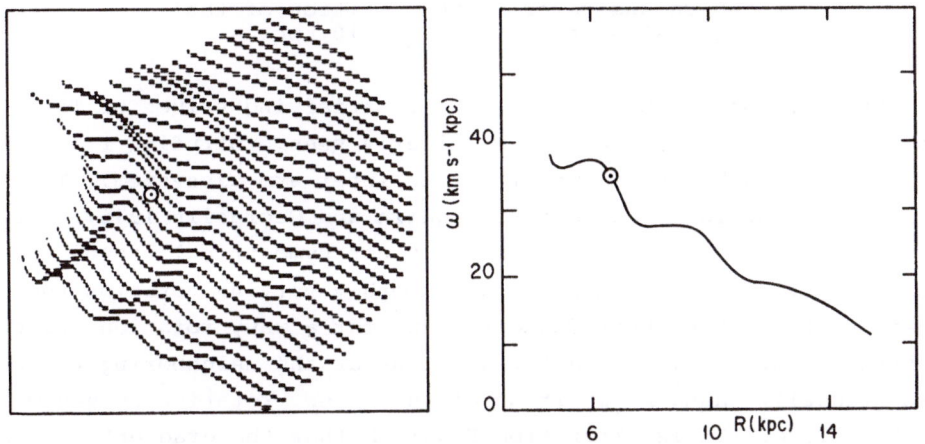

Figure 2. a) Surface representing the wavy rotational velocity field around the marked position of the Sun. Angular velocities, ω, are with respect to that of the outermost interarm population - namely 15.2 km s^{-1} kpc^{-1} at R=14.5 kpc (see Table 1). b) A meridional section of the surface passing through the Sun; this is the angular rotation curve. (ω vs. R).

The surface is constructed assuming that over a wide range of R ($\approx$7 to 15 kpc) the rotation velocity is constant both for arm (250 km s^{-1}) and as well as for interarm (220 km s^{-1}) regions (Pişmiş 1974) thus the average rotation curve is flat. Table 1 gives the pertinent ω values. The position of the arms is that of the double spiral (Fig. 1).

Table 1

Angular Velocities of Adopted Model at
distances of the spiral arms as in Fig. 2.

ω kms^{-1} kpc^{-1} R $\overset{*}{k}$pc	6.8	10.0	14.5
for V = 250 km s^{-1} (arm)	36.7	25	17.5
for V 220 km s^{-1} (smooth subsystem)	32.3	22	15.2
$\Delta\omega$ amplitude of wave	4.4	3.0	2.3

* R values refer to the position of the
spiral arms, with R$_{sun}$ = 10 kpc.

The portion of the surface $\omega(R,\phi)$ satisfying values of Table 1 is
shown in Figure 2. The waves are clearly seen protruding from the
surface defined by the slower rotating inter-arm subsystem. The sun
is slightly outside a maximum (Georgelin and Georgelin 1976). Due to
spiral structure the velocity field is not symmetrical around the sun-
center line. For the conventional smoothly varying rotation the determination of A may be satisfactory. But for a wavy rotation and the
adopted position of the sun the gradient at the neighboring regions,
with gradually changing position of the waves, rotation curves will
not be unique. It is clear from Figure 2 that the gradient of the non-
smooth velocity surface which is an average over velocity data say
1 - 2 kpc from the sun, will be direction-dependent and so will the
parameter A.

On the face of observational errors, random effects and the sim-
plifying assumptions involved, it may seem irrelevant to fuss about a
variation of A with direction. However, the diverse values of A (13 -
15 km s^{-1}) obtained up to the present may perhaps be explained in the
frame of our model. It may also be worthwhile to look into the north-
south asymmetry of galactic rotation in the light of a wavy rotation
law.

REFERENCES

Blitz, L. 1979, Ap. J. Letters 281, L115.

Bosma, A. 1978, Ph. D. Dissertation, University of Groningen.

Burbidge, M.E. and Burbidge, G.R. 1975, in Galaxies and the Universe, eds. A. Sandage, M. Sandage and J. Kristian (Chicago: The University of Chicago Press), p. 81.

Burton, W.B. and Gordon, M.A., 1978, Astron. and Astroph. 63, 7.

Cruz-González, C., Recillas-Cruz, E., Costero, R., Peimbert, M., Torres-Peimbert, S. 1974, Rev. Mexicana de Astron. Astrof., 1, 211.

Georgelin, Y.M. and Georgelin, Y.P. 1976, Astron. and Astrophys. 49, 57.

Kerr, F.J. 1962, Monthly Notices of the R.A.S. 123, 327.

Kerr, F.J. 1964, in The Galaxy and the Magellanic Clouds, IAU Symposium N° 207, eds. F.H. Kerr and A.W. Rodgers (Canberra Australian Academy of Sciences), p. 81.

Kraft, R. and Schmidt, M. 1963, Ap. J. 137, 249.

Pişmiş, P. 1960, Bol. Obs. Tonantzintla y Tacubaya 2, 3.

Pişmiş, P. 1965, Bol. Obs. Tonantzintla y Tacubaya 4, 3.

Pişmiş, P. 1966, in Non Stable Phenomena in Galaxies, IAU Symposium N° 29, ed. V. Ambartzumian (Yerevan: Academy of Sciences of Armenian SSR) p. 429.

Pişmiş, P. 1974, in Galaxies and Relativistic Astrophysics, Proceedings, First European Astronomical Meeting, eds. B. Barbanis and J. D. Hadjidemetriou (Berlin-Springer), p. 133.

Pişmiş, P. 1981, Rev. Mexicana de Astron. y Astrof. 6, 65.

Rubin, V.C. 1983, in Internal Kinematics and Dynamics of Galaxies. IAU Symposium N° 100 (ed. E. Athanassoula) p. 3.

Visser, H.C.D. 1978, Ph. D. Thesis. University of Groningen.

Rubin, V.C., Ford, W.K. and Thonnard, N. 1978, Ap. J. 225 L107.

Simkin, S.M. 1975, Ap. J. 195, 293.

MILKY WAY ROTATION
AND THE DISTANCE
TO THE GALACTIC CENTER
FROM CEPHEID VARIABLES

J.A.R. Caldwell
Mount Wilson and Las Campanas Observatories
813 Santa Barbara Street, Pasadena, CA 91101, U.S.A.

I.M. Coulson
South African Astronomical Observatory
P.O. Box 9, Observatory 7935, South Africa

Abstract: The existing data on Milky Way Cepheids are fitted with an axisymmetric Galactic rotation model. A new "standard candle" m_o is determined: the unreddened apparent magnitude of a hypothetical Cepheid with standard period, color, and metallicity at a distance of 1 R_o. It then follows that R_o = 7.8 ± 0.7 kpc from the cluster Cepheid zero point. Further important results are that $2AR_o$ = 228 ± 19 km s^{-1}, and that the observed Cepheid mean velocities average 3.1 ± 0.9 km s^{-1} too negative compared with the center-of-mass velocities.

1. Data on the Galactic Cepheids

We have compiled the existing data on Galactic Cepheids with regard to photoelectric magnitudes, reddenings, and mean radial velocities (CC87 = Caldwell and Coulson 1987). Nearly complete data exist for about 190 classical Cepheids, and partial data for an additional 80. Figure 1 shows the Galactic coordinates of the well-studied Cepheid sample (black dots) and of those lacking only velocities (crosses).

Good relative distances can be inferred for about 250 Milky Way Cepheids, and the radial velocity pattern can be investigated for about 190 of these. Figure 2 shows the locations of the same Cepheids as in Figure 1, but now projected onto the Galactic plane. The well-observed Cepheids are concentrated near the solar location, consistent with about a 12^m cutoff in past photometry and speedometry. Current observations by J.A.R. Caldwell and P.L. Schechter (MWLCO) are aimed at improving the data on the most distant Milky Way Cepheids and at discovering new members of this category.

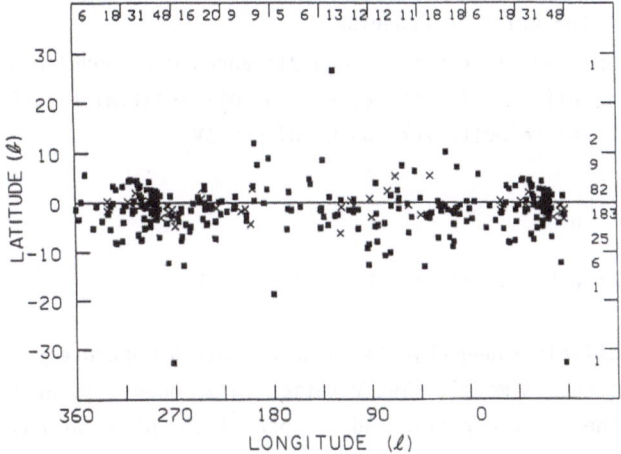

Fig. 1. The Galactic coordinates of 252 Milky Way Cepheids with good distances.

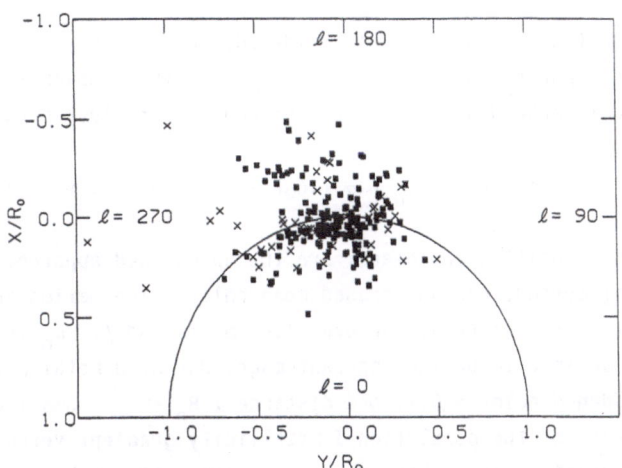

Fig. 2. The locations of 252 Milky Way Cepheids as projected onto the Galactic plane (distances in units of R_0).

Corresponding diagrams (CC87) in Galactocentric cylindrical coordinates show that the sun is located about 35 pc above the median z - plane of the Cepheids. It lies nearly exactly in the median plane of the nearest 15 Cepheids, those with distances less than 470 pc as projected onto the Galactic plane. The Cepheids as a whole suggest some degree of corrugation in the diagram of z versus Galactocentric azimuth.

2. Kinematics of the Galactic Cepheids

The observed (mean) radial velocities and distances were modeled by axisymmetric Galactic rotation, allowing for the effects of differential rotation, solar motion, and a possible radial velocity zero point offset δV_r:

$$V_r = (\omega - \omega_0) \bullet R_0 \sin\ell \bullet \cos b \qquad (1)$$

$$- u_0 \bullet \cos\ell \bullet \cos b - v_0 \bullet \sin\ell \bullet \cos b - w_0 \bullet \sin b - \delta V_r$$

Cepheids are relatively young objects which exhibit a moderately small scatter (11 km s^{-1}) about such a model. The rotation curve needed to evaluate (1) is modeled, within the Galactocentric radius span of the data, as possessing a degree of curvature n:

$$V(R) = V_0 \bullet (R/R_0)^n \qquad (2)$$

Notice that an arbitrary amount ($\pm$) of solid-body rotation may be added to V(R) in (2) with no effect upon V_r in (1). All distances in the problem are best scaled to R_0. The individual Cepheid distances are consequently evaluated as:

$$x = \text{distance}/R_0 = \text{dex}\{.2[\mathbf{m} - m_0 - \alpha \,(\mathbf{logP} - .9) - \beta \,(\mathbf{color} - .7)]\} \qquad (3)$$

Here the observed quantities (boldface) are the unreddened apparent mean magnitude, log of fundamental period, and unreddened mean color. The period and color coefficients of the Cepheid PL(C) law are given by α and β. m_0 is the distance scale parameter, defined to be the apparent magnitude of a metal-normal Cepheid with logP = 0.9, unreddened color = 0.7, and distance 1 R_0 away. Small adjustments to x were applied because of the population I metallicity gradient versus Galactocentric radius. This gradient is shown both by spectroscopy, and by broadband photoelectric colors (CC87) that become progressively bluer toward larger R. We estimate σ_x/x to be of order 20% or better.

A nonlinear least-squares method was used to solve for the parameters V_0, n, u_0, v_0, w_0, δV_r and m_0 in equations (1-3) by fitting the Cepheid data $V_r(x,\ell,b)$. We thank P.L. Schechter (MWLCO) for the use of his software and his extensive advice on this problem.

Table 1 gives the solution results from three alternative Cepheid luminosity laws (CC87): PLC(V,B-V), PL(V), and PL(I). V_0 and n are extremely highly correlated in the solution. Furthermore the solution is invariant to the addition of any amount of solid body rotation to the rotation curve. The consequence is that the data do not constrain V_0 itself, but instead strongly constrain the quantity

$2AR_O$. This is equal to V_O if rotation curve is flat at the sun. More generally $2AR_O$ is the value to which the rotation curve sampled by the Cepheids would linearly project if extrapolated to the Galactic center. The Cepheid data imply that $2AR_O$ = 228 ± 19 km s^{-1}, provided that the rotation curve is roughly linear over the range of the Cepheid sample.

Table 1. Model Results

MODEL:	PLC	PL-V	PL-I	AVG
no. in soln.	188	188	169	
$2AR_o$(km s^{-1})	238 ±20	218 ±18	228 ±18	228 ±19
u_o (km s^{-1})	8.5 ±1.4	8.1 ±1.4	7.6 ±1.4	8.1 ±1.4
v_o (km s^{-1})	13.6 ±1.2	12.7 ±1.2	12.1 ±1.2	12.8 ±1.2
w_o (km s^{-1})	(7.0)	(7.0)	(7.0)	(7.0)
δV_r (km s^{-1})	3.2 ±0.9	3.4 ±0.9	2.8 ±0.9	3.1 ±0.9
σV_r (km s^{-1})	11.6	11.6	10.8	11.3
m_o (mag)	10.84 ±.186	10.65 ±.182	9.91 ±.175	
M_o (mag)	-3.74 ±.03	-3.77 ±.03	-4.49 ±.03	
R_o (kpc)	8.28 +.75 −.69	7.66 +.68 −.62	7.58 +.65 −.60	7.83 +.69 −.63

The results for u_o and v_o are about standard. w_o has to be simply adopted and is not important to the problem. The radial velocity zero point offset δV_r is nonzero, about 3 ± 1 km s^{-1}, confirming what Kraft and Schmidt (1963) found. There is no trend of this offset with heliocentric distance or Galactic longitude, and its origin remains obscure. The clusters containing Cepheids ought to provide a check on the reality of nonzero δV_r, but very limited cluster velocity data exist at present. U Sgr and S Nor with their clusters yield an inconclusive δV_r = 1.4 ± 2.4 km s^{-1}. It may be worthwhile to note that the 47 Cepheids recently observed intensively with the SAAO radial velocity spectrometer yield a smaller δV_r consistent with zero: 1.24 ± 1.40.

m_o is probably our most important result. The Cepheid velocity and magnitude pattern collectively constrain this "standard candle" to an accuracy of better than $0.^m20$. m_o gives the numerical tie between R_o and the Cepheid luminosity zero point. We wish to emphasize that the kinematics of the 190 Milky Way Cepheids constrain only m_o, not R_o or the luminosity scale itself. The values obtained for R_o (last row) require a second distinct step of evaluating the Cepheid luminosity zero point from 23 cluster Cepheids (CC87) to give M_0. R_0 scales with the luminosity zero point (and vice versa); m_0 does not. The "best" R_0 obtained is 7.8 ± 0.7 kpc, which implies $A = 14.6$ km s^{-1} (all scaled to a Pleiades modulus of $5\overset{m}{.}57$).

These tabulated values assume that the rotation curve is roughly linear (not necessarily flat) over the range of the Cepheid sample. There is a particular combination of the kinematic and photometric data for each Cepheid which one can show to be linearly related if the rotation curve is linear. Figure 3 plots this regression and shows that a linear rotation curve is credible. If n is not precisely zero, a small change in R_0 and $2AR_0$ can be calculated following CC87.

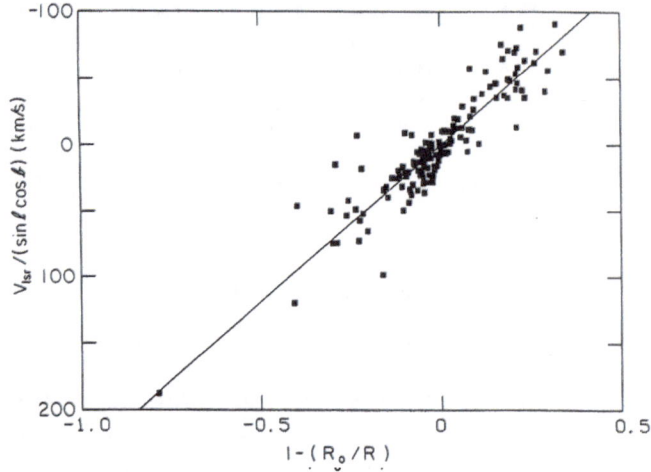

Fig. 3 Plot confirming that the rotation curve is approximately linear over the radius range of the available Cepheid data. Stars with $|\sin\ell\cdot\cos b| < 0.333$ are omitted.

The residuals from the solutions have been examined extensively for concealed trends. Little evidence of systematic deviations, e.g. of non-circular motions, was found. Monte Carlo tests showed that the biases in the solution parameters are small and that the parameter uncertainties are estimated satisfactorily.

This work was made possible by grant AST-8318504 from the National Science Foundation.

References

Caldwell, J.A.R., and Coulson, I.M. (1987). Astron.J. 93,1090.
Kraft, R.P., and Schmidt, M. (1963). Astrophys.J. 137,249.

THE DISTANCE TO THE GALACTIC CENTER
FROM OBSERVATIONS OF THE OUTER GALAXY

Leo Blitz (IAS and Maryland) and Jan Brand (MPI-Bonn)

Abstract

The method of Weaver (1954) to obtain the distance to the galactic center from observations on the solar circle is extended to include data at considerable distances from the solar circle. The circular velocity of the local standard of rest, Θ_0, is a free parameter in the solution, but because of the large number of data points, a relative uncertainty of only 6% is attained. The derived distance, 8.0 ± 0.5 kpc for $\Theta_0 = 220$ km s^{-1}, is sensitive only to streaming motions with scale sizes larger than ~ 10 kpc and to any zero point error in the stellar distance scale.

Introduction

Weaver (1954) recognized some time ago that the distance to the galactic center could be obtained from measurements of stars on the solar circle. Under the assumption of pure circular motion, distant objects with zero radial velocity should all lie on the solar circle. If the distance to such objects can be measured, one obtains an isosceles triangle where one side (the distance to the star) and one angle (the galactic longitude) are measured. One then simply solves for the remaining legs of the triangle, which gives the distance to the center.

Objects in the Milky Way do not, of course, move in pure circular orbits, and there is usually a considerable uncertainty associated with distance measurements. Because of the intrinsic velocity dispersion of stars and galactic streaming motions, some objects with zero radial velocity will be displaced from the solar circle, and some objects with non-zero velocities will be on it. Using the measured value of the Oort A constant, however, it is possible to derive a probability distribution for the location of objects near the solar circle given a measured radial velocity and distance. This technique was used by Feast and Shuttleworth (1965) to derive a distance to the galactic center from measurements of early type stars.

Now that a rotation curve has been measured to large galactocentric distance R, however, one can use the information in the rotation curve to determine R_0 for any object given its measured distance and velocity. We use the observations of star forming molecular clouds in the outer Galaxy (Blitz, Fich and Stark 1982; Brand 1986) to obtain a list of 109 objects suitable for analysis. In principle, if the mean error to any object is 25%, the uncertainty in the final result should be about 2.5%, excluding the effects of streaming. We find that the actual uncertainty, excluding zero point errors, is about 6%, largely because of the effects of streaming motions.

The Method

For an object in circular motion around the galactic center,

$$V_r = \left(\frac{R_0}{R}\Theta - \Theta_0\right)\sin l \cos b \tag{1}$$

where V_r is the radial velocity of the object relative to the local standard of rest, l, b are the galactic coordinates of the object and the subscript zero refers to values on the solar circle. For simplicity, assume a flat rotation curve ($\Theta = \Theta_0$). Then,

$$\frac{R_0}{R} = \frac{V_r}{\Theta_0 \sin l \cos b} + 1 \tag{2}$$

Now, the distance to any object is given by

$$R^2 - R_0^2 - (d\cos b)^2 + 2R_0 d \cos b \cos l = 0 \tag{3}$$

Setting $d_1 = d\cos b$, and $r = R/R_0$, we have

$$R_0 = \frac{-d_1 \cos l \pm d_1 \left(\cos^2 l + (r^2 - 1)\right)^{1/2}}{r^2 - 1} \tag{4}$$

For star-forming giant molecular clouds, r is determined from the mean V_r measured from CO profiles of the complex, and d_1 from the distance modules of the exciting stars. R_0 can then be determined by taking a weighted average of all of the individual determinations.

Caveats

Ironically, equation 4 blows up just for those objects for which Weaver introduced his method. That is, near the solar circle, $V_r \to 0$, $\frac{R}{R_0} \to 1$, and the denominator approaches zero. It is therefore important to determine the random uncertainties in the measurements to apply the proper weighting. In the final result quoted below, objects meeting the following criteria were excluded from the analysis: i) those in a $\pm 15°$ cone centered at $l = 0°$ and $l = 180°$, ii) those within a $5°$ cone centered at $l = 90°$ and $l = 270°$, iii) those within 1 kpc of the Sun, and iv) those with $V_r = 0$ within the uncertainties.

The circular velocity Θ_0 enters the distance determination through equation 2. However, by expanding equation 4 in powers of r, and substituting equation 2, it is easily seen that the solution for R_0 varies linearly with Θ_0 to first order. This was confirmed by finding solutions for R_0 using a range of values of Θ_0. The value quoted below, however, assumes a value of $\Theta_0 = 220$ km s^{-1}.

Streaming motions will cause errors because the measured radial velocity differs from the circular velocity appropriate to a given distance. However, if the sample covers a large enough surface area of the Milky Way, streaming motions smaller than the sample size will be averaged over galactic longitude. Our sample has a diameter of ~ 12 kpc, and streaming motions with scales smaller than this value should have no effect on the determination of R_0. The streaming motions will, however, act like an additional velocity dispersion, and will therefore add to the uncertainty in the derived value of R_0. Our error analysis indicates that streaming is probably the largest source of error in our derived value of R_0. We note in passing that the molecular clouds show a systematic motion in the Π direction of 4.2 km s^{-1} which has been subtracted from the measured radial velocities (Brand 1986).

The derivation of equation 4 assumes a flat rotation curve. Indeed, the best fit rotation curve to the CO and HI data is indeed very close to being flat (Brand 1986). However, the outermost points ($R > 1.65R_0$) deviate significantly from a flat curve and systematically lie above the curve $\Theta(R) = \Theta_0$. This introduces an additional error which can be compensated only with a more complex rotation curve. In this analysis, however, the outermost points are simply given a lower weight.

A systematic error is introduced by any zero point error in the stellar distance scale. The distances we have measured are based on Walraven photometry and the distances are thus tied to the local main sequence (Brand and Wouterloot 1988). Other distances come from a variety of sources (see Blitz, Fich and Stark 1982), and the overall systematic uncertainty due to zero point error is difficult to assess.

The Value of R_0

Excluding the 16 data points at $R > 1.65R_0$ gives a value of 7.5 ± 0.3 kpc for 93 objects; inclusion of these points raises the derived value to 8.2 ± 0.4 kpc. We therefore adopt a value of $R_0 = 8.0 \pm 0.5$ kpc based on a flat rotation curve and a value of $\Theta_0 = 220$ km s^{-1}. As a check to see if there are systematic deviations as a function of radius other than those of the most distant objects, we determine R_0 in various radial bins defined by $\Delta R' = R- < R_0 >$.

The results are shown in the table below.

Bin (kpc)	Number of objects	$< R_0 >$ (kpc)
$\Delta R' \geq 4$	16	12.3±1.3
$2 \leq \Delta R' < 4$	19	8.0 ±0.5
$1 \leq \Delta R' < 2$	29	7.1 ±0.7
$-1 \leq \Delta R' < 1$	39	7.5 ±0.6
$\Delta R' < -1$	6	8.2±0.3

Except for the most distant bin, where the deviations from a flat rotation curve are most pronounced, all of the derived values for R_0 determined from data in individual bins are within 1.5σ of 8.0 kpc.

Why the Method Works

It seems counter-intuitive to claim, as we do here, that one can obtain the distance to the galactic center by making observations away from the center, but consider the following. If the rotation curve is known, then the run of $\omega(R)$, the angular velocity of rotation is completely determined. $\omega(R)$, in turn, determines what the expected radial velocity will be for a measured distance from the Sun. That is, the rotation curve determines what the value of V_r will be for a given scale R_0 of the Galaxy. Our method simply inverts this procedure: we determine what the scale of the Galaxy is appropriate to the absolute distances and velocities that comprise the data set. The distances provide the anchor points for the velocity scale which is completely determined by the rotation curve and R_0. Observing the outer Galaxy allows us to sample a large fraction of the Milky Way because of the relatively small obscuration in the second and third galactic quadrants.

It may seem at first glance that there is circularity to the reasoning from which the value of R_0 is deduced: to derive a rotation curve, one needs to know the scale of the Galaxy: we then use the rotation curve to recover the scale. However, what one really does when a rotation curve is derived from a data set such as ours is to find $\omega(R) - \omega_0$ as a function of $R - R_0$; these quantities are directly measured by V_r and d in a circularly rotating galaxy. The derived rotation curve scales independently with R_0. However, once Θ_0 is fixed, the angular velocity rotation curve can only be fit with a particular value of R_0. It is for that reason that a value of R_0 is derived only for a set value of Θ_0.

Another way of seeing this is as follows: The solutions for R_0 and for the rotation curve are overdetermined because of the large number of data points. One could, for example, use one half of the data to derive a rotation curve, and the other half to determine R_0. The large number of data points assure that the statistical uncertainties will be little affected. A better way is to use all of the data in a maximum likelihood analysis to determine what rotation curve and what value of R_0 simultaneously provide the best fit to the data for a given Θ_0. Such an analysis is currently underway.

This work is partially supported by grant AST86-18763, and by the award of a fellowship to L. B. by the Alfred P. Sloan Foundation. L. B. also gratefully acknowledges the hospitality of the Institute for Advanced Study in Princeton where this paper was written. Most of the observations and analysis were done while J.B. was employed at the Sterrewacht Leiden.

References

Blitz, L., Fich, M., and Stark, A. A., 1982, *Ap. J. (Suppl.)*, **49**, 183.

Brand, J., 1986, Ph.D. Dissertation, University of Leiden.

Brand, J. and Wouterloot, J.G.A., 1988, *Astron. Ap. (Suppl.)*, in press.

Feast, M. W., and Shuttleworth, M., 1965, *M.N.R.A.S.*, **130**, 245.

Weaver, H. A., 1954, *A.J.*, **59**, 375.

II. LARGE SCALE SURVEYS AND OVERALL STRUCTURE

ROTATION AND THE OUTER GALAXY: COMMENTS ON TOPICS RAISED BY THE WORK OF FRANK KERR

Felix J. Lockman
National Radio Astronomy Observatory[1]
Edgemont Rd.
Charlottesville, Va. 22903

ABSTRACT

Several puzzles in galactic HI studies that were originally discussed by Kerr remain unsolved. The asymmetry in the rotation curve of the inner Galaxy is one of these. It cannot be explained solely as a peculiar motion of the LSR; it is important because small errors in the assumed galactic rotation curve can have tremendous consequences on kinematically derived quantities. Kinematic uncertainties alone make the surface density of HI in some places uncertain by at least a factor of four.

I. INTRODUCTION

I would like to use the occasion of this meeting, at which we honor Frank Kerr by talking about developments in the study of our Galaxy, to give a brief update on two of the many topics to which Frank has contributed: the symmetry of galactic rotation, and the interpretation of data from the outer Galaxy.

II. THE SYMMETRY OF GALACTIC ROTATION?

"Early checks indicated that there is approximate symmetry, showing that the simple circular orbit assumption is not too far in error. A closer look at the symmetry properties shows however that the apparent rotation curves are actually somewhat different on the two sides of the Sun-centre line" – Kerr 1962

This quotation is from one of Kerr's more well known papers. The issue is whether galactic rotation, as viewed with respect to our "Local Standard of Rest", is symmetric about the Sun-center line, and if not (since it is not) is it because the Galaxy is skewed, or because of an error in the adopted LSR? Kerr suggested that the LSR might need to be adjusted for a radial motion of 7 km/s.

There is a simple expression for the velocity with respect to the LSR of gas at some location in the Galaxy (R, θ, z), that has velocity components V_R, V_θ, V_z:

---

[1] The National Radio Astronomy Observatory is operated by Associated Universities Inc. under contract with the National Science Foundation.

$$V_{\text{LSR}} = \cos(b)[R_0 \sin(\ell)(\Omega - \Omega_0) - V_R \cos(\ell + \theta) + V_{R0} \cos(\ell)] + \sin(b)(V_z - V_{z0}). \qquad (1)$$

This expression is derived from geometry, and is entirely general since it includes every possible "noncircular" motion of the LSR through the terms V_{R0}, V_{z0}, and Ω_0, the last of which is the angular velocity of our adopted LSR, which may differ from the angular velocity of a reference frame in pure circular motion.

At each longitude in the inner Galaxy there is a unique "tangent point" where galactic rotation is projected totally along the line of sight while radial motions are projected orthogonally. In the galactic plane vertical motions are also orthogonal to the line of sight. We determine an HI rotation curve for the inner Galaxy from some measurable "terminal velocity" of an HI spectrum, V_t, which is related to V_θ at the tangent point. There are always two estimates of V_t for any distance from the galactic center: one derived from observations in the North (longitude quadrant I), and one from observations in the South (quadrant IV). The sum of the two terminal velocities is

$$\Delta V_t = V_t(North) + V_t(South) = \Delta V_\theta(R) + 2V_{R0} \cos(\ell). \qquad (2)$$

This equation is complete. Radial motions are entirely across the line of sight at a tangent point. Vertical motions, whether at the tangent point or of the LSR, make a negligible contribution to V_t at very low latitudes. Any peculiar motion of the LSR in the azimuthal direction (a difference between Ω_0 and the "true" circular velocity) cancels in the sum. It is clear that a radial motion of the LSR will have a unique $cos(\ell)$ signature in $\Delta V_t(\ell)$.

Figure 1 shows my recent determination of ΔV_t (according to the prescription of Shane and Bieger-Smith 1966) from the Weaver and Williams (1973) survey in the North and the Kerr et al. (1986) survey in the South. In most respects Figure 1 is in good agreement with what one would derive using the V_t values of Sinha (1978) or Rohlfs et al. (1986) who each used different data and a different procedure to obtain a V_t.

This figure, alas, cannot necessarily be taken at face value, for local conditions, not galactic rotation, can determine some part of ΔV_t. Near the Sun, for example, the V_{LSR} from galactic rotation is small and in the range of random motions of HI clouds; ΔV_t at $sin(\ell) \gtrsim 0.9$ does not necessarily contain any information about galactic rotation. And at small $sin(\ell)$ there is the galactic center, where again the terminal velocities are probably not determined by rotation but by motions with a non-circular component (e.g. Liszt and Burton 1980; Brown and Liszt 1984). The "3-kpc" arm, for example, probably perturbs ΔV_t near $sin(\ell) = 0.35$, although only over a few degrees of longitude (e.g. Bania 1986).

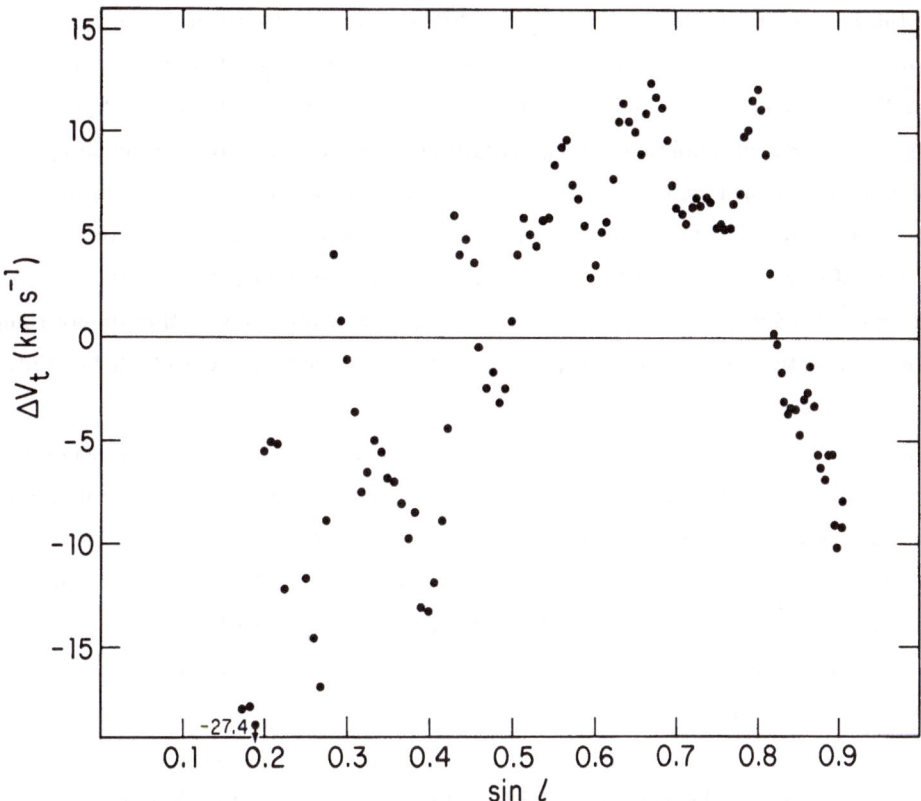

Figure 1. The difference in terminal velocity along the Galaxy's Northern and Southern tangent points. It is related to local motions through eq. 2.

Even if we ignore the possible influence of local conditions, it is clear from Fig. 1 that galactic motion is almost, but not quite, symmetric. Deviations are at most only 5% of the rotation. And the deviations are tantalizingly regular.

When Kerr (1962) postulated an outward motion of the LSR, it was in large part to see if this hypothesis would symmetrize the HI spiral pattern he derived for the Galaxy, because even then he knew that the solution to the slight rotational asymmetry *"...cannot be as simple as this, because there is no evidence for a systematic motion of the local centre of rest relative to the local gas"* (Kerr 1962). Nor is there any evidence in more recent data for a radial motion of the LSR relative to HI in absorption against the galactic center (Radhakrishnan and Sarma 1980; Liszt, Burton and van der Hulst 1985), and observers of HI emission are generally unconvinced or mildly agnostic about small radial motions (e.g. Sinha 1978 and Rohlfs *et al.* 1986). Furthermore, there is the fact that unless every negative value of ΔV_t is dismissed as a conspiracy between galactic nuclear phenomena (at low ℓ) and local effects (at high ℓ), the curve of ΔV_t just doesn't have the $cos(\ell)$ form that signifies radial motion of the LSR.

Still, hints of an outward motion of the LSR, or some variant on it, seems to pop up over and over again, e.g. in a study of molecular clouds associated with optical HII regions (Blitz and Fich 1983), and in an analysis of the velocities of young, nearby stars (Ovenden, Pryce and Shuter 1983). The suggested amplitude for V_{R0} is typically 2-4 km/s, and the sign is always positive. Resolution of the issue requires observations over a large range of longitude (i.e. distance from the galactic center) to aviod bias. For example, bright radio HII regions show a velocity asymmetry suggestive of a radial motion of the LSR (Lockman 1979), but that is because they lie mainly at $0.4 < sin(l) < 0.8$ where ΔV_t is indeed positive. Species or observations that do not sample all radii give a selective view of ΔV_t; it is possible that HI is the only species suitable for study of the entire phenomenon.

Several other possible explanations for the form of Fig. 1 are unconvincing when examined in detail. Streaming along spiral arms produces high frequency ripples in ΔV_t, not a broad pattern. While there certainly are high frequency variations in ΔV_t, they are not the dominant effect; the variation in the rotation curve appears to have a much longer radial wave length. A kinematic model that gives the Galaxy a slight ellipticity does not reproduce Figure 1 either, unless the orientation of the ellipse is allowed to change with distance from the galactic center.

But is it reasonable to expect there to be a single explanation for every feature in Figure 1? There is actually some evidence suggesting that part of the phenomenon may be caused by a "local" rather than a global structure, and I want to digress just a bit to discuss it.

Recently Clemens (1985) claimed that his analysis of CO observations revealed a "local" motion of the LSR with a magnitude of about 7 km/s in the direction of galactic rotation. I do not think that this claim is very convincing for it is based solely on Northern observations, and one would not reach the same conclusion from the Southern data. But the effect that he noticed is interesting. It is is caused by a band of gas (see the HI and CO maps in Appendices A and B of Burton and Israel 1983) which starts at the terminal velocity around $\ell = 58°$, but which appears to cross V=0 at $\ell = 75°$ rather than at $\ell = 90°$ where it should if the LSR were not "racing ahead". Unfortunately for the Clemens hypothesis, there is no corresponding band of HI (or of CO) in the South which is being left behind.

This asymmetry shows up in Figure 1 as the negative-going ΔV_t at $sin(\ell) > 0.85$, but, as shown in eq. (2), ΔV_t does not notice the magnitude of Ω_0. Rather, we are seeing something like a local "compression" of galactic rotation relative to the Sun-center line. This effect is probably the one found by Ovenden, Pryce and Shuter (1983) in their analysis of the kinematics of ~ 1000 young stars within 5 kpc of the Sun (in which, incidentally, they saw no hint that the LSR needed adjusting by more than about 2 km/s). However, a true compression of rotation about the Sun should show up equally in terminal velocities of Northern and Southern HI (or CO), while the disturbance that causes ΔV_t to go negative (and Clemens to advocate an extra motion of the LSR) happens only in the North. If stars do really show this effect then it can't be due to gas just being blown around, but until the issue is resolved it is not clear how to interpret $\Delta V_t(\ell)$ at the higher longitudes. The low velocities at the northern tangent-point, whatever their origin, persist over several kpc.

The tangent points are the perfect reference for viewing galactic motion. The situation there can be summarized as follows: 1) galactic rotation in the inner Galaxy *is* remarkably symmetric. But there are systematic, large-scale departures from circular rotation that appear sometimes like a "compression" and sometimes like an "expansion" relative to the Sun-center line. 2) the LSR does not have a large radial motion relative to gas along the tangent points. It has no radial motion relative to gas toward the galactic center. In addition, the circular motion of the LSR seems to be approximately correct. Deviations are at most 2% of the local rotational velocity, or 2-4 km/s. 3) There is no simple kinematic model, like density-wave streaming, or simple elliptical motion, which reproduces the main aspects of Figure 1.

II. INTERPRETATION OF THE OUTER GALAXY

"It has not been sufficiently emphasized in the past that every spiral diagram which has been derived from 21-cm results is an *interpretation, rather than* **the** *interpretation, of the available data."*

"The form of the finally-derived diagram will therefore depend on the kinematic picture which has been adopted." – Kerr 1962 (original emphasis)

These sentences are from a paper which demonstrates the malleability of derived spiral patterns to slight perturbations in kinematic assumptions. It is this sensitivity which makes the question of rotational asymmetries so important (see also Burton 1971; 1976). While Kerr was talking specifically about the problem of deriving a galactic spiral pattern from HI data, his warning applies in general, and to quantities that naively might seem quite robust. Let me give a single example.

Figure 2 shows two interpretations of the HI surface density in the outer Galaxy. Both curves were derived from **identical** data (the Weaver and Williams 1973 HI survey) and were interpreted identically except for the rotation curve that was used to convert velocity into distance. The surface density graph labeled FLAT is what you get by assuming that $V_{\theta 0} = 250$ km/s at all $R > 10$ kpc (in this section I am using the "old" values of R0 and $V_{\theta 0}$). The other wildly different surface density pattern results if one adopts a rotation curve which is mildly rising from 250 km/s at R=10 kpc to 285 km/s at R=19 kpc then is flat thereafter (this is the rotation curve used by Kulkarni, Blitz and Heiles 1982). The extreme difference in amplitude and shape between 10 and

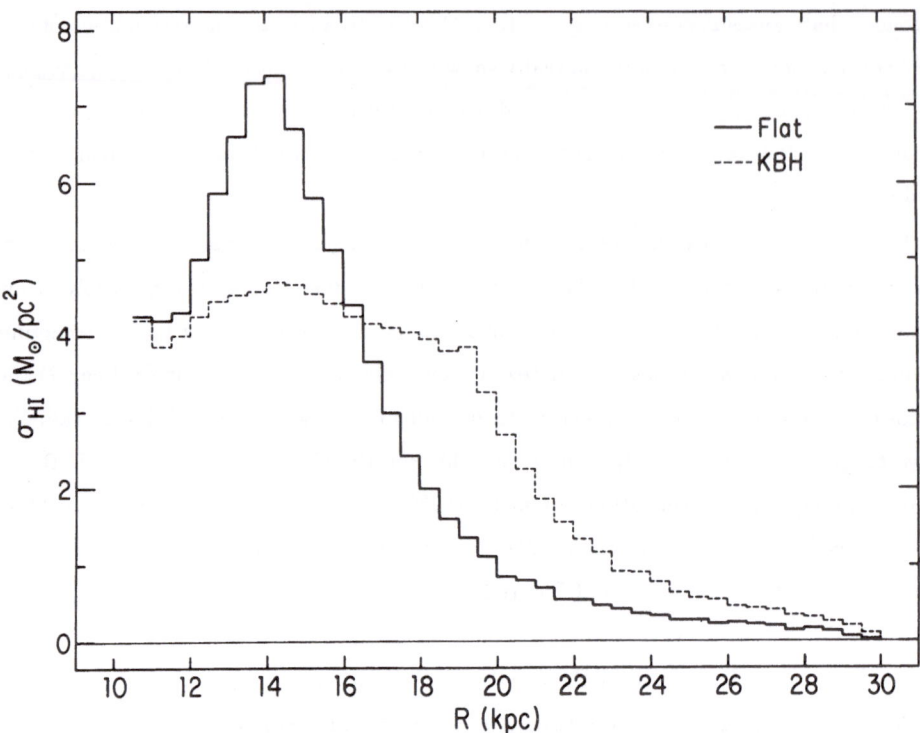

Figure 2. The Weaver and Williams (1973) HI survey analyzed with a FLAT rotation curve gives a surface density vs. R diagram shown by the solid line; if the same data are analyzed using the rotation curve of Kulkarni, Blitz and Heiles (1982), which differs from a flat curve by $< 14\%$ at any point, one gets the surface density curve shown by the dashed line. The rotation curves differ by only 2% at $R < 14$ kpc. Note that this figure is on a scale where $R0 = 10$ kpc and $V_{\theta 0} = 250$ km/s.

14 kpc is caused by a difference in rotational velocity of < 5 km/s, i.e. only 2%. At R=19 kpc the surface density estimates differ by a factor of 4, while the rotation curves differ by only 14%. In both interpretations the surface density falls approximately exponentially at large R, but this is only because both rotation curves were assumed to be flat in the outermost parts of the Galaxy.

We should use the difference between the two estimates of surface density as a measure of *minimum* uncertainty in kinematic analyses of the outer Galaxy, for there are other ways to get things wrong (some of which are discussed in Lockman 1986). Unfortunately, this source of error is rarely admitted. In a nice turnaround of the whole business, however, Knapp, Tremaine and Gunn (1978; see also Knapp, this volume) *assumed* that the HI surface density must decrease exponentially with R (as it does, on average, in many other galaxies) then found the rotation curve which would give us that result for galactic HI. The curve was flat.

The best conclusion to this section would be to repeat Kerr's words which opened it; in the outer Galaxy one should not push the accuracy of HI "interpretations" too far.

III. TWO PROBLEMS

"The calculations for the regions of the Galaxy at the greatest distance are affected by HI emission from nearby regions, either as 'stray radiation', or through local material at noncircular velocities being attributed to greater distances" – Kerr 1983

This may seem like a trivial concern, but I think that it is vital, for it strikes at the fundamental assumption of our kinematic interpretations: that HI emission occurs at a given velocity because galactic rotation put it there. The operational side of this assumption is shown in Figure 3. Here is an HI spectrum in the direction of the galactic warp that has been marked to show how a rotation curve maps observed velocity into distance from the galactic center and distance from the plane (note: this is for a flat rotation curve with $R0 = 8.5$ kpc and $V_{\theta 0} = 220$ km/s). This spectrum does not have the best signal-to-noise ratio (it is only a 40^s integration), but it shows detectable emission out to perhaps -150 km/s, and this is verified in longer integrations. Kerr's comments quoted above, when applied to this spectrum, lead to two questions: 1) is the emission around -85 km/s, which purports to be at R=15 kpc and z=4 kpc, really coming from the sky, or is it merely an instrumental artifact, and 2) if real, is our interpretation of its velocity correct?

Let's take the second question first, for it just reminds us that local, low velocity, HI is very bright compared to the weak features under study here. The low velocity portion of this particular spectrum (off the top of Figure 3) has a brightness temperature of 10 K. Can we be certain that local gas doesn't have a faint extension to -85 km/s or contain a small high-velocity component? Perhaps the gas around -130 km/s contributes a bit of a wing at -85 km/s also. If so, we are misinterpreting the emission at -85 km/s and thereby erroneously boosting the scale-height of the outer Galaxy. Caution is called for, and duly noted.

The first question though, raises a different and difficult issue: stray radiation. Every radio telescope, in addition to its main beam, has weak broad sidelobes which also pick up emission and convey it into the receiver. A one percent sidelobe lying on the 100 K brightness of HI emission in the galactic plane will produce a 1 K spurious signal in an HI spectrum. The situation is somewhat analogous to scattered light in an optical telescope, but it is generally worse, because the most villainous sidelobes of radio telescopes lie many tens of degrees away from the direction the telescope is pointing. A typical radio telescope pointed at the north galactic pole receives about equal amounts of HI from the main beam and from the far sidelobes (Lockman, Jahoda and McCammon 1986). This effect was discovered 25 years ago (van Woerden, Takakubo and Braes 1962), and modern 21cm receivers reach the stray-radiation limit in a few seconds (recent discussions of the problem, and three different approaches to its cure, are in Kalberla, Mebold and Reich 1980, Heiles, Stark and Kulkarni 1981, and Lockman, Jahoda and McCammon 1986).

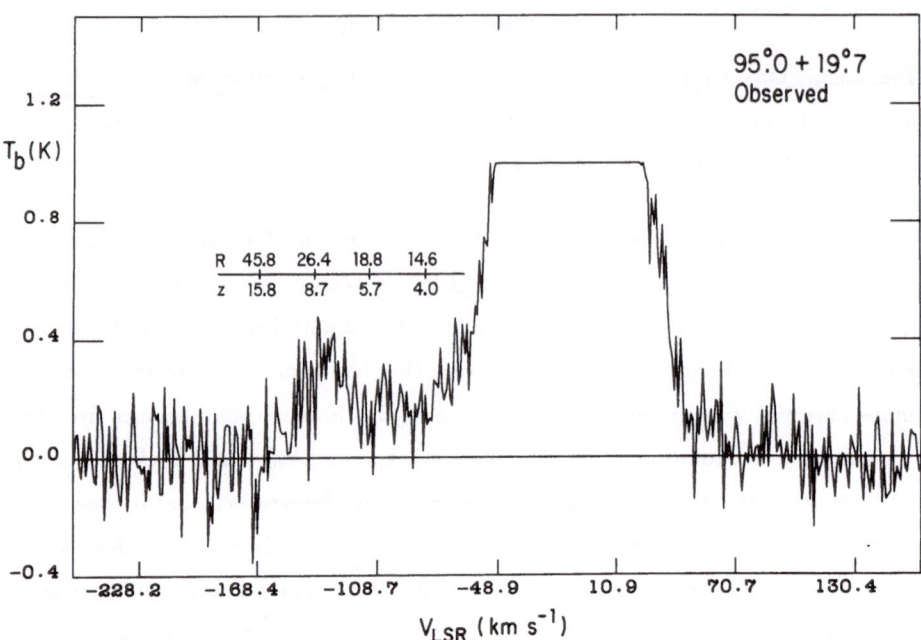

Figure 3. The HI spectrum towards $\ell = 95°$, $b = 19.7°$ as observed with the 43m telescope of the NRAO at Green Bank. A flat rotation curve with $V_{\theta 0} = 220$ km/s at all $R > R0 = 8.5$ kpc relates velocity to distance from the galactic center (R) and distance from the galactic plane (z) as shown by the fiducial marks.

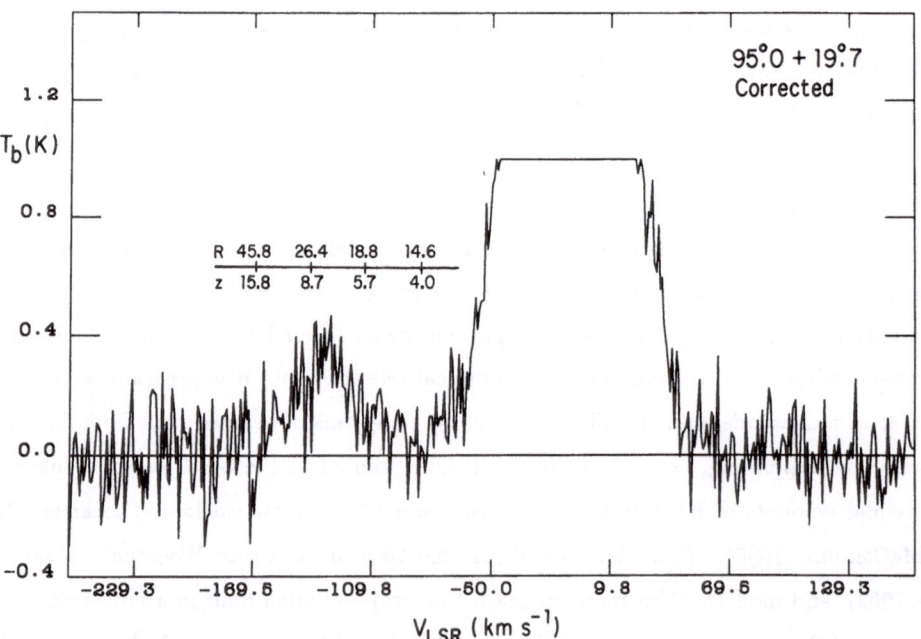

Figure 4. The HI spectrum of Figure 3 after correction for stray radiation. The emission that was at -85 km/s in the uncorrected spectrum does not appear here. It was an instrumental artifact.

The observed spectrum of Figure 3, when corrected for stray radiation, becomes the spectrum of Figure 4. The HI at -85 km/s has disappeared. It was due to stray radiation. It is quite likely that our descriptions of HI in the outermost Galaxy have been distorted by stray radiation, and possibly also by velocity dispersion, as Kerr suggested. But note that the most negative velocity emission is real. It comes from the galactic warp, which is **not** an instrumental artifact.

It is cumbersome at present to obtain HI spectra with reasonable angular resolution and minimal stray radiation, but some recent advances in receiver and feed design at Green Bank should soon eliminate this menace from galactic HI studies.

IV. FINAL THOUGHTS

Frank Kerr's research has touched on fundamental issues for the study of our Galaxy – its inner as well as outer parts. I have spent this time talking about small effects, effects that Kerr noticed and discussed. They can have large consequences, especially for studies of the Galaxy's outermost regions. Yet it is a tribute to Frank Kerr that he and his colleagues understood the importance of subtle effects many years ago, even as they were drawing a picture, using the rudest methods of analysis on only the brightest HI, that has remained our basic model of the Galaxy to this day.

REFERENCES

Bania, T.M. 1986, *Ap. J.*, **308**, 868.

Blitz, L. and Fich, M. 1983, in *Kinematics, Dynamics and Structure of the Milky Way*, ed. W. Shuter, Dordrecht:Reidel, p. 143.

Brown, R.L. and Liszt, H.S. 1984, *Ann. Rev. Astr. Ap.*, **22**, 223.

Burton, W.B. 1971, *Astr. Ap.*, **10**, 76.

Burton, W.B. 1976, *Ann. Rev. Astr. Ap.*, **14**, 275.

Burton, W.B. and Israel, F.P. 1983, in *Surveys of the Southern Sky*, ed. W.B. Burton and F.P. Israel, Dordrecht: Reidel.

Clemens, D.P. 1985, *Ap. J.*, **295**, 422.

Heiles, C., Stark, A. A. and Kulkarni, S. 1981, *Ap.J. Lett.*, **247**, L73.

Kalberla, P.M.W., Mebold, U. and Reich, W. 1980, *Astr. Ap.*, **82**, 275.

Kerr, F. J. 1962, *MNRAS*, **123**, 327.

Kerr, F.J. 1969, *Ann. Rev. Astr. Ap.*, **7**, 39.

Kerr, F.J. 1983, in *Surveys of the Southern Galaxy*, ed. W.B. Burton and F.P. Israel, Dordrecht: Reidel, p. 113.

Kerr, F.J., Bowers, P.F., Jackson, P.D. and Kerr, M. 1986, *Astr. Ap. Suppl.*, **66**, 373.

Knapp, G.R., Tremaine, S.D. and Gunn, J.E. 1978, *A.J.*, **83**, 1585.

Kulkarni, S., Blitz, L. and Heiles, C. 1982, *Ap. J. (Letters)*, **259**, L63.

Liszt, H.S., Burton, W.B. and van der Hulst, J.M. 1985, *Astr. Ap.*, **142**, 245.

Liszt, H.S. and Burton, W.B. 1980, *Ap. J.*, **236**, 779.

Lockman. F.J. 1979, *Ap. J.*, **232**, 761.

Lockman, F.J., Jahoda, K. and McCammon, D. 1986, *Ap.J.*, **302**, 432.

Lockman, F.J. 1986, *in Gaseous Halos of Galaxies*, ed. J.N. Bregman and F.J. Lockman, NRAO: Green Bank, p. 63.

Ovenden, M.W., Pryce, M.H.L. and Shuter, W.L.H. 1983, in *Kinematics, Dynamics and Structure of the Milky Way*, ed. W. Shuter, Dordrecht:Reidel, p. 67.

Radhakrishnan, V. and Sarma, N.V.G. 1980, *Astr. Ap.*, **85**, 249.

Rohlfs, K., Chini, R., Wink, J.E. and Böhme, R. 1986, *Astr. Ap.*, **158**, 181.

Shane, W.W. and Bieger-Smith, G.P. 1966, *B.A.N.*, **18**, 263.

Sinha, R.P. 1978, *Astr. Ap.*, **69**, 227.

Weaver, H. and Williams, D.R.W. 1973, *Astr. Ap. Suppl.*, **8**, 1.

van Woerden, H., Takakubo, K. and Braes, L. L. E. 1962, *B. A. N.*, **16**, 321.

MAPPING THE GALAXY AT 21 CM WAVELENGTH: THE BOSTON UNIVERSITY-ARECIBO GALACTIC H I SURVEY

T. M. Bania
Astronomy Department
Boston University
725 Commonwealth Ave.
Boston, MA 02215

Over its operating frequency range Arecibo provides the highest angular resolution attainable by filled aperture techniques. For most purposes filled apertures are the instruments of choice for galactic scale spectral line surveys because the measurement is unambiguous: no spatial Fourier components are lost up to the limit set by the illumination pattern. Although far higher resolutions are available with interferometers, no truly galactic scale surveys have ever been made with these devices, probably due to logistical, political, psychological, and physiological limitations. Thus large scale surveys of various spectral species will continue to be the province of filled aperture antennas.

Because of its size Arecibo's suitability for such projects in terms of angular resolution and sensitivity is unique. However, although mechanical constraints limit Arecibo's view to δ's between roughly $0°$ and $+40°$, it still has over 13,000 square degrees of sky accessible to observation. This corresponds to over 5 million individual 4 arcmin beams at 21 cm wavelength. Even with Arecibo, time constraints severely limit the scope of the survey projects that are feasible, especially if one attempts to account for the confusion produced by the high sidelobe levels of the line feeds. At present, only the 1407 MHz flat feed has both high gain and $\leq 1\%$ sidelobes and hence a low confusion level for spectral surveys of extended sources. Nonetheless, there are already two large-scale galactic surveys in progress at Arecibo: the Cornell-Arecibo OH survey and the Boston University-Arecibo HI survey. Here we focus on the BU-Arecibo galactic HI survey and the reason why we are conducting yet another galactic-scale 21 cm survey.

In the inner galaxy, that part which lies within the solar orbit, cool foreground HI clouds can be seen absorbing background 21 cm line radiation. The Arecibo survey of the vertical distribution of atomic hydrogen (Bania and Lockman 1984, hereafter the BL survey) produced a catalogue of 177 such objects whose kinematic distances are fixed by the special geometry required to produce self-absorption. Self-absorption can occur within the solar orbit where galactic rotation makes the LSR velocity–distance from the sun relationship double valued. Thus cold foreground clouds can absorb 21 cm line radiation from hotter gas residing at the far kinematic distance. Detection of self-absorption resolves the distance ambiguity. These kinematic distances can be quite accurate: the $\sim15\%$ uncertainty found by BL results from a combination of random cloud motions and streaming motions which introduce ±5 km s^{-1} perturbations to a cloud's velocity.

These cool HI clouds appear in the BL survey with a surface density of ~20 clouds per square degree and with an apparent cloud mean free path of ≤0.7 kpc. As a population, the BL clouds resemble galactic molecular clouds in their vertical extent and angular size. There is little doubt that we are observing cool HI associated directly with molecular clouds in many cases (Liszt, Burton, and Bania 1981; Bania 1983; BL). The 4 arcmin resolution of the Arecibo data is comparable to the angular resolutions of present large scale galactic CO surveys. In fact, many of the BL survey objects are quite small: approximately a third of the sources have angular sizes ≤8 arcmin. As the figure below shows, sources this size would be nearly invisible in HI surveys made with coarser angular resolution. At 21 cm Arecibo is capable of resolving a 10 pc diameter cloud at a distance of 8 kpc. This is just the distance to the tangent point at a galactic longitude of 35°. Arecibo can thus determine the kinematic distances to self-absorbing 'standard' HI clouds anywhere in the accessible portion of the inner galaxy.

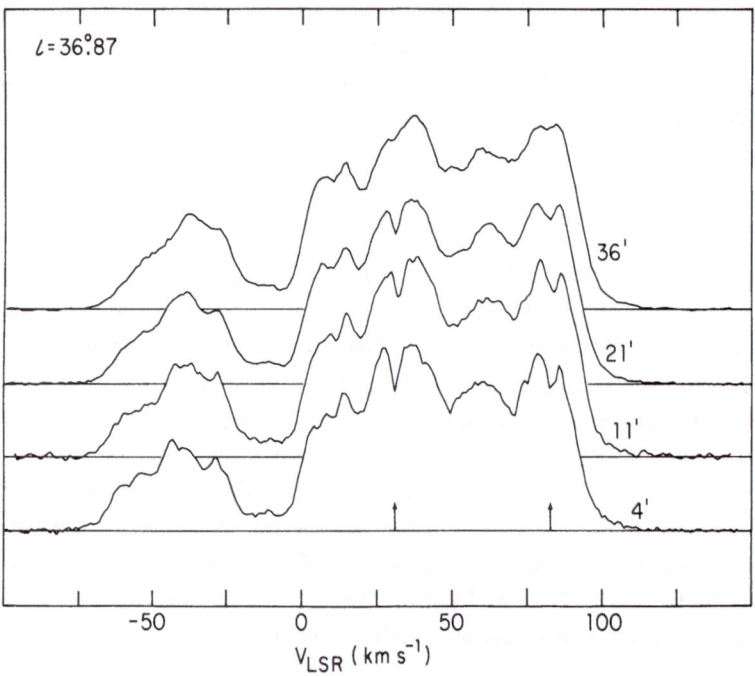

H I spectra for the same direction made at 4 different angular resolutions: Hat Creek (HPBW=36');
Green Bank 43 m (HPBW=21'); Green Bank 91 m (HPBW=11'); and Arecibo 305 m (HPBW=4').
Arrows indicate the positions of H I self-absorption features. Figure reproduced from the BL survey.

The BU-Arecibo HI survey will map 70 square degrees of the first galactic quadrant at 2 arcmin spacings with the 4 arcmin beam of the 1407 MHz flat feed. It is expected to discover and to establish the distances to over 1,000 cold HI clouds that will be detected by their HI self-absorption spectral signatures. It will also search for self-absorbed HI in the direction of star clusters, giant molecular cloud complexes, bipolar flows, HII regions and supernova remnants. A major goal of the survey is to establish Arecibo onto an international standard brightness temperature scale–the Berkeley Hat Creek scale. It will also improve upon this standard by studying the sidelobe response of the flat feed by mapping standard fields at different epochs. All measurements will be ultimately referred to the Bell Labs galactic HI survey made with the horn telescope that discovered the 3 K cosmic background radiation.

Correlations between these survey clouds and other astronomical sources can provide many astrophysically interesting results:

1. The Structure of the Inner Milky Way – The majority of the HII regions, supernova remnants, and molecular clouds in the Milky Way reside within the inner $|l| \leq +60°$ of the galactic plane. Arecibo can provide many new kinematic distances and thus a better determination of the structure of the galaxy for the region which is exterior to 5 kpc galactic radius. Since self-absorption can be used to corroborate or revise previous distance estimates to presumed spiral tracers, the accuracy of spiral structure patterns could be assessed. For example, the BU-Arecibo survey will be able to check the distances to half of the large molecular complexes identified by Dame *et al.* (1986).

2. The Infrared Luminosity Function for the Inner Milky Way – The BU-Arecibo survey can in principle be used to establish the distances to galactic IRAS sources, especially those radiating principally at 60 and 100 μm wavelength. Such objects have IRAS color temperatures between 10–300 K. Sources with $T_{color} \geq 25$ K require internal heat sources, possibly even protostellar ones. Any identification of an IRAS source with an Arecibo cloud provides the distance to, and hence the luminosity of, the infrared source. Even a modest identification rate will therefore provide the first determination of the infrared luminosity function for the inner galaxy.

3. Studies of Thermal and Nonthermal Continuum Sources – The 5 GHz survey of the galactic plane by Altenhoff *et al.* (1978) lists over 300 discrete continuum sources in the survey region. The resolution of the kinematic distance ambiguity to these objects using HI self-absorption would be more reliable than distances set by 6 cm H_2CO absorption because H_2CO can absorb against the 2.7 K cosmic background radiation and yet be seen through a foreground optically thin HII region (Whiteoak and Gardner 1974).

4. Interstellar Cloud Properties and the Mass Spectrum – Many HI self-absorption features correlate with molecular clouds. Any coincidence between an Arecibo HI self-absorption cloud and one emitting in ^{12}CO or ^{13}CO gives a distance and size for the molecular cloud. These HI /CO clouds can be observed in several molecular species that are collisionally excited at higher densities than CO in order to derive the column density structure of these objects, giving their total atomic and molecular mass. The properties of these clouds will provide an independent determination of the interstellar cloud size and mass spectrum. These distributions may help to constrain agglomeration theories for cloud growth (Kwan 1979; Cowie 1980).

5. The H I Structure of Molecular Clouds – The BU-Arecibo survey will make unbiased maps of the HI structure of giant molecular clouds (GMCs) and GMC complexes. The HI distribution around many first galactic quadrant GMCs, including 50% of the largest molecular complexes, will be completely mapped. When compared with CO emission surveys, these maps can be expected to show whether or not HI halos are ubiquitous features of GMC complexes.

6. <u>Atomic and Molecular Gas in Star Forming Regions</u> – The distribution of atomic and molecular gas around HII regions and star clusters of varying ages gives information regarding the nature, duration, and efficiency of the star formation process, together with insight into mechanisms which destroy GMCs. These include the shear of differential galactic rotation, stellar winds and bipolar flows, ionization fronts, and supernovae. Mapping the gas distribution surrounding galactic HII regions and open star clusters provides the data necessary to study this interaction.

Arecibo can map the HI distribution about 34 of the open clusters in Leisawitz's catalog (1983) in order to search for cold atomic gas clouds in the vicinity. This sample spans a range of distance, richness, and main-sequence turnoff age and many of these clusters have known distance modulii (Lyngå 1981; Janes and Adler 1982; Leisawitz 1983). The detection of HI self-absorption in star clusters with known optical distance modulii would provide a check on the kinematic distances.

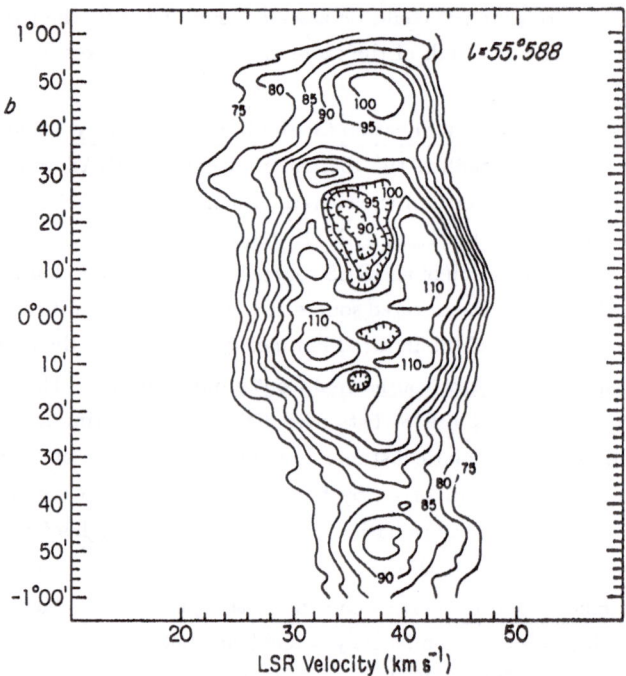

H I brightness temperature contours are shown in this figure from Bania (1983).

The BL and BU-Arecibo surveys have already found cold self-absorbing HI in several star clusters. For example, Bania (1983) has analyzed the self-absorption at $(l,v)=(56°, 36 \ km \ s^{-1})$ shown above. The position, velocity, and distance of the HI self-absorption seen in this direction suggest that it is physically associated with an anonymous OB association that has the Cepheid GY Sagittae as a member (Forbes 1982). The kinematic distance of 3.25 kpc is confirmed to within 5% (!) by the spectroscopic parallax of the OB association which is spatially and kinematically coincident with the HI /CO cloud. It is thus likely that the cold atomic clouds are the remnants of the original molecular cloud from which the $\sim 10^7$ year old cluster formed. The HI is distributed in two clouds, each at least 15 pc in diameter. If their temperature is ~ 60 K, the cloud HI masses are each $\sim 10^3$ M$_\odot$.

REFERENCES

Altenhoff, W. J., Downes, D., Pauls, T., and Schraml, J. 1978, *Astr. Ap. Suppl.*, **35**, 23.

Bania, T. M. 1983, *A.J.*, **88**, 1222.

Bania, T. M., and Lockman, F. J. 1984, *Ap. J. Suppl.*, **54**, 513. (**BL**)

Cowie, L. L. 1980, *Ap. J.*, **236**, 868.

Dame, T. M., Elmegreen, B. G., Cohen, R. S., and Thaddeus, P. 1986, *Ap. J.*, submitted.

Forbes, D. 1982, *A.J.*, **87**, 1022.

Janes, K. A., and Adler, D. 1982, *Ap. J. Suppl.*, **49**, 425.

Kwan, J. 1979, *Ap. J.*, **229**, 567.

Leisawitz, D. 1983, Open Cluster Database , private communication.

Liszt, H. S., Burton, W. B., and Bania, T. M. 1981, *Ap. J.*, **246**, 74.

Lyngå, G. 1981, Catalogue of Open Cluster Data , Obs. de Strasbourg, Centre de Donnees Stellaires.

Whiteoak, J. B., and Gardner, F. F. 1974, *Astr. Ap.*, **37**, 389.

THE SHAPE OF THE OUTER-GALAXY HI LAYER

W.B. Burton
Sterrewacht Leiden
P.O.Box 9513, 2300 RA Leiden
The Netherlands

Frank Kerr and others first recognized the globally warped nature of the gas layer in the outer parts of the Galaxy in the early Leiden and Sydney 21-cm surveys. Their work showed that the layer is warped above the b=0° plane in the northern data and below it in the southern, that the radial extent of the layer is greater in the south than in the north but that the warp amplitude is more extreme in the north, and that the thickness of the gas layer increases at larger radii (Kerr 1957, Burke 1957, Oort et al. 1958, Kerr 1962).

Motivation for more recent investigations of the global warp has been provided by a number of factors. It has become clear that systematic deviations from a flat disk are a common characteristic of the HI morphology of many spiral galaxies. Furthermore HI is the only accessible tracer of the mass in the outer parts of many galaxies; it is plausible that the shape and motions of the warped gas layer are governed by dark matter. Additional motivation is provided by the realization that the physical circumstances in the outer Galaxy differ from those in the inner, for reasons not yet fully understood.

Our embedded viewing perspective complicates many Milky Way studies, but is ideal for study of the morphology of the warped layer. Two advances facilitate such work. One advance concerns the improved accuracy with which the outer-Galaxy rotation curve is known and the consequently more accurate distances (see Brand 1986, and Kulkarni et al. 1982). Another advance concerns the improved data for the southern hemisphere (Kerr et al. 1986) as well as for the northern one (Burton, 1985); the Dwingeloo 25-m telescope is currently undertaking a deep northern survey. This report describes briefly and largely pictorially work towards a quantitative description of the HI layer in the Galactic outskirts. Some of the figures reproduced here will appear in the Verschuur/Kellermann compendium (Burton, 1988); a complete report is in preparation by Burton and Kwee.

References

Brand, J. 1986, PhD dissertation, University of Leiden

Brinks, E., and Burton, W.B. 1984, Astron. Astrophys., 141, 195

Burke, B.F. 1957, Astron. J., 62, 90

Burton, W.B. 1985, Astron. Astrophys. Suppl., 62, 365

Burton, W.B. 1988, in "Galactic and Extragalactic Radio Astronomy",
    G. Verschuur and K.I. Kellermann, eds., Reidel Pub. Co. (in press)

Henderson, A.P., Jackson, P.D., and Kerr, F.J. 1982, Ap. J., 263, 116

Kerr, F.J. 1957, Astron. J., 62, 93

Kerr, F.J. 1962, Monthly Not. Roy. Astron. Soc., 123, 327

Kerr, F.J., Bowers, P.F., Jackson, P.D., and Kerr, M. 1986, Astron.
    Astrophys. Suppl., 66, 373

Kulkarni, S.R., Blitz, L., and Heiles, C. 1982, Ap. J., 259, L63

Oort, J.H., Kerr, F.J., and Westerhout, G. 1958, Monthly Not. Roy.
    Astron. Soc., 118, 379

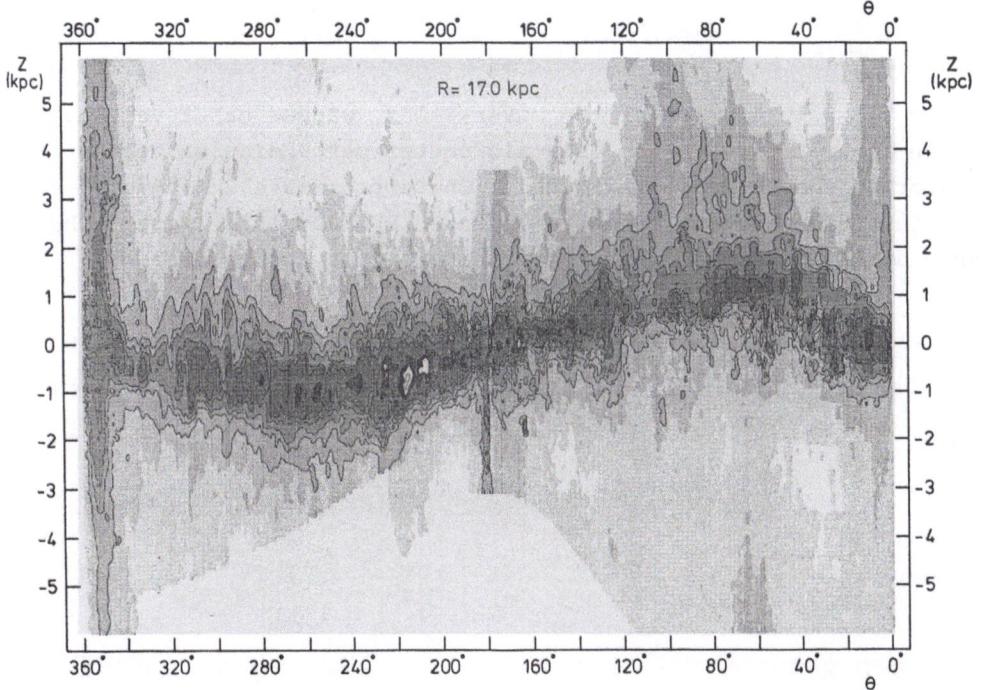

Figure 1. Illustration of the outer-Galaxy warped gas layer showing
n_{HI} in a cylinder centered at R=17 kpc. At R<17 kpc the azimuthal
variation of the warp is approximately sinusoidal; at larger R, the
amplitude of the warp is greater at $\Theta<180°$ than at $\Theta>180°$.

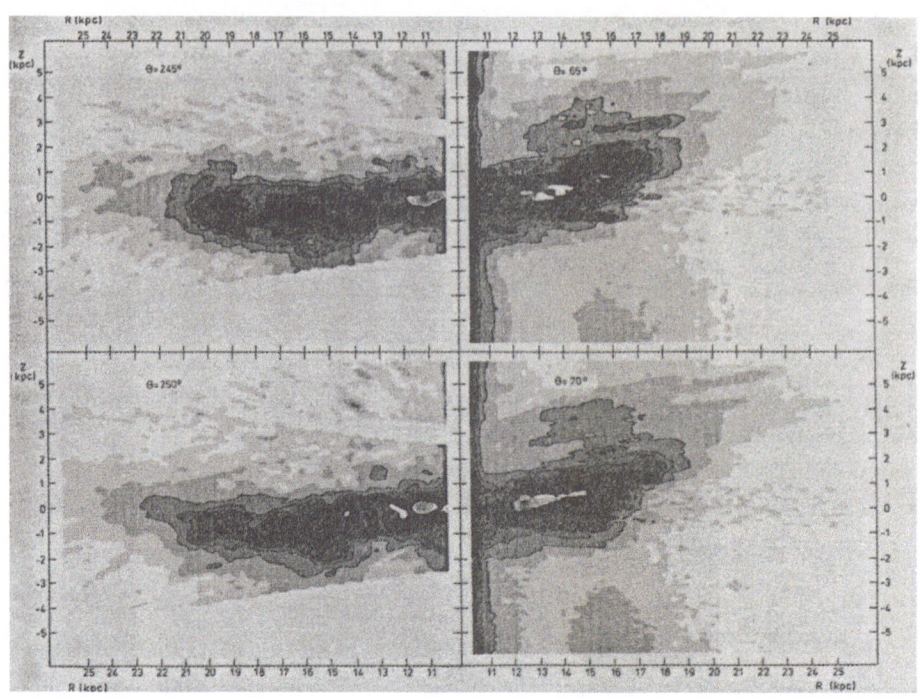

Figure 2. Illustration of the outer-Galaxy warped gas layer showing n_{HI} in a sheet through the Galactic center perpendicular to the equator at the indicated azimuths. The approximate 2π-symmetry of the warp justifies pairing of maps separated by 180° in azimuth. The bands at R≤11 kpc, caused by the finite σ_v, may be ignored.

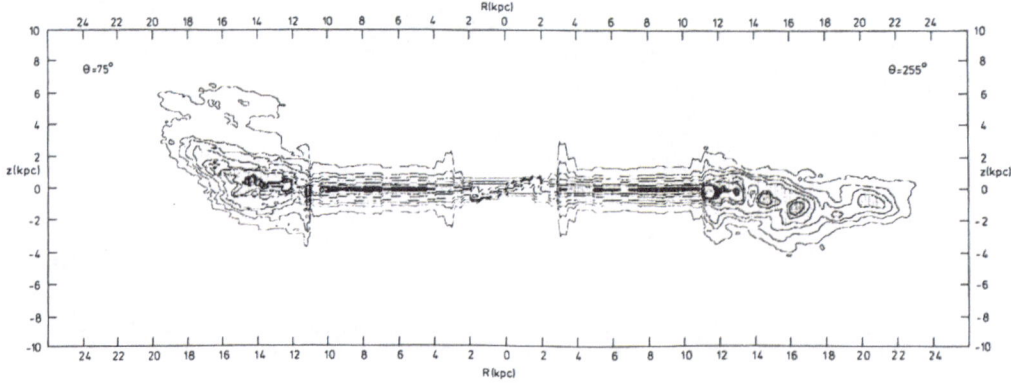

Figure 3. Schematic representation of the Galactic HI layer. The cross section represents a cut through the Galactic center perpendicular to b=0° in a sheet oriented in azimuth approximately along the line of nodes of the Galactic warp.

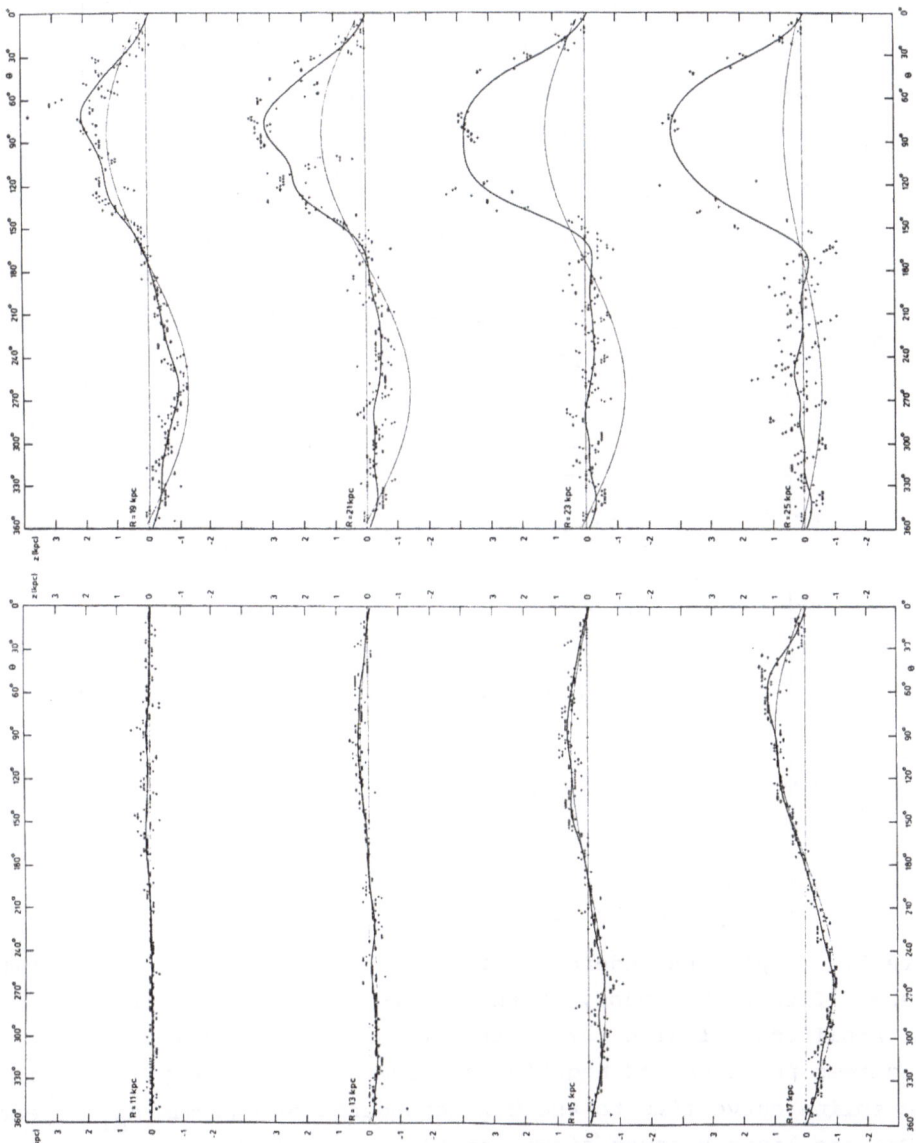

Figure 4. Distance from the plane b=0° of the maximum gas density at different Galactocentric distances. The + symbols give the z-height variation with Galactocentric azimuth of the maximum n_{HI} on the walls of cylinders of radius R. The more smoothly varying lines show least-squares fits of sine curves to the symbols; the less regular curves show fits by a cubic-spline algorithm. From such fits, carried out at different R, follows information on the amplitude and line of nodes of the Galactic warp.

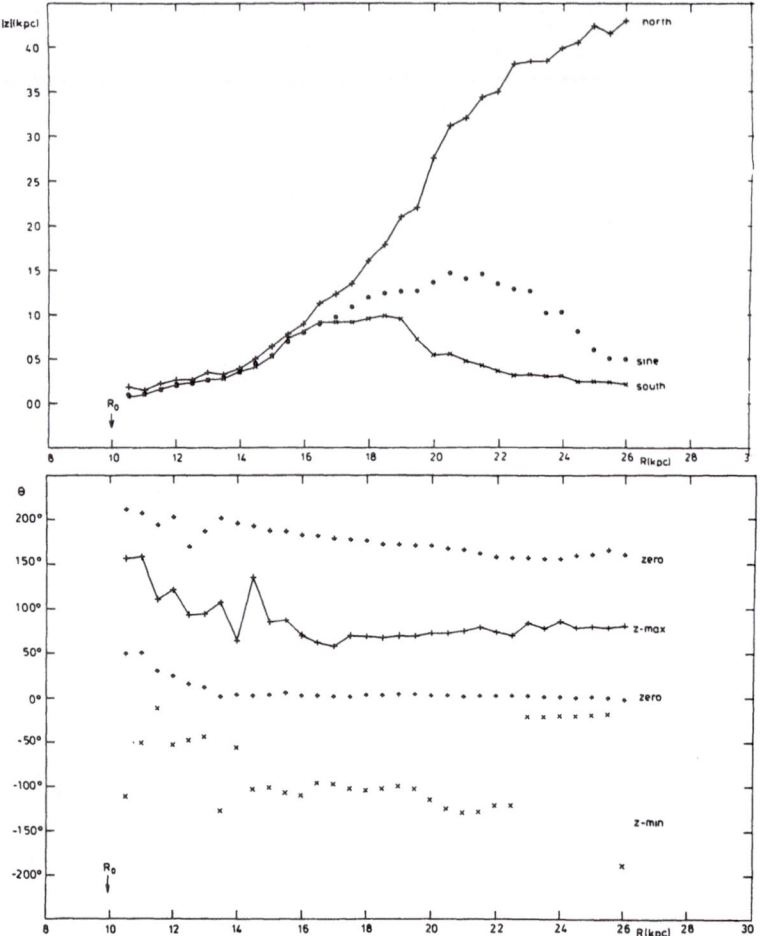

Figure 5. (upper) Dependence of the warp amplitude on R. The symbols represent fits to Θ,z maps, of the Figure 1 sort, at intervals of 0.5 kpc. Amplitudes of sine-curve fits are shown by o's; +'s and x's show the extreme positive and negative z values, respectively, found in the cubic-spline curve fits to the Θ,z dependence of maximum n_{HI} at each R. The gas layer becomes significantly warped at R>13 kpc. It grows approximately linearly, and equally in the two Galactic hemispheres, until about 16 kpc; at R≥17 kpc the amplitude increases strongly in the northern hemisphere but decreases in the southern one.
(lower) Radial dependence on azimuth showing positions where the gas layer reaches its maximum height above (+ symbols) the plane z=0 kpc and its maximum height below (x symbols) this plane, as well as the positions where the gas layer crosses z=0 kpc. The zero crossings, separated by approximately 180°, give the position angle of the line of nodes of the warp. This angle varies little with R: evidently the line of nodes of the Galactic warp does not precess.

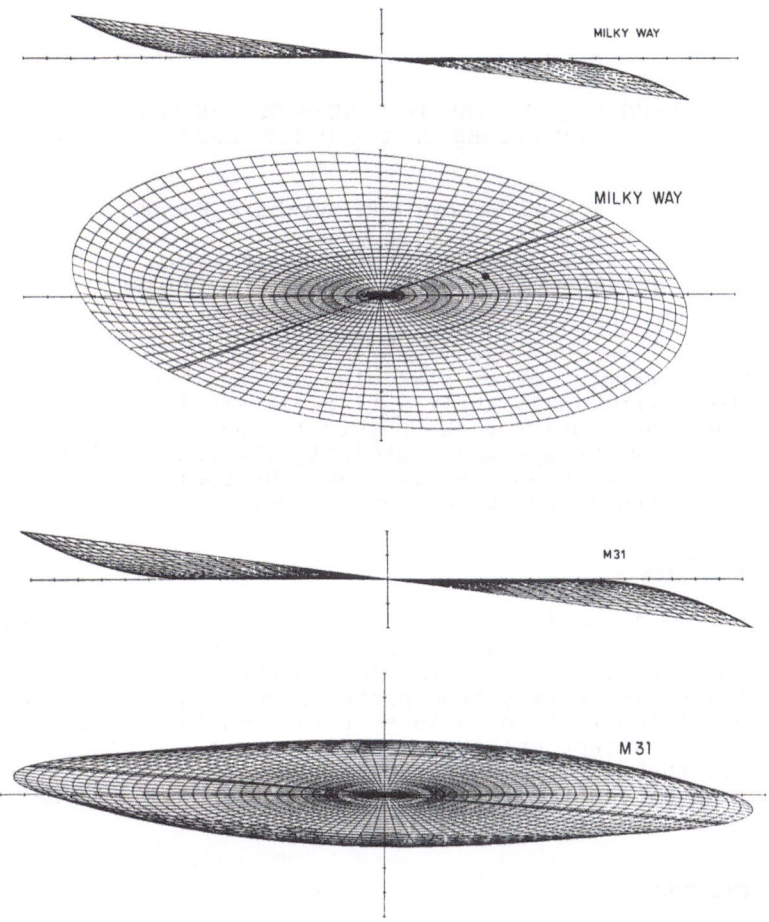

Figure 6. Simple geometrical models of the warped Galaxy, and of M31, drawn to the same scale. The plots show the midplane of the gas layers. Galactocentric rings are drawn at 1 kpc intervals. Rays are drawn at intervals of azimuth; a heavy line shows the line of nodes. In the edge-on situation both galaxies are viewed looking along the lines of nodes. The panel second from top shows the warped Milky Way oriented as it would be for a viewer in M31; the dot shows the Sun's position. The bottom panel shows the warped midplane of M31, drawn in accordance with the model parameters of Brinks and Burton (1985) determined from kinematic characteristics of HI observations. The warp of M31 is quite similar to that of our Galaxy, in terms of extent, amplitude, flare, and non-precession of the line of nodes.

STAR CLUSTERS AND THE THICKNESS OF THE GALACTIC DISK
AS PROBES OF THE OUTER GALAXY

Roland Wielen and Burkhard Fuchs

Astronomisches Rechen-Institut
Heidelberg, Federal Republic of Germany

ABSTRACT: Properties of star clusters and the thickness of the galactic disk, especially their variation with galactocentric distance, provide tests for the physical conditions in different parts of a galaxy. We discuss especially the radial variation of the dissolution times of open clusters, and the radial variation of the disk thickness under various circumstances.

1. INTRODUCTION

The physical conditions in the outer Galaxy (and in other parts of the Galaxy) should be mainly investigated by direct observations. In some cases, additional information on these physical conditions may be obtained indirectly from probes such as star clusters. Such probes can, at least, be used to test the conclusions which have been drawn from more direct evidence. Aside from star clusters, the thickness of the galactic disk can also provide some insight into the conditions at different parts of our Galaxy and in external galaxies.

2. STAR CLUSTERS

Star clusters depend in two ways significantly on their environment: firstly, the galactic gravitational field produces a finite tidal radius r_t for a cluster; secondly, the dissolution of star clusters depends both on the overall tidal field of the host galaxy and on encounters with massive objects, such as giant molecular clouds (GMCs) or massive black holes (MBHs).

2.1 Tidal Radii of Star Clusters

The tidal radius r_t of a star cluster which moves in a nearly circular orbit of radius R around the center of its host galaxy, is given by

$$r_t = (\, Gm_c/(4\omega^2 - \kappa^2)\,)^{1/3} \, , \tag{1}$$

where G is the gravitational constant, and m_c the total mass of the cluster. The rotational frequency $\omega(R)$ and the epicyclic frequency $\kappa(R)$ for orbits of stars or clusters in the galaxy are determined by the galactic rotation curve. For example, for a flat rotation curve with a circular velocity of

$$v_o = \omega(R)\ R = const., \tag{2}$$

we have

$$\kappa(R) = 2^{1/2} \, \omega(R) \; , \tag{3}$$

and hence

$$r_{t,flat}^3 = Gm_c/2\omega^2 \; , \tag{4}$$

or

$$r_{t,flat} = (Gm_c/2v_0^2)^{1/3} \, R^{2/3} \; . \tag{5}$$

For our Galaxy, we derive, using the new IAU values of $v_0 = 220$ km/s and $R_0 = 8.5$ kpc,

$$r_{t,flat} = 9.3 \; pc \; (m_c/250 \; m_\odot)^{1/3} \; (R/R_0)^{2/3} \; . \tag{6}$$

Table 1 illustrates the corresponding increase of $r_{t,flat}$ with R for a typical open star cluster with a total mass of 250 solar masses. For the (unrealistic) case of a Keplerian rotation curve,

$$v_{circ}(R) = (Gm_{gal}/R)^{1/2} \; , \tag{7}$$

where $v_{circ}(R)$ is the circular velocity at galactocentric distance R and m_{gal} is the total mass of the galaxy, we obtain for the tidal radius of a cluster:

$$r_{t,Kepler}^3 = Gm_c/3\omega^2 \; , \tag{8}$$

or

$$r_{t,Kepler} = (m_c/3m_{gal})^{1/3} \; R \; . \tag{9}$$

The dependence of the tidal radius r_t of a cluster on the galactocentric distance R differs only slightly, by a power of $R^{1/3}$, in the two cases of a flat or a Keplerian rotation curve. For example, even at $R = 2R_0$, the tidal radii, if normalized at $R = R_0$, would differ by a factor of 2/1.59 = 1.26 only. Hence, a discrimination between a flat or a Keplerian rotation curve for the outer Galaxy from observations of tidal radii of star clusters should be very difficult. Furthermore, there are severe difficulties, firstly in measuring accurate values of tidal radii of remote clusters, and secondly in interpreting these (exact meaning of r_t uncertain; individual determinations of the cluster masses m_c difficult; in the case of using average values for m_c: possible variations of m_c and r_c with R; only clusters older than about one galactic orbital period ($2\pi/\omega(R)$) had time to adjust their outer radius to r_t; non-circular orbits in the case of globular clusters).

2.2 Dissolution Times of Star Clusters

The influence of the galactic environment on the dissolution time of a star cluster has been discussed in detail by Wielen (1985, 1987). In the limit of large median cluster radii r_c, say for r_c larger than 0.1 r_t, the evaporation time of the cluster due to the internal relaxation in the cluster and due to the tidal effect of the galactic gravitational field, is given (Wielen 1987, Eq.9) by:

$$T_{lim} = NT_t/\phi_0 \; , \tag{10}$$

where

$$T_t = 0.022 \ (4\omega^2 - \kappa^2)^{-1/2} \tag{11}$$

is the 'tidal time', and

$$\phi_0 = 0.56 \tag{12}$$

is a constant describing the number of escaping stars per crossing time in the case of an isolated cluster; N is the total number of stars in the cluster.

In the case of a flat rotation curve (Eqs.2 and 3), we obtain

$$T_{lim,flat}(R) = N(0.022/(2^{1/2}\phi_0 \ \omega)) = 0.028 \ N \ v_0^{-1} \ R \ . \tag{13}$$

We can rewrite T_{lim} in terms of the galactic orbital period, $T_{circ}(R) = 2\pi/\omega(R)$:

$$T_{lim,flat}(R) = 2.2 \ (N/500) \ T_{circ}(R) \ . \tag{14}$$

This means that a typical open cluster with N = 500 stars would survive over slightly more than two galactic revolutions. From Eq.13, we see that the evaporation time $T_{lim,flat}$ increases proportional to R.

Let us now consider the additional dissolution of clusters by the gravitational tidal fields of passing giant molecular clouds (GMCs). We shall use here the formulae and notation of Wielen (1987), and shall limit ourself to the important case of $p_1 > p_0$ (this means that the cluster can be disrupted completely by a single close encounter with a GMC). The dissolution time $T_{n,1}$ (Eq.12 of Wielen, 1987) of a cluster, due to GMCs alone, is then given by:

$$T_{n,1} = 0.053 \ (Gm_c/r_c^3)^{1/2}/(G\rho_{an}) \ . \tag{15}$$

ρ_{an} is the overall mass density of GMCs. Since ρ_{an} decreases strongly with R, especially outside the solar circle, the dissolution time of open clusters due to GMCs increases very strongly with R.

The total dissolution time T_{tot} of an open cluster, due to both the internal evaporation and the external disruption by passing GMCs, is approximately derived by adding the corresponding escape rates:

$$T_{tot} = 1/((1/T_{lim}) + (1/T_{n,1})) \ . \tag{16}$$

In Table 2, we list the dissolution times $T_{lim,flat}$, $T_{n,1}$ and T_{tot} as functions of the galactocentric distance R for a typical open cluster with N = 500 stars, total mass $m_c = 250 \ m_\odot$, and median cluster radius $r_c = 1$ pc. From Table 2, we predict a strong increase of $T_{tot}(R)$ with increasing R, especially in the range $R_0 < R < R_0 + 3$ kpc. This steep increase is mainly due to the low frequency or absence of GMCs in the outer Galaxy. This was first pointed out by van den Bergh and McClure (1980), who used this interpretation in order to explain the relatively larger number of old open clusters in the outer Galaxy.

The observational data on the age distribution of open clusters as a function of R are in good agreement with our predictions listed in Table 2. For example, Figure 4 of Lynga (1980) shows that there are, in his sample, 37 'outer' clusters (with 90° < l < 270°, corresponding to $R > R_0$), but only 17 'inner' clusters (0° < l < 90° and 270° < l < 360°, corresponding to $R < R_0$) with ages larger than about $3 \cdot 10^8$ years (log age > 8.5), while the number of inner and outer clusters is essentially equal for younger clusters.

Table 1. Tidal radius r_t of a star
cluster of total mass $m_C = 250\ m_\odot$
as a function of galactocentric
distance R.

R(kpc)	r_t(pc)
4	5.6
6	7.4
R_0 = 8.5	9.3
10	10.4
12	11.7
14	13.0
16	14.2
18	15.3
20	16.4

Table 2. Dissolution times T of a typical open star cluster
($m_C = 250\ m_\odot$, $r_C = 1$ pc) as a function of
galactocentric distance R.

R (kpc)	$R-R_0$ (kpc)	ρ_{an} $(m_\odot/pc^3)$	$T_{lim,flat}$	$T_{n,1}$ (10^8 years)	T_{tot}
4.5	−4	0.13	2.8	1.1	0.8
8.5	0	0.02	5.3	6.2	2.9
9.5	+1	0.01	5.9	12.4	4.0
10.5	+2	0.004	6.5	31.0	5.4
11.5	+3	0	7.1	∞	7.1
12.5	+4	0	7.7	∞	7.7
13.5	+5	0	8.4	∞	8.4

Our Table 2 predicts essentially the same value (2.9·10[8] years) for this critical age limit. The agreement may be, however, somewhat fortuitous: the observational data may suffer from severe selection effects (older clusters are probably more easily detected in the outer part of the Galaxy than in the inner part); furthermore, an increase in the average total mass of open clusters with increasing R would also produce an increase of the dissolution time T with increasing R.

3. THICKNESS OF THE GALACTIC STELLAR DISK

The thickness of the galactic stellar disk, H, is determined by various properties of the Galaxy: (a) by the compressive K_z force, (b) by the initial velocity dispersion of the stars at birth (probably roughly equal to the turbulent velocity of the interstellar gas), (c) by the diffusion of stellar orbits due to the random acceleration of stars by the irregular gravitational field in the Galaxy, and (d) perhaps, in the outer Galaxy, by the effects of the warp. Therefore, the thickness H of the disk and its radial variation, H(R), provide in principle important information on the physical conditions in the relevant parts of the Galaxy through the effects (a) to (d).

For illustration, we shall use here some simplifications: (1) We assume that the disk is made up of only one stellar generation with a velocity dispersion σ_W in the z direction independent of z (isothermal case). This assumption is very unrealistic, but may be sufficient for discussing the main trend of the R dependence of H. (2) With respect to the K_z force, we consider two limiting cases only: (2a) The disk is completely self-gravitating (i.e. the disk produces the whole K_z force; no other component contributes to K_z). (2b) The disk is completely non-self-gravitating (i.e. disk stars move as test particles in a linear K_z force, produced by other, much thicker components such as the halo or the dark corona). A more realistic description of the galactic stellar disk at the solar galactic distance R_o has been given in earlier papers (Wielen and Fuchs, 1983, 1985; Fuchs and Wielen, 1987).

We define the thickness H of the disk by

$$H = \mu_d/2\rho_d \ , \tag{17}$$

where μ_d is the surface density of the disk, and ρ_d is the central volume density of the disk (at z=0; an index 0 is supressed for simplicity !). In the self-gravitating case (2a), $H = H_{sg}$ is given by

$$H_{sg} = \sigma_W/(2\pi G\rho_d)^{1/2} = \sigma_W^2/(\pi G\mu_d) \ . \tag{18}$$

For the non-self-gravitating case (2b), we have

$$H_{nsg} = \sigma_W/(8G\rho_{tot})^{1/2} \ , \tag{19}$$

where ρ_{tot} is the total mass density at z=0. For the ratio H_{nsg}/H_{sg}, we find

$$H_{nsg}/H_{sg} = 0.88 \ (\rho_d/\rho_{tot})^{1/2} \ . \tag{20}$$

A convenient and rather accurate interpolation formula between the self-gravitating and non-self-gravitating case is given by:

$$H = \sigma_W/(8G(\rho_{nd} + (\pi/4)\rho_d))^{1/2} \ , \tag{21}$$

Table 3. The radial variation of the disk thickness H(R) for various cases.

Initial velocity dispersion $\sigma_{W,o}$ neglected:

Source of diffusion	Diffusion produces	Resulting thickness H and its trend with R			
		Self-gravitating disk		Non-self-gravitating disk	
		H_{sg}	Trend	H_{nsg}	Trend
Massive black holes	$\sigma_W^2 \propto \rho_{MBH}$	$H \propto \rho_{MBH}/\mu_d$ $\propto (1/R^2)\, e^{\alpha R}$	↑	$H \propto (\rho_{MBH}/\rho_{tot})^{1/2}$	↑
Instabilities in the disk	$\sigma_W^2 \propto \mu_d$	$H = const.$	→	$H \propto (\mu_d/\rho_{tot})^{1/2}$	↓
Giant molecular clouds	$\sigma_W^3 \propto \mu_{GMC}$	$H \propto \mu_{GMC}^{2/3}/\mu_d$	↓	$H \propto \mu_{GMC}^{1/3}/\rho_{tot}^{1/2}$	↓
Extreme case of no diffusion	$\sigma_W = \sigma_{W,o}$	$H \propto \sigma_{W,o}^2/\mu_d$ $\propto \sigma_{W,o}^2\, e^{\alpha R}$	↑	$H \propto \sigma_{W,o}/\rho_{tot}^{1/2}$	↑

where the 'non-disk' density ρ_{nd} is given by

$$\rho_{nd} = \rho_{tot} - \rho_d . \tag{22}$$

The velocity dispersion σ_W of the disk stars in the z direction is partly determined by the initial velocity dispersion, $\sigma_{W,0}$, but mainly by the diffusion process. Possible sources of the diffusion of stellar orbits are (1) giant molecular clouds (GMCs), (2) density variations in the galactic disk due to transient instabilities of the disk, and (3) massive black holes as basic constituents of the dark corona (proposed by Lacey and Ostriker (1985) and Ipser and Semenzato (1985)).

In this paper, we are mainly interested in the possible radial variation of the disk thickness, H(R), in the outer Galaxy. A general overview of the various possible cases is given in Table 3. We assume an exponential disk, $\mu_d \propto e^{-\alpha R}$, and an isothermal corona, $\rho_{corona} = \rho_{MBH} \propto 1/R^2$. The probable variation of H with increasing R is indicated by the direction of the arrows. We should add that the warp phenomenon may have the tendency to produce an additional increase of H with R. It is clear from Table 3, that an observed tendency for the radial variation of H would give important hints on the physical conditions in the (outer) Galaxy, such as the source of orbital diffusion, degree of self-gravitation of the disk, and relative importance of the initial velocity dispersion. For example, GMCs may be ruled out as an important source of orbital diffusion, because the much smaller number of GMCs in the outer Galaxy would produce a steeply decreasing thickness H of the disk with increasing R, in contrast to observations in external galaxies and probably also in conflict with the situation in our Galaxy.

4. CONCLUSIONS

It would certainly not be convincing to derive physical properties of the outer Galaxy directly from star cluster properties or from the thickness of the galactic disk as a function of R. Such information provides, however, valuable tests for models of the outer Galaxy derived from more direct evidence. Of course, the dependence on R of star cluster properties and of the disk thickness is an interesting problem in itself.

REFERENCES

Fuchs, B., Wielen, R.: 1987, in The Galaxy, Eds. G. Gilmore and R. Carswell, NATO ASI Series, D. Reidel Publ. Comp., Dordrecht (in press).
Ipser, J.R., Semenzato, R.: 1985, Astron. Astrophys. 149, 408.
Lacey, C.G., Ostriker, J.P.: 1985, Astrophys. J. 299, 633.
Lynga, G.: 1980, in IAU Symposium No.85, Star Clusters, Ed. J.E. Hesser, D. Reidel Publ. Comp., Dordrecht, p.13.
van den Bergh, S., McClure, R.D.: 1980, Astron. Astrophys. 88, 360.
Wielen, R.: 1985, in IAU Symposium No.113, Dynamics of Star Clusters, Eds. J. Goodmann and P. Hut, D. Reidel Publ. Comp., Dordrecht, p.449.
Wielen, R.: 1987, in IAU Symposium No.126, The Harlow Shapley Symposium on Globular Cluster Systems in Galaxies, Eds. J.E. Grindlay and A.G.D. Philip, D. Reidel Publ. Comp., Dordrecht (in press).
Wielen, R., Fuchs, B.: 1983, in Kinematics, Dynamics and Structure of the Milky Way, Ed. W.L. Shuter, D. Reidel Publ. Comp., Dordrecht, p.81.
Wielen, R., Fuchs, B.: 1985, in The Milky Way Galaxy, Eds. H. van Woerden, R.J. Allen and W.B. Burton, D. Reidel Publ. Comp.,Dordrecht, p.481.

THE MILKY WAY IN HIGH RESOLUTION U PHOTOMETRY AND INFERENCES ON ITS STRUCTURE

Th. Schmidt-Kaler, W. Schlosser
Fakultät für Physik und Astronomie, Astronomisches Institut
Ruhr-Universität Bochum
Universitätsstr. 150, 4630 Bochum 1

A high resolution ultraviolet photometry of the Milky Way is presented (Schlosser, Schmidt-Kaler, Schneider, in prep.). The resolution is $0°\!.25$ for the southern part $\ell = 235°...360°...15°$. Including the data after Winkler, Schmidt-Kaler, Pfleiderer (1984), it is $1°$ for the whole Milky Way (fig. 1). The half width of the brightness distribution perpendicular to the equator is $8°\!.5$. This is definitely narrower than in the visual.

The central bulge is visible south of the plane. With a flattening of B/A = 0.59 in the ultraviolet, the Galaxy is classified as Sb $(-Sb^+)$ I - II. This result confirms the earlier classification obtained by visual inspection from very-wide-angle photographs in the optical regions UBVR (Schmidt-Kaler, Schlosser 1973) and from various lines of evidence (van den Bergh 1968, 1972).

The next inner spiral arm -I is seen most conspicuously between $\ell = 280° - 310°$ and between $\ell = 320° - 345°$. Both fragments of the spiral arm are inclined to the galactic equator. The fragment $320° - 345°$ (shingle 2) is extremely sharp ($z_{1/2} = 1°\!.3 = 30$ pc at a distance of 1.3 kpc). The question arises if this filament is really so narrow, or if it appears so by the effect of foreground dark clouds. To decide this question we used the Atlas of Dark Clouds by Feitzinger and Stüwe (1984). It turns out that the narrow spiral fragment is not an illusion due to the particular distribution of nearby dark clouds but is real.

The typical velocity dispersion of interstellar matter and young early type stars is $\pm$ 6 km/s. This means that the narrow fragment originated from a very thin sheet of interstellar matter, and has an age of at most $5 \cdot 10^6$ years.

The surface brightness of the Milky Way is strongly affected by interstellar extinction, especially in the ultraviolet. Feitzinger and Stüwe's (1984) Dark Cloud Atlas is used to remove this influence to a certain degree in order to reconstruct the Milky Way as it would appear to us without those nearby dark clouds: a "Milky Way without dark clouds".

The absorption A_V of a dark cloud of given opacity class C was found to depend on the background brightness, i.e. the latitude b, according to $A_V = 0\overset{m}{.}55 \cdot \exp\,(|b|/5^\circ) \cdot C$.

Fig. 2 shows the extinction-corrected brightness distribution. The run of the Milky Way appears now much smoother and is more concentrated towards the Galactic equator. The maxima are south of the plane with $\bar{b} = -0\overset{\circ}{.}37 \pm 0\overset{\circ}{.}16$. With 2 kpc as characteristic distance, the Sun is positioned 13 ± 6 pc north of the Galactic plane. For comparison, the midplane of CO emission exhibits warps of amplitude $\pm$ 20 pc about $z = -20$ pc below the $b = 0^\circ$ plane (Sanders et al. 1984). More than 50 % of the visual light of the surroundings of the Sun is produced by stars with $M_V \lesseqgtr 1$ as can be inferred from the luminosity function (see, e.g., Scheffler and Elsässer 1982), i.e. by stars of spectral types O-A 2. Constructing the ultraviolet luminosity function from $\phi\,(M_V)$ by means of the colour index U - V it can be seen that more than 50 % of the U intensities is produced by O - B9 stars with $M_V \lesseqgtr 0$. Bahcall and Soneira (1980) and Scheffler (1982) have summarized the observations of the z-distribution and agree in giving a nearly constant scale height of about $z_0 = 110$ pc for all $M_V \lesseqgtr 2$. In the case of an exponential distribution the scale height is equal to 0.7 x full width at half maximum intensity: $z_0 = \overline{|z|} = 1.4\ z_{1/2}$. The scale height of the obscuring dust is somewhat controversial. Bahcall and Soneira (1980) assume the same value as for the young population I. This is confirmed by Sanders et al. (1984) for the molecular clouds. From a study of the reddenings of more than 10^4 stars and clusters Neckel and Klare (1982) determined, within 0.5 kpc from the Sun, $z_0 = 40$ pc. The width of Feitzinger and Stüwe's (1986) dark cloud distribution in galactic latitude gives (for 0.5 kpc as

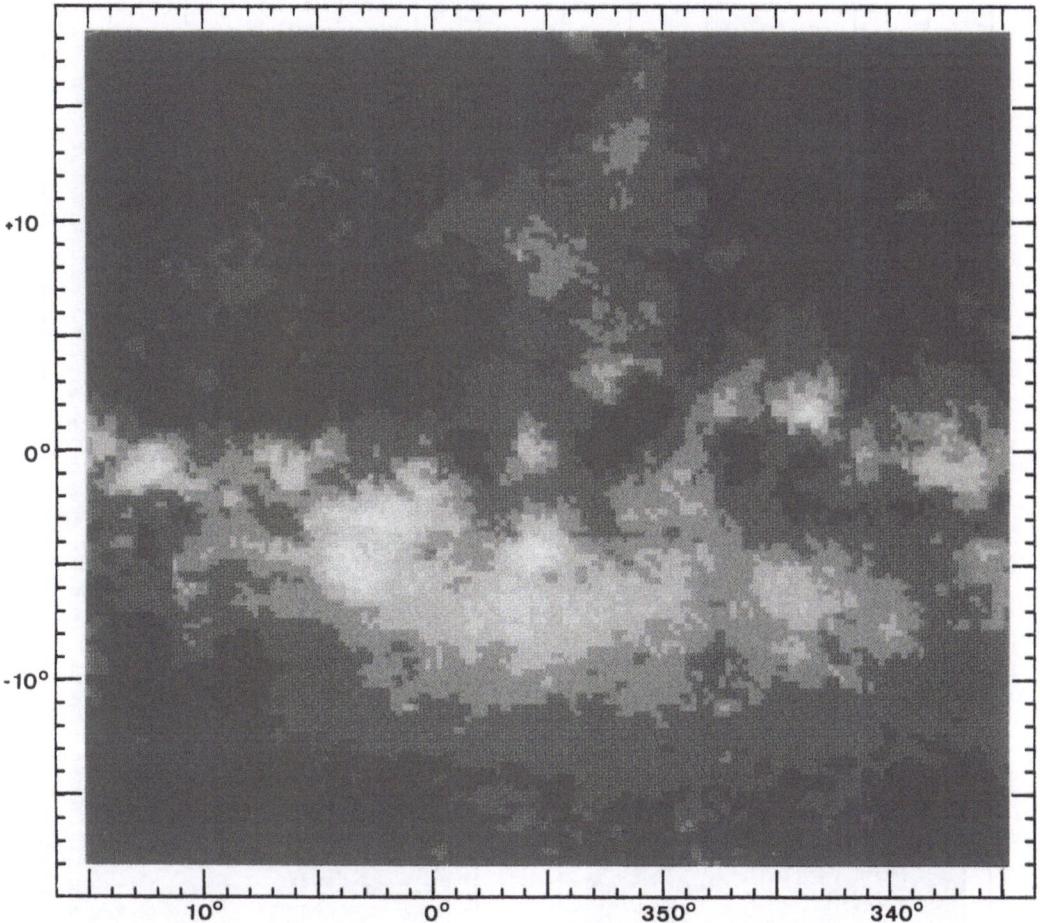

Fig. 1. The Galactic Center region in ultraviolet. This part of
the high-resolution photometric map of the Southern Milky Way
(0.25 degrees resolution) displays the bulge of our Galaxy, and
at l = 337°, b = -1° the eastern part of shingle 2.

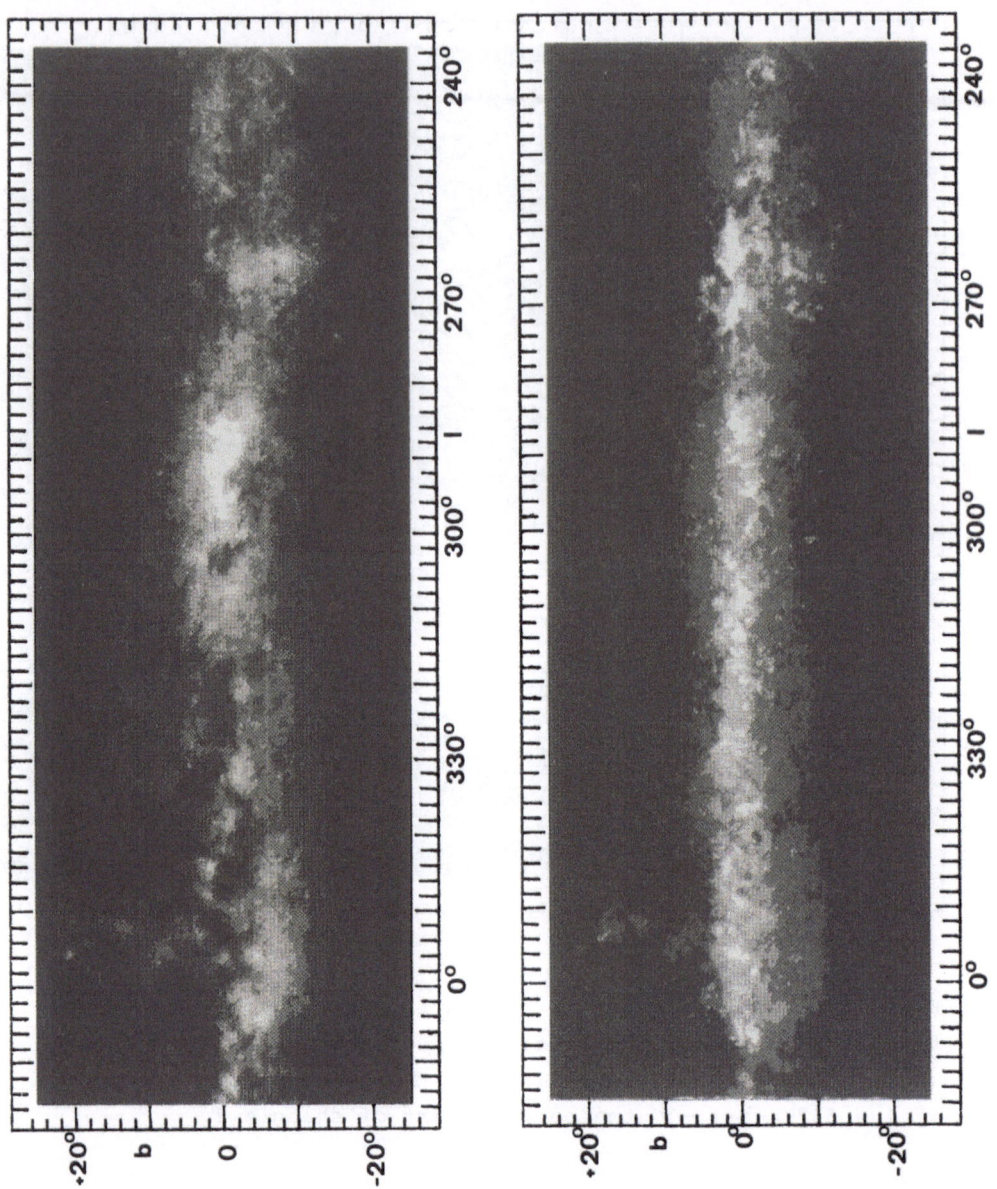

Fig. 2. Ultraviolet surface brightness of the Southern Milky Way
(left) and corrected for interstellar absorption (right). This
zero-order reconstruction of a "Milky Way without Dark Clouds"
shows a surprisingly smooth brightness distribution.

the mean distance) z_o = 75 pc. Feitzinger and Stüwe's survey reaches up to distances of about 1 kpc; they do not recognize clouds with $A_v \lesssim 0.^m5$ nor the continuously distributed obscuration. If we eliminate the influence of these dark clouds on the light distribution of the Milky Way then the Milky Way intensities will be only slightly affected by dark clouds for $|z| > z_o$ or $|b| > 4.^\circ5$. Indeed, the uncorrected and the corrected distribution for $|b| \gtrsim 5^\circ$ are very similar.

The whole Milky Way Photometry in U, B, V, R will be completed in the next two years and used to construct a quantitative model of our Galaxy.

References:

Bahcall, J.V., Soneira, R.M.: Astrophys. J. Suppl. 44 (1980), 73
Feitzinger, J.V., Stüwe, J.A.: Astron. Astrophys. Supp. 58 (1984), 365; corrig. 63 (1986), 207
Feitzinger, J.V., Stüwe, J.A.: Astrophys. J. 305 (1986), 534
Neckel, Th., Klare, G.: Mitt. Astr. Ges. 57 (1982), 249
Sanders, D.B., Scoville, N.Z.: Astrophys. J. s76 (1984), 182
Scheffler, H.: Landolt-Börnstein, Numerical Data Vol. 6 c (1982), 189
Scheffler, H., Elsässer, H.: Bau und Physik der Galaxis, Bibliograph. Institut Mannheim 1982, 121
Schmidt-Kaler, Th., Schlosser, W.: Astron. Astrophys. 29 (1973), 409
van den Bergh, S.: J. R. Astron. Soc. Canada 62 (1968), 149; Astron. Astrophys. 20 (1972), 469
Winkler, Chr., Schmidt-Kaler, Th., Pfleiderer, J.: Astron. Astrophys. Supp. 58 (1984), 705

FIRST RESULTS OF A MILKY WAY CONTINUUM SURVEY AT 45 MHz

H. Alvarez, J. Aparici, J. May and F. Olmos

Departamento de Astronomía, Universidad de Chile
Casilla 36-D, Santiago, Chile

INTRODUCTION

One of the important areas of research in modern astronomy is the study of the nonthermal continuum radio emission from our Galaxy since this is produced by the interaction of cosmic ray electrons with galactic magnetic fields. Because of the spectrum of the synchrotron mechanism, the emission is better studied at frequencies below about 100 MHz. However, since most of the world radio observatories are in the Northern Hemisphere and only very few operate at low frequencies, the Southern Hemisphere has not been well covered by ample continuum surveys. Partly to fill this need the University of Chile Radio Observatory at Maipú built during the 1970's a large array operating at 45 MHz (May et al. 1984) and with which the exploration of the southern sky has been practically finished. Table 1 shows the surveys covering the whole Southern Hemisphere below 100 MHz. This list does not include all-sky surveys prepared by combining observations at different frequencies (Yates, 1968; Cane, 1978; and Landecker and Wielebinski, 1970). The fourth column lists the temperature steps between continuous isophotes while the fifth column shows the percentage of the background temperature corresponding to that step. This number is intended to give an idea of the sensitivity of the system, and was computed for the arbitrary position α=05h, δ=-60°. In Table 1 it is seen that, in general, there are few southern surveys and that only two, at 85 and 45 MHz, were made with a good combination of angular resolution and sensitivity. The most recent all-sky survey and closest to 45 MHz is that by Haslam et al. (1982) at 408 MHz. It has an angular resolution of 0°.85 x 0°.85, ΔT=2 K and $\Delta T/T$=10%.

OBSERVATIONS

The observations were made with the 45-MHz transit array at the Maipú Radio Astronomy Observatory. The array is made up of 528 full-wavelength dipoles oriented E-W and distributed into six groups that feed a Butler matrix. This matrix produces several beams, each 4°.6(α) x 2°.4(δ) at the zenith, staggered along the meridian and sepa-

rated approximately by 1°8, The beams observe independent and simultaneously and can be steered, all together, in declination by electric
phasing. Since only the four central beams were used, the antenna at
a given position scans the sky in four strips covering a band about
5°4-wide at a time. The instrument behaves well approximately within
50° from the zenith, therefore the declination coverage of the data

TABLE 1

Continuum surveys covering the whole Southern
Hemisphere below 100 MHz

f (MHz)	Declination coverage	Angular resolution ($\alpha° \times \delta°$)	ΔT (K)	$\Delta T/T$ (%)	Instr.	Ref.
100	30° to -90°	17x17	50	9	array	(1)
85	-20 to -90	3.5x3.8	100	10	dish	(2)
55	20 to -90	14x14	2000	50	dish	(3)
45	20 to -86	4.6x2.4	100	4	array	(4)
30	0 to -90	11x11	1800	25	dish	(5)
16.5	0 to -90	1.5x1.5	2x10	25	array	(6)
to		to				
2.1		7.5x7.5	10	25	array	

(1) Bolton and Westfold (1950), (2) Yates et al. (1967), (3) Rohan and
Soden (1970), (4) This paper, (5) Mathewson et al. (1965), (6) Ellis
(1982).

ranges from practically the South Pole up to about +20°. The antenna
was moved through nineteen selected positions in such a way that the
end beams of adjacent positions overlapped, producing the anchorage of
the declination strips. Since low frequency observations are very sensitive to climatic and ionospheric conditions, and to man-made interference, the antenna was kept in each position for several weeks in order
to obtain good data. Also, to help eliminating diurnal and seasonal
effects, each positions was observed in two epochs separated by about
six months.

Calibrations signal from a noise generator were injected through
the preamplifiers every hour and their temporal drift in gain was corrected by linear interpolation. To achieve a highly stable gain the
receiver was installed in a temperature-controlled box. To make an
overall calibration of the system, whose temperature is dominated by
the sky noise, we observed a region in the sky (α=19h 00mh, δ=+5°),
well covered at low frequencies by northern observers, and measured
an antenna temperature of 29500 K, in good agreement with them. The

pointing was constantly checked with strong radio sources. The observations were made with 1 MHz bandwidth, and from 1982 throughout 1985.

THE MAP

The data consist of a large number of antenna temperature profiles 24-h long for declinations which are integer multiples of 1°8. Since, for the reasons described earlier, the profiles obtained are not exactly the same for a given declination, we have adopted as the true profile the envelope of all good data available for that declination. In right ascension there are data points every minute. In declination the data are separated by 1°8 and the antenna temperature was linearly interpolated to generate profiles at intermediate declinations. In the matching of the data by overlapping end beams of adjacent positions the criterion adopted is that the highest temperature prevails, and any rise in temperature is distributed linearly across the four beams. Actually we found that the angular distance between beams is not constant at 1°8, and that it increases with zenith distance. Since the software was prepared to accept antenna temperature profiles at the selected declinations separated by multiples of 1°8, we reduced the actual declination to the corresponding selected values by linear interpolation.

The map was corrected for errors in the azimuth orientation of the array and in the levelling of its reflector plane. The map has not been corrected for either beam smoothing or sidelobes effects and the extragalactic sources have not been removed. To make the map more useful the presentation will follow the format of the 408-MHz survey by Haslam et al. (1982). The declination coverage will range from -86° up to +20°.

Figure 1 shows the South Pole region at 45 MHz. For comparison it shows also the map at 85 MHz by Yates et al. (1967), the best available at low frequencies. Both maps have comparable angular resolution and the contour separation is the same, 100 K. The 45-MHz sensitivity is better as it is evidenced by the presence of the Small Magellanic Clouds (SMC) about 01h and -72°, and of the strong radio source 0842-75, both of which are absent in the 85-MHz map. The Large Magellanic Cloud (LMC) is prominent in both maps, however the center of its distribution is different because the 45-MHz map is dominated by the synchrotron emission while the 85-MHz is predominantly thermal emission from 30 Dor (Alvarez et al. 1987). A peculiar feature, seen in both maps, is a tongue of sharp contour laying between the SMC and the pole. Another interesting feature is a spur that points away from the tongue and towards the LMC; we are currently investigating the possibility that

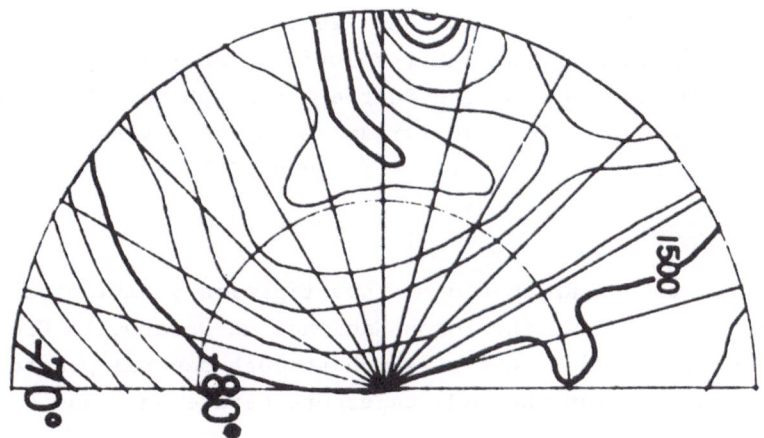

85 MHz   Yates et al.(1967)   3°5 × 3°8

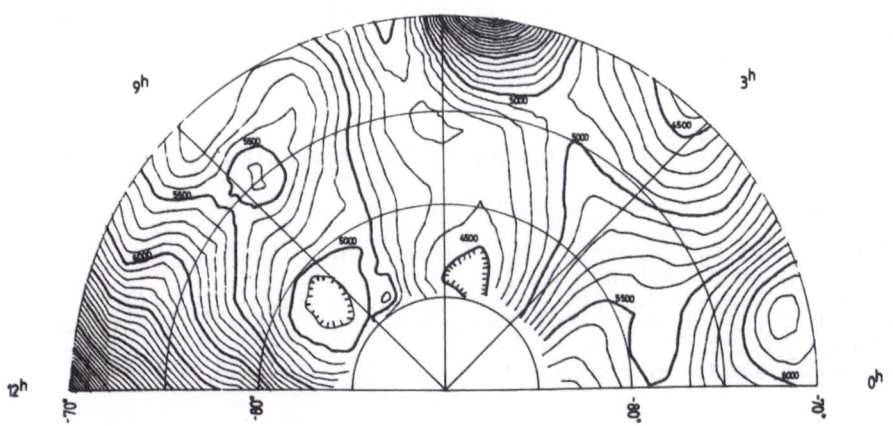

Maipu Survey at 45 MHz 4°6 × 2°4

Figure 1. South Pole region. Contours represent antenna temperature,
in K, in steps of 100 K. Shown for comparison is the 85-MHz
map by Yates et al. (1967). Several interesting features discussed in the text are seen on the Maipú Survey.

this spur is a bridge that connects both Clouds. The intense radiosource
0410-75   is at the tip of the 5000 K isophote and it probably sharpens
it a bit, however the source does not appear obviously on the map be_
cause it scintillates strongly. The minima at about right ascensions
04h and 08h, and declination -82° could be due to absorption by local
HII regions. A special study of the whole region surrounding the two
Magellanic Clouds is being conducted and preliminary results on the SMC
will be reported elsewhere.

CONCLUSIONS

Over four years of observing we have covered 65% of the sky between approximately declinations -86° and +20°. Our 45-MHz survey will have the best combination of angular resolution, sensitivity and sky coverage among low frequency maps of the Southern Hemisphere. To facilitate comparisons we are using the same format as the map by Haslam et al. (1982) which is the all-sky survey at the frequency closest to 45 MHz.

ACKNOWLEDGEMENTS

We thank the Departamento de Investigación y Bibliotecas of the University of Chile and the Fondo Nacional de Desarrollo Científico y Tecnológico for funding this research through several grants. A. Gallardo was responsible for the radiotelescope maintenance and C. Monsalve for the photographic work.

REFERENCES

Alvarez, H., Aparici, J., May, J.: 1987, Astron. Astrophys. 176, 25
Bolton, J.G., Westfold, K.C.: 1950, Australian J. Sci. Res. A 3, 19
Cane, H.V.: 1978, Australian J. Phys. 31, 561
Ellis, G.R.A.: 1982, Australian J. Phys. 35, 91
Haslam, C.G.T., Salter, C.J., Stoffel, H., Wilson,: 1982, E.W. Astron. Astroph. Suppl. 47, 1
Landecker, T.L., Wielebinski, R.: 1970, Australian J. Phys. Astrophys. Suppl. 16
Mathewson, D.S., Broten, N.W., Cole, D.J.: 1965, Australian J. Phys. 18, 665
May, J., Reyes, F., Aparici, J., Bitran, M., Alvarez, H., Olmos, F.: 1984, Astron. Astrophys. 140, 377
Rohan, P., Soden, L.B.: 1967, Australian J. Phys. 23, 223
Yates, K.W., Wielebinski, R., Landecker, T.L.: 1967, Australian J. Phys. 20, 595
Yates, K.W.: 1968, Australian J. Phys. 21, 167

Galaxian Structure and X-ray Astronomy

Herbert Gursky
E. O. Hulburt Center for Space Research
Naval Research Laboratory
Washington, D. C. 20375

DEDICATION AND FOREWORD

It is a pleasure for me to acknowledge my 20 year personal and professional association with Frank Kerr and to contribute this short paper in his honor. My association with Professor Kerr began when Paul Gorenstein and I approached him to discuss the possibility of conducting radio observations in order to elucidate the properties of certain x-ray sources. Out of that emerged a collaboration, with Edward Grayzeck, that resulted in several papers relating the line of sight absorption in x-rays to the column density at 21 cm.

At the Symposium, and by way of introducing me, Gart Westerhout, in his inimitable style, told the story of how I appeared at the University of Maryland in 1966 just at the time the National Science Foundation was informing him that funding for radio astronomy was being reduced to allow for an increase in support of astronomers conducting optical observations of x-ray sources. I dimly recall a cool reception on the part of Westerhout; however, he did send me on to Kerr. Westerhout went on to state that several years later the National Science Foundation did increase its support for 21 cm observations. I'm not sure Westerhout realized how appropriate his introduction would be, since the events described in this paper could have contributed to both of these changes in funding.

This paper will be a very brief summary of our understanding of the relation of observed x-ray emission to the structure of normal galaxies, both how the subject began and where it now stands. As is frequently the case we, or at least I, thought we had a better understanding early in the discipline compared to what has presently emerged.

The observational basis for our understanding of the relation of x-ray emission to galaxian structure lies with three distinct kinds of information:

1. The distribution of discrete sources in the Milky Way.
2. The presence in our Galaxy of a soft, diffuse background.
3. **The observation of discrete sources and integrated emission in external galaxies.**

There are substantial limitations to using this information. There are no distance indicators generally available for discrete sources, meaning that sources can only be localized in the Galaxy on an ensemble basis. The opacity of the Galaxy is very high for photons below about a kilovolt so that the diffuse background is only seen locally and its interpretation is also limited by the lack of a distance indicator. And we are still in our infancy in terms of observing external galaxies in x-rays.

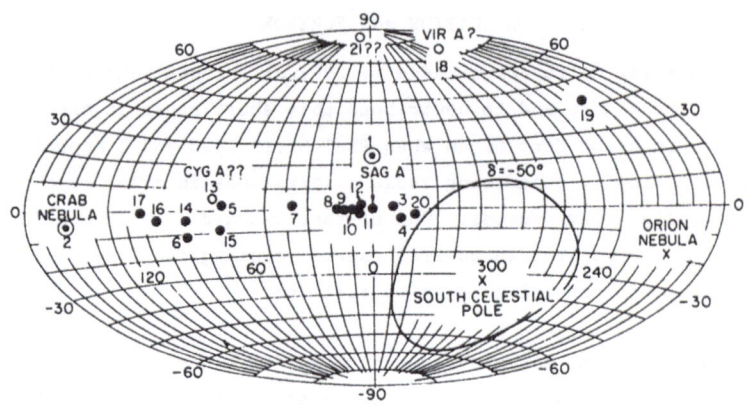

FIGURE 1. THE COSMIC X-RAY SOURCES AS KNOWN IN 1966, BASED ON
SOUNDING ROCKET OBSERVATIONS.

The first positive evidence for celestial x-ray emission was obtained from a rocket experiment conducted in 1962 (1). By 1966 a large number of discrete sources had been found. As shown in Figure 1, there was observed a strong concentration of sources at low galactic longitude. A second concentration of sources was seen in the Cygnus region and single sources were seen elsewhere near the galactic plane. Several sources were seen at high galactic latitude that could have been extragalactic. I and my colleagues at American Science and Engineering in Cambridge conducted two rocket experiments in 1966 that contributed to the information in Figure 1. The first of these was carried out in April in collaboration with the group at MIT and made use of modulation collimators that allowed for arc minute accuracy and resulted in the localization and identification of Sco X-1(2,3). As a second experiment that year, we could not simply refly the modulation collimator payload since there was no single source of sufficient brightness and interest to which we were willing to dedicate a single rocket flight. Instead we developed an entirely new payload, comprising slat collimators and proportional counters and decided to concentrate on localizing the sources in Cygnus. In addition we performed a single scan at the beginning of the rocket flight extending from Sagittarius to Cygnus. Incidentally, that payload contained the largest area of proportional counters flown up to that time, about 800 cm2, and represented the prototype of the Uhuru satellite instrumentation. The Cygnus region was chosen as the primary target because it was known to contain the bright sources Cyg X-1 and Cyg X-2 which were among the earliest observed sources. Also the NRL group had reported Cyg-A to be an X-ray source(4). Finally, Cygnus being at high declination was more favorably located for American optical astronomers and subject to less obscuration than the sources near the galactic center. The rocket experiment was conducted during October 1966 and resulted in a number of important results. We obtained accurate locations (10 arc min) for Cyg X-1, Cyg X-2 and the previously unreported Cyg X-3(5). An optical study of source locations did yield the optical counterpart of Cyg X-2, but no credible candidates for the other two(6). We did not see Cyg-A. Two other results emerged of special importance to the question of galactic structure. The scan along the Milky Way yielded positions of sources precise in galactic longitude and we observed a deficiency in low energy x-rays for several sources which could be interperted as absorption in the interstellar medium along the line of sight. It was this last result that sent us to Frank Kerr.

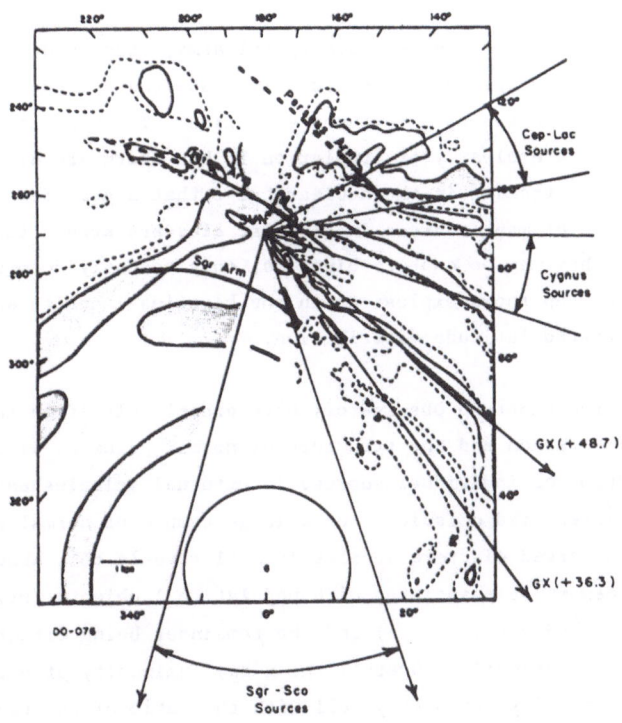

FIGURE 2. OVERLAY OF REGIONS OF LONGITUDE CONTAINING X-RAY SOURCES
ON SHARPLESS' MAP OF THE SPIRAL ARMS.

The distribution of these sources with respect to the then picture of the spiral arm structure of the galaxy is shown in Figure 2 which is an overlay of the longitude regions of the x-ray sources on the spiral arms as found in 21 cm radiation. We concluded that there was a strong correlation between the x-ray sources and the spiral arms based on the decrease in source density beyond l=20 degrees and the apparent correlation between sources and the Serpens, Cygnus and Perseus spiral arms(7).

Subsequent observations, especially with the Uhuru satellite, showed that many of our interpretations of the Cygnus rocket results were incorrect. The low energy deficiency in x-ray spectra can be intrinsic to the sources rather than being caused by the interstellar medium. Cygnus X-3, whose line of position is directly along the Cygnus spiral arm, is located well beyond that arm based on the observation of 21 cm absorption in its radio emission. Also we missed what could have been a spectacular result; namely, the optical identification of Cyg X-1, which was in fact the brightest star in our error box. We were looking for a faint, blue object based on the Sco X-1 identification.

Regarding the correlation of sources with spiral arms, studies of individual sources and models of the origin of the sources indicate the presence of a mix of populations that is complicated by a tortuous evolution. The high-mass binaries, as exemplified by Cyg X-1, are clearly of population I, but there are also numerous low mass systems and a concentration in the nuclear bulge that argues for population II objects. The existence of many sources in globular clusters argues for very old systems, except that they may be capture binaries, only recently formed. Also the possible occurrence of supernova explosions in the binaries leads to anomolous velocities and a distorted latitude distribution.

Observations with the Einstein Observatory have greatly clarified the relation between x-ray source emission and the structure of normal galaxies since it has provided the observation of individual sources in external galaxies and of the distribution of the integrated emission from a large sample of normal galaxies. The distribution of the observed discrete sources in M-31 reveals that about half the objects do in fact seem to be associated with population I objects (bright visible objects, dust, neutral hydrogen. . . .) and the remainder being either globular cluster or inner bulge sources(8). Overall the x-ray luminosity of normal galaxies tracks the visible luminosity reasonably well with the ratio of the two being roughly 2×10^{-4}, independent of galaxy type.

Returning to our own galaxy, the soft-x-ray background has been mapped out in a series of sounding experiments by the Wisconsin group (9). The striking feature of the background distribution is that the emission is actually fainter along the equator than it is at mid-latitudes. This immediately points out the effect of the high opacity of the interstellar medium to soft x-rays and vitiates against all but local studies being carried out with these data. Attempts to map this soft x-ray emission in external galaxies have not been particularly successful.

There has been remarkable progress in x-ray astronomy and important information regarding the galaxy has emerged from the synthesis of x-ray, radio and optical observations. To go further, we need to discuss prospects. Unhappily, nothing will happen soon, given the state of the space sciences and the necessity for major observatories. In order for the X-ray sources to be useful tracers of galactic structure it will be necessary to find means to distinguish the various categories of sources and to measure their distance, both of which should be possible. Also it may be possible to record the soft X-ray background at a high enough energy (1 kev) to avoid the worst of the opacity effects and to allow its mapping in external galaxies. It is still the case that the X-ray sources, being very bright. are observable to large distances and that the soft X-ray background carries unique information on the interstellar medium. These and other factors are bound to be important in developing a better understanding of the structure of galaxies.

REFERENCES

1. Giacconi, R., Gursky, H., Paolini, F. R., and Rossi, B., Physics Review Letters, 9, 439, 1962.

2. Gursky, H., et al., Astrophysical Journal, 146, 310, 1966.

3. Sandage, A., et al., Astrophysical Journal, 146, 316, 1966.

4. Byram, G.T., Chubb, T. A., and Friedman, H., Science, 152, 166, 1966.

5. Giacconi, R., Gorenstein, P., Gursky, H., and Waters, J. R., Astrophysical Journal, 148, L119, 1967.

6. Giacconi, R., et al., Astrophysical Journal, 148, L129, 1967.

7. Gursky, H., Gorenstein, P., and Giacconi, R., Astrophysical Journal, 150, L75, 1967.

8. Van Speybroek, L., et al., Astrophysial Journal (Letters), 234, L45, 1970.

9. Burstein, P., et al., Astrophysical Journal, 213, 405, 1977.

III. SPIRAL STRUCTURE

THE CLOUDY INTERSTELLAR MEDIUM:
AGGREGATION OF GIANT MOLECULAR CLOUDS
IN SPIRAL STRUCTURES

William W. Roberts, Jr. and David S. Adler
University of Virginia
Charlottesville, Virginia

Physical mechanisms and dynamical processes underlying the clumpy, cloudy ISMs of spiral galaxies are investigated. Through computational studies, we focus on the important roles played by the orbital dynamics of ISM clouds, dissipative cloud-cloud collisions, and self gravitational effects in the aggregation of giant molecular clouds in the global spiral structures of such galactic systems.

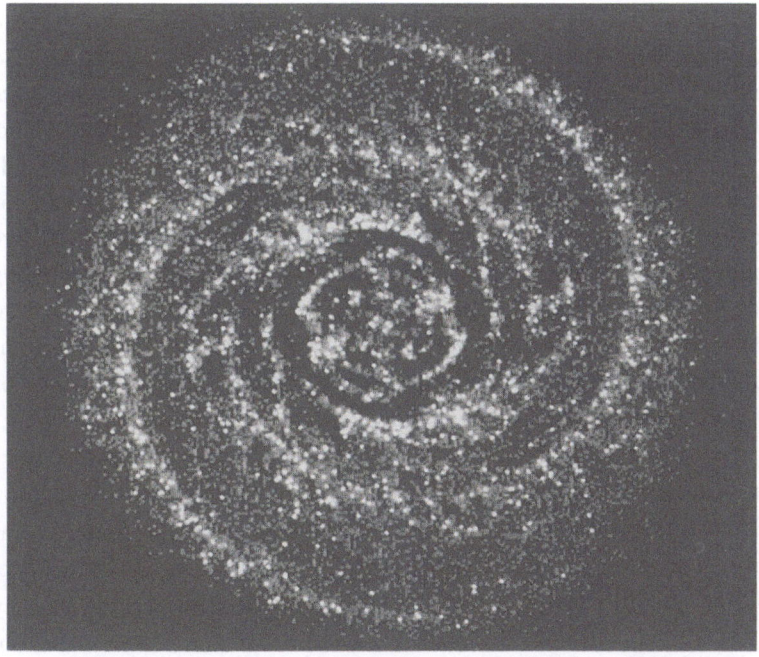

Figure 1. Photographic intensity map of the gas (patches) and young to middle-aged stellar association (white dots) distributions in a representative simulation at a sample epoch [t = 800 Myr] during the computations.

In the "N-body," cloud-particle computational code developed to study cloudy galactic disks (Roberts and Hausman, 1984; Hausman and Roberts, 1984; Roberts and Stewart, 1987; and Adler, 1987), the gaseous interstellar medium is simulated by a system of particles, representing clouds, which orbit in a spiral-perturbed galactic gravitational field. Self gravitational effects of the clouds are included via Fourier Transform techniques, adapted from those developed by Miller (1976) and Miller and Smith (1979a, 1979b). The "cloud-particles" undergo dissipative collisions with other clouds and experience velocity-boosting interactions with expanding supernova remnants. Associations of protostars form in clouds following such "collisions" and "supernova interactions" but take finite times before becoming active themselves and undergoing their own supernovae events.

Figure 1 shows the results of one representative simulation in which the gas mass to stellar disk mass is 10%. Plotted here, in a photographic intensity map at one sample time epoch [800 Myr] during the computations, is the computed global distribution of gas clouds, represented by patches, and young to middle-aged stellar associations active (with supernova events) during the past 60 Myr, represented by white dots. The gas clouds and the stellar associations triggered from the clouds exhibit aggregations of giant complexes along the global, spiral-wave-arm structures. The stellar associations are still strongly correlated with the gas, with few associations which are not adjacent to clouds. Self gravitational effects of the gas cloud system are playing an important role in the formation of the local aggregations and the massive structures exhibited on local scales.

In order to better view and study the gas by itself, we display in Figure 2 at the same sample time epoch [800 Myr] a photographic intensity map of the gas cloud system alone, without the stellar associations superposed. The raggedness and patchiness of the global gaseous spiral structure is indeed evident, with massive aggregations of clouds and giant cloud complexes appearing throughout the arm regions. Giant gaseous knots and some spur-like and feather-like features can be traced. The assembly of the giant cloud complexes from smaller clouds is found to be remarkably efficient in view of the fact that collisional coalescence of individual clouds into larger entities is not a requirement of the model. Dissipative collisions among clouds are not the primary mechanism which determines the global coherence and the strong arm-to-interarm density contrasts observed in these spiral arm structures, but are important for the growth of individual complexes on local scales (Roberts and Stewart, 1987). Of prime importance are the gravitational effects driving the low velocity dispersion gas, including its own self gravity. Gravitationally driven crowding and temporary trapping of cloud orbits in spiral arms, along with dissipative cloud-cloud collisions, underlie the aggregation of the clouds into the giant cloud complexes as well as the organization of the complexes along the ragged global spiral structures. Likewise, the dispersal of giant cloud complexes is not restricted in dependence on rapid and highly disruptive star formation episodes. Indeed, the loosely-associated aggregations and giant complexes of clouds, followed in their evolution in the computational studies, are found to continually disassemble and reassemble rather naturally over time under the influence of the various physical mechanisms and dynamical processes underlying the cloudy ISMs in these model spiral galaxies. It may be that substantially fewer GMCs than heretofore thought are gravitationally bound entities.

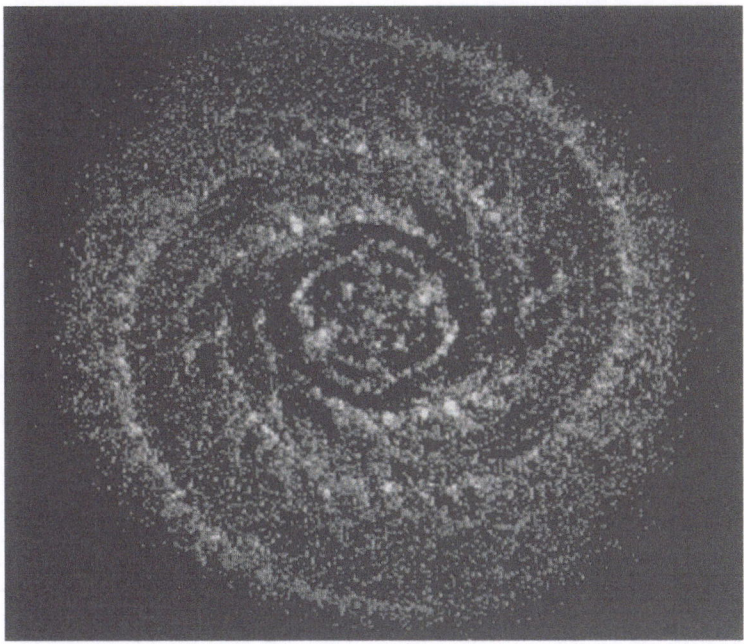

Figure 2.   Photographic intensity map of the gas (patches) alone, without the stellar associations superposed.

One primary motivation for formulating and developing this type of computational approach is the ability to simulate and study such ragged, realistically-disorderly model "snapshots" of galaxies. What is most striking is the strength with which gas cloud aggregations and complexes interact on local scales and strongly perturb the global two-armed gaseous spiral structure. These perturbations take the form of multitudes of spurs, feathers, and secondary features which continually break apart and reform as the loosely-associated aggregations and giant complexes of clouds continually disassemble and reassemble over time.

Figure 3 displays variations of selected physical quantities with respect to spiral phase around a representative annulus in the disk at two sample time epochs [800 Myr and 900 Myr]: cloud number density, velocity components perpendicular and parallel to spiral equipotential loci, velocity dispersion among gas clouds, and distribution of young to middle aged stellar associations active during the past 60 Myr. First and foremost it is important to note how strong the enhancements are in number density of the gas cloud system. The density distribution of the self gravitating cloud system is strongly-peaked with peak-to-mean values $[n_{max}/<n>]$ typically 3:1 and arm-to-interarm contrasts $[n_{max}/n_{min}]$ typically 6:1, with arm thicknesses typically on the order of one kpc. Gas clouds move into the gaseous density wave arm (from left to right) where they pile up. Note the sharp deceleration in the $u_\perp$ velocity component from supersonic to subsonic near 180^0 spiral phase, with a much more gradual rise downstream. This charac-

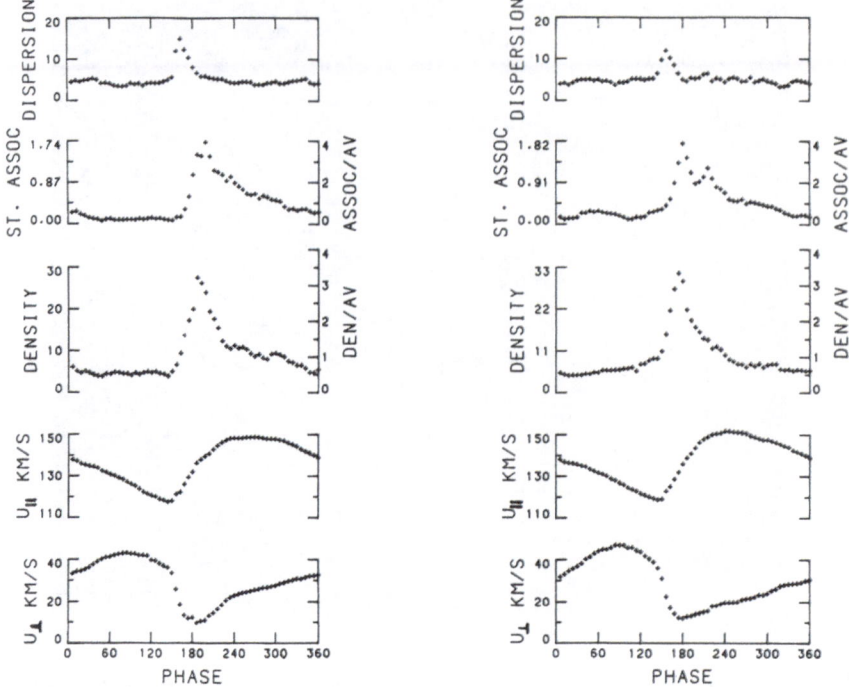

Figure 3. Spatial variation, at two sample time epochs [t = 800 and 900 Myr], of young to middle-aged stellar associations (active during the past 60 Myr) and of the number density, velocity dispersion, and velocity components (perpendicular and parallel to spiral equipotential contours) of the gas cloud system, plotted versus spiral phase about a representative annulus (500 pc wide) at 8 kpc. Comparison of the distributions at the two epochs demonstrates that a steady state prevails despite transient (stochastic) variations on local scales.

teristic skewness in the $u_\perp$ velocity component as well as the corresponding characteristic asymmetry in the parallel velocity component, both representative of a galactic shock, is as strong here as in non self gravitating simulations. Such skewness is less apparent in the density distribution, with the density rise occurring over a broad shock width of 370 - 510 pc (five to seven cells). If interstellar dust travels with and is aggregated with the cold cloudy interstellar gas, then such gas density ridges might also have strong dark dust concentrations and be traceable by dark dust lanes. The enhancement in the distribution of triggered stellar associations, active during the past 60 Myr, is broad and covers a region shifted downstream by as much as several hundred parsecs on average from the gas density ridge and by as much as 500 pc from the middle of the shock region.

Figure 4 shows the velocity field for the cloud system at the sample time epoch: t = 800 Myr. Instantaneous velocity vectors of 1000 randomly selected clouds (per half disk) are displayed. Dots mark the current positions of clouds; line segments point along velocity directions. Note the strong convergence of the flow in the regions of the spiral arms, evidenced through the systematic difference between orbital directions of clouds entering the arms and those of clouds leaving the arms. An equally strong divergence of the flow is evident in the interarm regions. Such convergence of the flow field, followed by divergence, constitutes <u>orbit crowding</u> of the collective system of cloud-particle orbital trajectories in the regions of the global spiral arms. Indeed the strong pile-up of gas clouds in the arms is interpreted here to be a manifestation of the collective orbit crowding phenomenon, driven by the spiral perturbed gravitational field and gas self gravity. In the presence of a 5% to 10% spiral perturbing force field, the low dispersion cloud system is capable of participating in strong orbit crowding, leading to strong pile-up of gas clouds in global spiral arms. Dissipative cloud-cloud collisions aid gravitationally driven orbit crowding in the assembly of massive cloud complexes from smaller clouds. It is possible that this leads to the subsequent gravitational collapse of massive cloud agglomerations, providing suitable environments for active star formation.

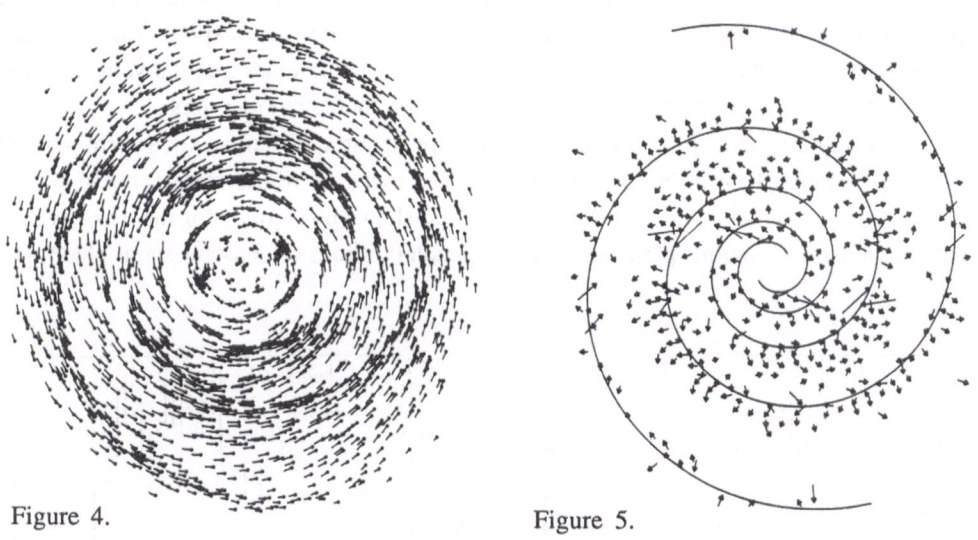

Figure 4.   Figure 5.

Figure 4.   Instantaneous velocity vectors of 1000 gas clouds (per half disk) randomly selected. Dots are at current positions of clouds; line segments point along velocity directions. Note that all velocities are shown with respect to the frame corotating with the spiral perturbation.

Figure 5.   Local velocity differences between postsupernova associations and clouds. Only differences significant at better than the 1 σ level are shown.

The intriguing nature of the cloud system's orbit crowding leads us to inquire whether there are any significant differences between the stellar and cloud velocity fields. Figure 5 examines this by displaying the <u>differences</u> between the average stellar and cloud velocities. Each arrow is at the center of one of the 576 bins and points in the direction of the (stellar association - cloud) velocity difference. The stellar associations considered here are those young to middle aged stellar associations which have undergone their supernova events up to 60 Myr in the past. Only those bins in which the velocity difference is statistically significant to the one-sigma level are represented, assuming an intracell velocity dispersion of 6 km s^{-1} (one dimension).

We can see that across much of the disk, especially the interarm regions, the difference between the average stellar and cloud velocities is indeed small. However near the arms the differences are significant and of order of 10-20 km s^{-1}. In particular, stellar associations just outside (i.e. downstream) of the spiral arm region tend to have a positive radial component relative to the clouds. It is natural that the differences should be greatest in the arms, since this is where the higher frequency of cloud collisions and SNR impulses will have the greatest influence on the clouds' otherwise unperturbed trajectories. The positive radial velocity difference of the stellar associations in the arms apparently reflects the loss of kinetic energy which collisions inflict upon the cloud orbits. Note that because the stellar associations considered here were all formed within clouds no more than 80 Myr (and active no more than 60 Myr) in the past, about two-thirds the time required for a single passage between spiral arms, the stellar associations are still rather strongly correlated with their parent clouds and the velocity differences shown in Figure 5 are lower than is typical for older, more common stars.

In our model the velocities of clouds and newborn stellar associations are determined self-consistently by their physical interactions, so we need make no restrictive assumptions about initial velocities of the newborn associations. Figure 6 shows histograms for the velocity components of stellar associations perpendicular and parallel to lines of constant spiral phase. The

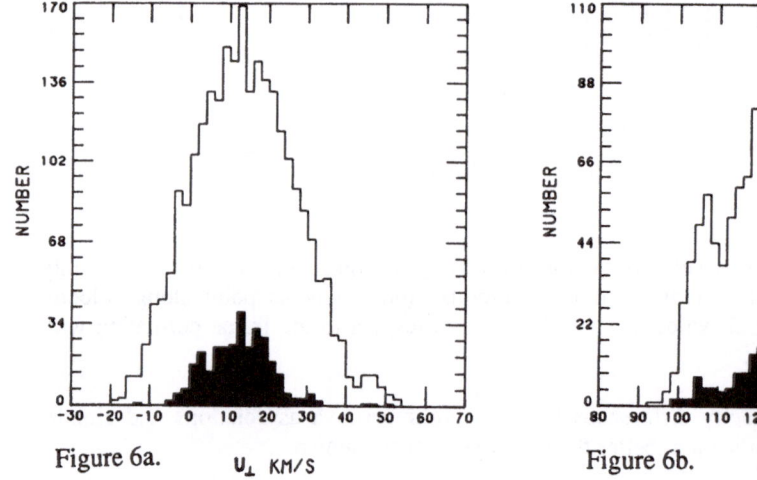

Figure 6a. $U_\perp$ KM/S   Figure 6b. $U_\parallel$ KM/S

Figure 6. Histograms of the components of the stellar association velocities perpendicular and parallel to loci of constant spiral phase. The blackened sample is presupernova associations; the white sample is postsupernova associations. Both samples are restricted to galactocentric radius 4 - 10 kpc and spiral phase 150^0 - 240^0.

black-shaded sample denotes the presupernova (not yet active) associations; the white sample denotes the postsupernova associations active during the past 60 Myr. Both samples are restricted to galactocentric radius 4 - 10 kpc and spiral phase 150^0 - 240^0. Because the velocity changes in our cloud-particle models' arms are not discontinuous, we find initial velocities spanning the full range between pre-arm and post-arm values. In Figure 6a, we see that both presupernova and postsupernova associations are peaked strongly in the range 10 - 20 km s^{-1} with only small tails in the negative region. Very few associations recross the potential minimum inwards; most associations penetrate the downstream interarm region to varying distances.

Acknowledgements

This work was supported in part by the National Science Foundation under grants AST-82-04256 and MCS-83-04459 and the National Aeronautics and Space Administration under grant NAGW-929.

References

Adler, D.S., 1987, Ph.D. Dissertation, University of Virginia, in preparation.

Hausman, M. A. and Roberts, W. W. 1984, Ap. J., 282, 106.

Miller, R.H. 1976, J. Comput. Phys., 21, 400.

Miller, R.H. and Smith, B.F. 1979a, Ap. J., 227, 407.

Miller, R.H. and Smith, B.F. 1979b, Ap. J., 227, 785.

Roberts, W. W. and Hausman, M. A. 1984, Ap. J., 277, 744.

Roberts, W.W., and Stewart, G.R. 1987, Ap. J., 314, 10.

CO EMISSION FROM THE SOUTHERN GALACTIC PLANE AND GALACTIC STRUCTURE

W.H. McCutcheon, Department of Physics
University of British Columbia, Vancouver, B.C.
and
B.J. Robinson, R.N. Manchester, J.B. Whiteoak
Division of Radiophysics, CSIRO, Sydney

ABSTRACT

The original southern galactic plane CO survey (294° < 358° and 2° < ℓ < 13° at -0°.075 < b < 0°0.75) has been extended to cover the longitude range 279° < ℓ < 300° at -0°.95 < b < -0.70. The Carina arm is well defined in CO emission over a length of eleven kiloparsecs. The tangent point, and emission inside and outside the solar circle are clearly distinguished in the ℓ-v plane as a characteristic loop structure. This transforms to part of a spiral segment in the galactic plane with a pitch angle of about 11°.

CO and HI terminal velocities measured over the complete longitude range above are in close agreement. Our fourth quadrant CO terminal velocities are also compared with CO velocities from the first quadrant.

INTRODUCTION

The initial CO survey of the southern galactic plane, made with the 4-m telescope at Epping, NSW, covered 294° < ℓ < 358° with -0°.075 < b < 0°.075, and subsequently was extended to cover 2°< ℓ < 13° (Robinson et al. 1982, 1984). This survey was combined with northern surveys (Robinson et al. 1983, 1984) and analyzed for evidence of spiral structure in our own Galaxy. Emission from the CO molecule is the best tracer of the giant molecular clouds which have been identified as the regions where massive stars form in the Galaxy. The warm molecular clouds (T_k>10 K) have been shown by the observations of Sanders et al. (1985) to be a spiral arm population closely associated with HII regions, whereas the cool clouds (T_k<10 K) appear to be widespread in the Galaxy, both in and out of spiral arms.

The southern CO survey (Fig. 1 frontispiece) has well defined clumps of emission along the run of terminal velocities and these were interpreted as being tangential points to spiral segments. Robinson et al. (1983) presented evidence for four spiral segments with pitch angles in the range 11° - 13°. One of the segments from the fitted model predicted a tangential point at $\ell = 282°$ (the Carina arm), but this longitude was beyond the range of the original survey. Subsequently Cohen et al. (1985) reported CO observations of the Carina arm.

The southern survey was extended in 1984 to cover 279° < ℓ < 300° at -0°.95 < b < -0°.70 and CO emission from the Carina arm, extending over 11 kpc from the inner to the outer Galaxy and having a tangential point at $\ell = 282°$, is clearly detected. In this paper we show that the new data are consistent with the earlier analysis on spiral structure. We also compare the CO and HI southern terminal velocities, and the CO southern and northern terminal velocities.

THE CARINA ARM

CO emission, which defines the Carina arm, is easily identified in Fig. 1 (frontispiece) at both positive and negative velocities in the range 280° < ℓ < 300°. Since an arm crosses the line of sight very slowly around the tangential point, a change of only a few degrees in longitude covers many kiloparsecs of arm length. This results in a high density of sources in a given beam width which makes the tangent point a prominent feature. On the near side of the arm (the negative velocity side), clouds will be nearer and presumably more easily detected, but the number in the beam width is low because the telescope beamwidth subtends only a small length of the arm. On the far side, the number in the beamwidth is higher because the telescope beamwidth subtends a larger length of arm, but the clouds are more distant and hence weaker. These features are evident in the loop of emission in Fig. 1, with emission being most intense around the tangential point, and becoming much weaker on the negative velocity side of the arm for $\ell > 300°$, and on the positive velocity side at $\ell \simeq 300°$.

To determine the pitch angle of the segment of arm shown in Fig. 1, it is convenient to transform that part of the ℓ-v plane into (ℓnR, ϕ) co-ordinates, shown in Fig. 2. A logarithmic spiral (Robinson et al. 1983) will appear straight on this plot, with a slope that depends on the pitch angle. The number of clouds and their positions were determined from the observed spectra, and the distances were determined from the association with HII regions from the survey of Caswell and Haynes (1987). Only the molecular clouds associated with HII regions having unambiguous distances are shown. A cloud was assumed to be associated with one or more HII regions if there was a high correspondence among their ℓ, b, and v values. In constructing Fig. 2, we use the value $R_0 = 10$ kpc, in order to use the distances derived by Caswell and Haynes. The solid line is the best fit to the data points. However, actual galactic features are not lines, but have a finite radial thickness. The dashed lines indicate a radial thickness of 1 kpc. As seen in Fig. 2, much of the scatter in the data points must be due to the finite thickness, a point also stressed by Caswell and Haynes (1987).

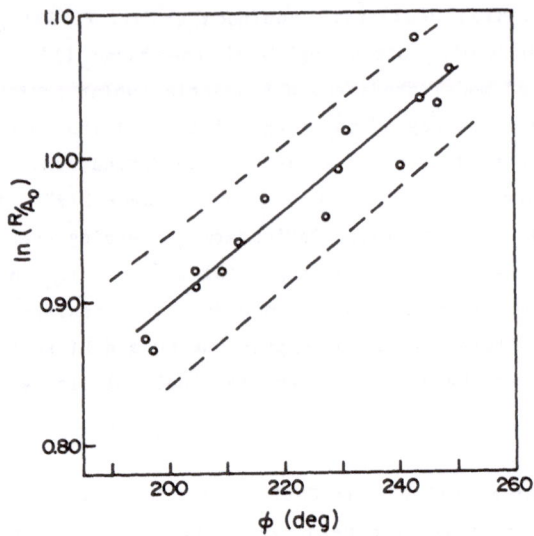

Figure 2. A ($\ln R$, ϕ) plot for the clouds of the Carina arm in Fig. 1. R is the distance to a cloud and was determined from the association with HII regions from the survey of Caswell and Haynes (1987). Only the molecular clouds associated with the HII regions having unambiguous distances are shown. A_0 = 4 kpc is the constant used by Robinson et al. (1983). ϕ is the azimuthal angle of a cloud measured in the direction of increasing longitude starting from ℓ = 0°. The dashed lines indicate an arm thickness of 1 kpc. The slope of the solid line gives a pitch angle of 10°.5.

The pitch angle, determined from the solid line is 10.5°, consistent with our earlier analysis on spiral structure (Robinson et al. 1983). Our results on the Carina arm are in agreement with the measurements of Grabelsky et al. (1987), and Cohen et al. (1985).

TERMINAL VELOCITIES

At most longitudes, the fourth quadrant data (Fig. 1 frontispiece) show well defined maximum velocities which can be associated with the tangential velocities of gas at galactocentric radii R/R_0 = sin ℓ. The terminal velocities, V_T, were measured in a manner similar to that used by Burton and Gordon (1978). Further details are given in Robinson et al. (1987).

The values of V_T are plotted as a function of sin ℓ in Fig. 3(a). The run of terminal velocities is approximately linear over a large range of sin ℓ (-0.955 to -0.25). There are, however, a number of deviations from this linear variation. Significant "holes" where the terminal velocity is more positive than the general trend can be seen at several longitudes, eg. sin ℓ = -0.64 or ℓ = 320°. These holes, which are readily visible in Fig. 1, are not due to the limited latitude coverage of the survey (Robinson et al. 1987) but are regions which are genuinely

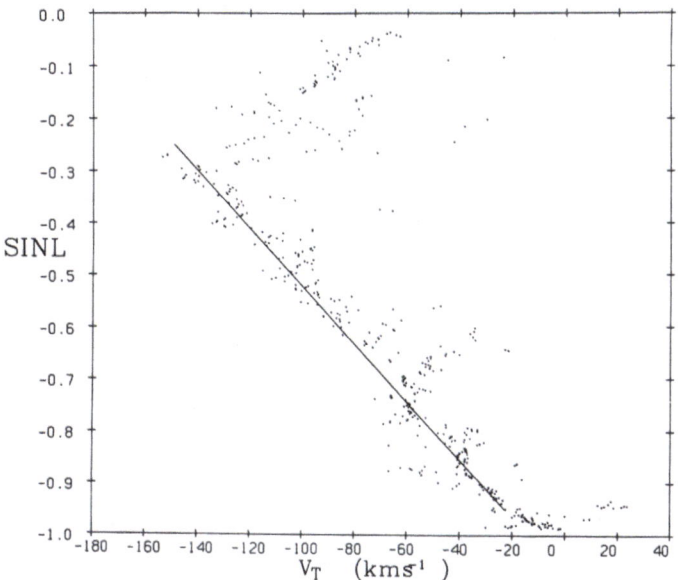

Figure 3(a). CO terminal velocities plotted against sin ℓ. The straight line represents the equation $V_T = -180 \sin \ell -194$ kms^{-1}.

deficient in gas. At other longitudes, eg. sin $\ell \simeq -0.4$, -0.5, and -0.78, or $\ell = 336°$, $330°$, and $309°$, the terminal velocities are more negative than the general trend. These kinematic irregularities may indicate streaming motions of ~ 10 kms^{-1} (eg. Burton and Gordon 1978). The points near sin $\ell = -0.88$ ($\ell = 298°$) may be spurious as any emission at these more negative velocities is weak.

In Fig. 3(b), the fourth quadrant HI terminal velocities, derived from the data of Kerr et al. (1987) are compared with the CO velocities from Fig. 1. There is excellent agreement between the two sets of data, both in the general linear trend and in departures from the trend, indicating that the kinematics of the atomic and molecular gas are very similar.

In Fig. 3(c), the fourth quadrant CO terminal velocities are compared with CO terminal velocities for the first quadrant (Burton and Gordon 1978, Clemens 1985). The overall trend of the CO velocities from the two quadrants is very similar, indicting that the dominant large scale motion of the Galaxy is circular rotation.

For $|\sin \ell| \sim 1$, the slope of the terminal velocity line gives AR_0 where A is Oort's first constant. In Fig. 3(a), the slope of the line yields the value AR_0 = 90 ± 3 kms^{-1}, considerably lower than Northern Hemisphere values such as 110 kms^{-1} (Gunn, Knap and Tremaine 1979), and 126 kms^{-1} (Clemens 1985), although we note that the data for sin $\ell < -0.95$ depart significantly from the straight line. This may be a result of streaming motions in the Carina arm which crosses the solar circle close to its tangent point at $\ell = 282°$ (see Fig. 1). Because of this, it

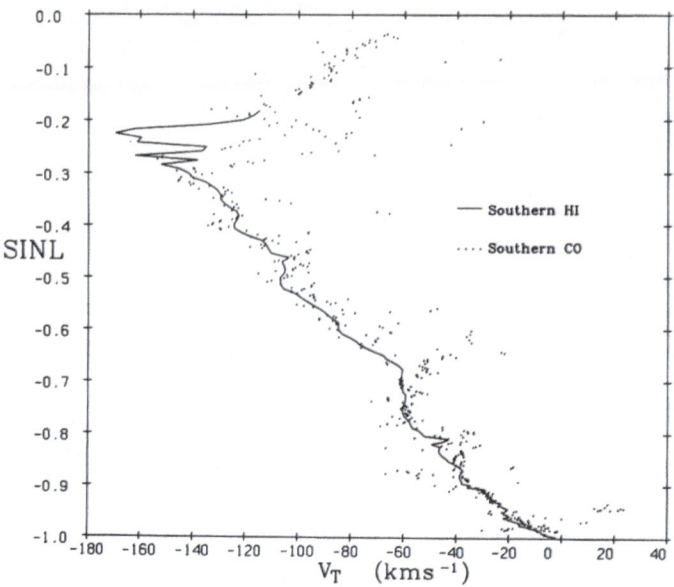

Figure 3(b). A comparison of CO and HI terminal velocities. The HI terminal velocities were derived from the data of Kerr et al. (1986).

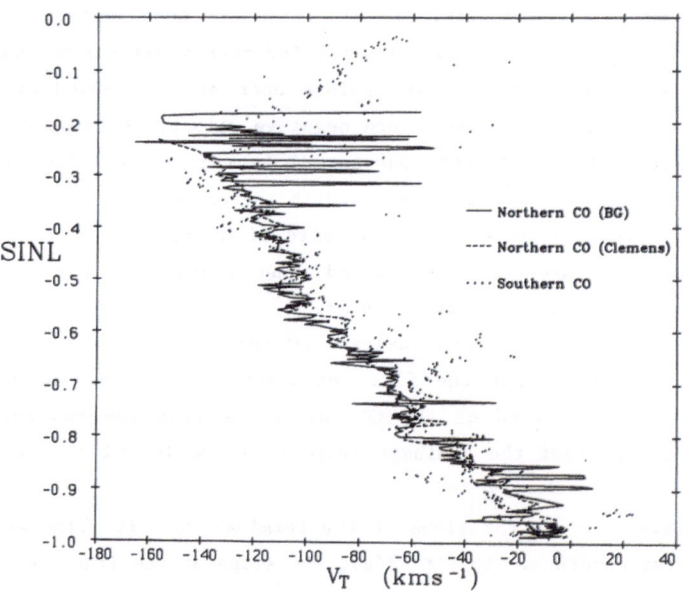

Figure 3(c). Comparison of CO terminal velocities for the first and fourth galactic quadrants. The northern curves are from Burton and Gordon (1978), and from Clemens (1985). Both of these first-quadrant curves have the signs of the axes reversed.

may not be possible to obtain a reliable value for AR_0 from these data.

Any peculiar velocity of the l.s.r. about the galactic centre can be inferred from terminal velocities at $R/R_0 = |\sin \ell| = 1$. Shuter (1982) deduced a peculiar azimuthal l.s.r. motion of 7 kms^{-1} toward $\ell = 90°$ and Clemens (1985) obtained a similar value. Fitting a straight line to our data over the range $-0.995 > \sin \ell > 0.988$ yields the intercept $V_T(R_0) = 4 \pm 2$ kms^{-1}, which is consistent with the previously obtained value. However, because of possible Carina arm streaming motions mentioned above, our value may not be that reliable.

Fig. 3(c) shows a systematic tendency for the northern values of $|V_T|$ to be less than the southern values in the ranges $0.25 < |\sin \ell| < 0.45$ and $0.85 < |\sin \ell| < 0.98$, and greater in the range $0.5 < |\sin \ell| < 0.8$. This type of variation with longitude is not consistent with a radial motion of l.s.r., a conclusion also reached by Clemens (1985).

SUMMARY

Our Southern Hemisphere CO survey, extended to cover $279° < \ell < 300°$ at $-0°.95 < b < -0°.70$, clearly shows CO emission from the Carina arm, extending over 11 kpc from the inner to the outer Galaxy and having a tangential point at $\ell = 282°$. The loop structure in the ℓ-v plane transforms to part of a spiral segment in the galactic plane with a pitch angle of about 11°.

CO and HI terminal velocities, measured over the range $279° < \ell < 350°$ are in very close agreement. The overall trend of the fourth quadrant CO terminal velocities is very similar to that of the first quadrant data, indicating that the dominant large scale motion of the Galaxy is circular rotation. Small differences of ~ 8 kms^{-1} between the two data sets form a variation with longitude which is not consistent with a radial motion of the l.s.r.

REFERENCES

Burton, W.B., and Gordon, M.A. 1978, Astron. Astrophys. 63, 7.
Caswell, J.L., and Haynes, R.F. 1987, Astron. Astrophys. 171, 261.
Clemens, D.P. 1985, Astrophys. J. 295, 422.
Cohen, R.S., Grabelsky, D.A., May, J., Bronfman, L., Alvarez, H., and Thaddeus, P. 1985, Astrophys. J. 290, L15.
Grabelsky, D.A., Cohen, R.S., Bronfman, L., and Thaddeus, P. 1987, Astrophys. J. 315, 122.
Gunn, J.E., Knapp, G.R., and Tremaine, S.D. 1979, Astron. J. 84, 1181.
Kerr, F.J., Bowers, P.F., Jackson, P.D., and Kerr, M. 1986, Astron. Astrophys. Suppl. Ser. 66, 373.
Robinson, B.J., Manchester, R.N., Whiteoak, J.B., Otrupcek, R.E., and McCutcheon, W.H. 1987, Astron. Astrophys. (in preparation).
Robinson, B.J., Manchester, R.N., Whiteoak, J.B., Sanders, D.B., Scoville, N.Z., Robinson, B.J., McCutcheon, W.H., Manchester, R.N., and Whiteoak, J.B. 1983, in Surveys of the Southern Galaxy, eds. W.B. Burton, and F.P. Israel (Dordrecht: Reidel), p.1.
Robinson, B.J., McCutcheon, W.H., and Whiteoak, J.B. 1982, Int. J. Infrared and Millimeter Waves, 3, 63.
Sanders, D.B., Scoville, N.Z., and Solomon, P.M. 1985, Astrophys. J. 289, 373.
Sanders, D.B., Solomon, P.M., and Scoville, N.Z. 1984, Astrophys. J. 276, 182.
Shuter, W.L.H. 1982, Mon. Not. R. Astron. Soc., 199, 109.

SUBSTRUCTURE IN SPIRAL ARMS

by

Steven A. Balbus

Virginia Institute for Theoretical Astronomy

I. Introduction

The classical work of Goldreich and Lynden-Bell (1965) and Julian and Toomre (1966) has taught us that differentially rotating gas (and stellar) disks can be far more unstable than rigidly rotating disks. The underlying physical mechanism is lucidly set forth with characteristic verve in the review article of Toomre (1981). The shearing flow in a galactic disks locally drags the wave crests of an embedded plane wave perturbation in the same direction that its constituent particles naturally oscillate in their epicyclic excursions. This kinematic "mini-resonance" prolongs the compressive phase of a perturbative oscillation and self-gravitational forces are thereby aided in preventing the subsequent reëxpansion of the wavelet. In its full global manifestation, this "swing amplification" process may be responsible for much of the large scale spiral structure seen in disk galaxies.

In this report, I would like to look rather carefully at how the amplifier behaves on considerably smaller scales. In particular, the local conditions in the spiral arm regions of disk galaxies will be the setting of interest here. I shall assume that the potential of the arm rigidly rotates (at least on scales small compared with the galactic radius) relative to the gas flow, and that the resulting highly nonlinear density compression in the interstellar medium near the arm forms the background flow in which the perturbations develop. The perturbation length scales are thus considered to be small compared with the galactic radius and the interarm spacing.

With the added constraint that the spiral arm potential be tightly-wound, the picture presented here for the background medium will be recognized as essentially the scheme developed by Fujimoto (1968) and Roberts (1969) in which the notion of large scale galactic shocks gained favor. Two comments are of some relevance. First, we will be looking at scales *small* compared with the interarm spacing length, so the periodic boundary conditions employed by Roberts (1969) need not constrain us here. Second, the essential, and hitherto neglected, role played by self-gravity on the structure of the background flow has recently been emphasized by Lubow, Balbus, and Cowie (1986). Probably the most novel element found in this investigation is that under conditions when the self-gravity becomes important, the classical shock wave structure can disappear completely! The combination of disk geometry, self-gravity, and Coriolis forces introduces a kind of interflow communications network that can prevent the fluid from being caught by surprise and developing a discontinuous shock structure. (Mathematically, the equations become elliptic rather than hyperbolic; the same thing happens in planetary rings [Shu *et al.* 1985].) The possible disappearance of shocks does *not* mean the disappearance of large scale gas compressions along spiral arms however, and the question of shocks vs. continuous structure is not a crucial one for present purposes.

II. Evolution of Density Perturbations.

On its Galactic circumnavigation, a fluid element is compressed and then rarified as it passes through an arm. In this work, we focus in on an element just after the point of maximum compression. In a Lagrangian coordinate system tailored to follow the element, we examine the local stability of the fluid to a large wavenumber embedded spatial plane wave with an arbitrary **k** vector. The functional form of the time dependence is then solved for, and the nature of the response as a function of **k** is examined.

In a frame rotating at the angular velocity of the spiral arm Ω_p, consider the motion of the undisturbed gas. The basic coordinate system is shown in fig. (1). The inclination angle i satisfies $\sin i \ll 1$. We introduce coordinates x and y which form a quasi-cartesian system as shown in fig. (1). The x-coordinate points nearly radially outward and is defined to be locally perpendicular to the spiral arm. The y-coordinate points along the arm in the direction of the azimuthal velocity. Let r be the local galactic radius. We denote by u_T the total x-component of the velocity vector as viewed in the rotating frame, and let u be only that component induced by the presence of the spiral potential; i.e. $u_T - u \equiv u_c$ is the x-component of the circular velocity in this frame. Similarly, v_T and v are the respective y-components of the velocity. The arm-to-arm spacing L is assumed to satisfy $L \ll r$, and the circular velocity in the inertial frame $r\Omega$ is much larger than u, u_T, or v. The latter are all comparable. The background flow is assumed to be one-dimensional in x, and the finite thickness of the Galactic disk is ignored here. Under these conditions, the general equations of motion are (Roberts 1969, Spitzer 1978):

$$\frac{du}{dt} + \frac{\partial}{\partial x}(\chi + \Phi) - 2\Omega v = 0, \tag{1a}$$

$$\frac{dv}{dt} + \frac{\kappa^2 u}{2\Omega} + \frac{\partial}{\partial y}(\chi + \Phi) = 0, \tag{1b}$$

$$\frac{d\ln \Sigma}{dt} + \frac{\partial u_T}{\partial x} + \frac{\partial v_T}{\partial y} = 0, \tag{1c}$$

$$\nabla^2 \phi_g - 4\pi G\Sigma\delta(z) = 0. \tag{1d}$$

In the above, Φ represents an external (stellar) potential, and

$$d\chi \equiv \frac{dP}{\Sigma} + d\phi_g, \tag{2}$$

where P and ϕ_g are the two-dimensional gas pressure and the self-gravitating gas potential, respectively. Σ is the gas column density through the disk, κ is the stellar epicyclic frequency (the gaseous epicycles are modified by local compressions), and G is the usual gravitational constant. The pressure and density are related by a polytropic formula:

$$P = K\Sigma^\gamma \tag{3}$$

where K is a dimensional constant, and γ is related to the polytropic index n by $\gamma = 1 + 1/n$. Finally, d/dt denotes the Lagrangian co-moving time derivative.

To discuss the local stability of the flow satisfying equations (1), the oxymoron of an "infinite small patch" is useful. That is, we take the local flow properties of density, pressure, etc. , of a

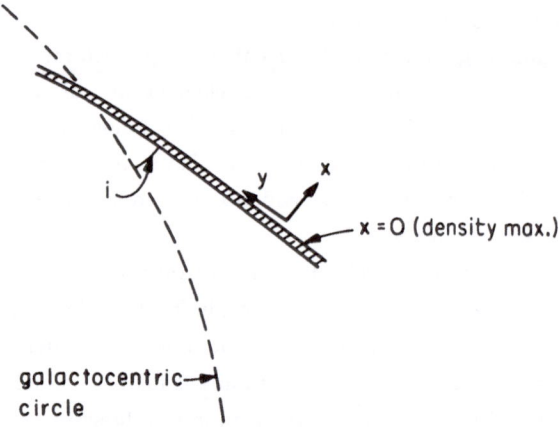

Fig. 1.

infintesimal fluid element, and assume that they extend infinitely in a plane. The time evolution of the element is then descibed by choosing coordinates that comove with the unperturbed flow, and that allow the flow properties to change homogeneously with time. Thus, each fluid element is described by its own "cosmological" evolution. As in the classical treatments of gravitational instability in cosmology, (Lifschitz 1946, Bonnor 1957) we consider Eulerian perturbations of plane waves $\exp(i\mathbf{k}\cdot\mathbf{r})$, but using Lagrangian coordinates. Unlike Friedmann cosmologies, the volume distortions of the background are anisotropic, and this leads to results of considerable significance. The Goldreich–Lynden-Bell swing amplifier is one such result, and the generalization here to include the effects of expansion of the fluid downstream from a spiral arm further enriches the behavior of density perturbations.

The technical details of switching coordinates to the Lagrangian frame are a little complicated, and seem best left to a more leisurely development in a paper that will be published in *The Astrophysical Journal* (Balbus 1988). Here, I will simply write down the final equation for the evolution of density perturbations, and describe some of its most important features. First, a few more intermediate results are needed.

Denote by T as $-k_x/k_y$ as viewed in an Eulerian frame. That is, $-T$ is the x-wavenumber to y-wavenumber ratio as measured by coordinates which do not distort with the flow. (The minus sign is included because in the limit of pure shear flow, T becomes directly proportional to the time t, and is identical to the Goldreich–Lynden-Bell τ.) As mentioned in the Introduction, the direction toward which the perturbation wave crests are dragged determines the nature of the amplification. Retrograde dragging, as occurs in swing amplification, enhances growth, but prograde dragging (*i.e.* in the opposite sense of the retrograde epicyclic motion) damps growth. Thus, the sign of dT/dt is an important quantity. If the Galactic potential gives rise to a flat rotation curve (in the absence of spiral structure), then it is not difficult to show (Balbus 1988):

$$\frac{dT}{dt} \propto \frac{T}{\Omega}\frac{d\ln\Sigma}{dt} + \frac{\sigma'}{\mathcal{R}} - 2, \tag{4}$$

where σ' denotes the ratio of the density at the point of maximum compression (Σ') to its azimuthally averaged value, Σ_0, and $\mathcal{R}$ is a volume expansion factor which is unity when $\Sigma = \Sigma_0$.

A uniformly sheared sheet has a constant negative value for dT/dt, but density compressions and an expanding background can result in more complicated behavior for this quantity.

The first term in equation (4) is a consequence of anisotropic volume distortions. An embedded pattern stretched along one axis in an elastic sheet would clearly be distorted and elongated parallel to this axis. This stretching causes its own type of wave crest rotation, and initially leading crests are dragged retrograde in an expanding post-compression background. This is in the same sense as the shear flow would drag wave crests in a disk in which the angular velocity decreased outward with radius. But the final two terms in equation (4) indicate that for sufficiently high values (in excess of 2 for a flat rotation curve) of the compression ratio σ', the local angular velocity may actually increase outwards. This is nothing more than local specific angular momentum conservation. But the consequences of this "reverse shear" flow may be quite far-reaching.

We are now ready to return to the equation for the evolution of the perturbations. It may be written

$$\left[\frac{d^2}{dt^2} + S(t)\right] \widetilde{\delta\psi} = 0, \tag{5}$$

where,

$$\widetilde{\delta\psi} = \frac{\delta\rho}{\rho} \left(1 + T^2\right)^{-1/2} \tag{6a}$$

and,

$$S(t) = \left(k_y a^2\right) - 2\pi G \frac{\Sigma'}{\mathcal{R}} |k_y| \left(1 + T^2\right)^{1/2} + \frac{\kappa^2 \Sigma'}{\mathcal{R}\Sigma_0}$$

$$- \left(1 + T^2\right)^{1/2} \frac{d^2}{dt^2} \left(1 + T^2\right)^{-1/2} + \frac{\kappa^2}{\Omega} \frac{\Sigma'}{\mathcal{R}\Sigma_0} \frac{dT/dt}{1 + T^2}. \tag{6b}$$

The first three terms in equation (6b) are the familiar Jeans group familiar to all practitioners of disk dynamics from the radial stabilty calculations going back to Safronov (1960) and Toomre (1964). The penultimate term, which it will be noticed is independent of dynamical content, arises from the curvature of the quantity in equation (6a). The final term in equation (6b) is the heart of the amplification/damping mechanism. It is basically a product of the local κ^2 of the gas × the rate of change of the angle made by the wave crests and the x-axis. Thus its presence demands both "shaking" and shearing (Toomre 1981). But shearing need not be due to Galactic differential rotation alone.

From equation (4), we see that in regions in which $\sigma'/\mathcal{R} > 2$ the reverse shear flow disrupts the swing amplification mechanism. The compressive phase of an oscillation is kinematically shortened when the reverse shear conditions prevail. Were it not for the first term in equation (4), compressed flow in spiral arms would behave in a manner entirely different from standard swing amplification: the most rapidly growing instabilities would be the unsheared Jeans-like disturbances. But the presence of the first term complicates the situation. Instead, a competition is set up between the tendency of the reverse shear to damp non-radial directions, and the tendency of flow expansion to twist leading wave crests retrograde, thereby causing a *divergence* (as opposed to shear) swing amplifier. The result of this competion is shown in fig. (2).

Fig. (2) is a polar plot. The angular variable is $\tan^{-1}(k_x/k_y)$ and the radial variable is proportional to the peak amplitude reached by a perturbation while passing through a spiral arm

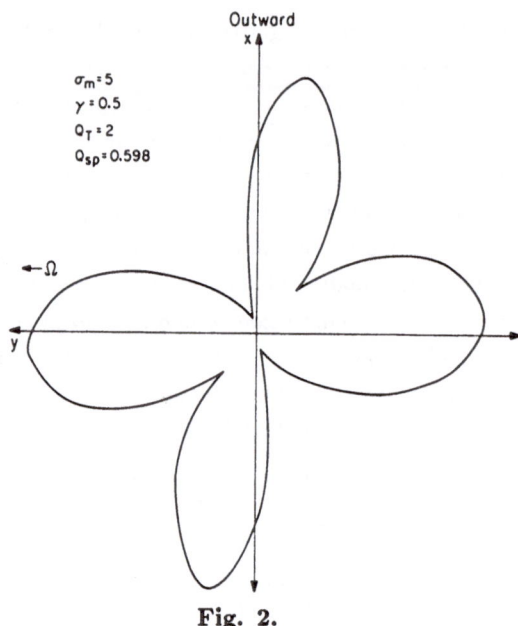

Outward
x

$\sigma_m = 5$
$\gamma = 0.5$
$Q_T = 2$
$Q_{sp} = 0.598$

$\leftarrow \Omega$

y

Fig. 2.

region. (The radial scale is set by the maximum linear amplitude attained, which happens to be 84 for this plot.) Thus, fig. (2) measures how sensitive the growth rate is as a function of the initial wavenumber direction of the perturbation. The *magnitude* of the wavenumber for peak growth must also be found, it is not represented on this plot. Naturally, some sort of Galactic model is needed for the background flow. The results shown are for a background density enhancement peak of $\sigma' = 5$ with a sech^2 gas distribution in the spiral arm zone. (σ' is denoted as σ_m in the figure.) The equation of state for the gas was modeled as a polytrope with $\gamma = 0.5$; the sech^2 distribution should not be regarded as anything more than a model fit. The initial conditions correspond to $|\delta\rho/\rho| = 1$ and to a vanishing initial time derivative. The parameter Q_T refers to the Toomre (1964) Q-parameter of the unenhanced, azimuthally averaged disk gas. The parameter Q_{sp} is essentially the local value of Q_T at the density enhancement peak; it takes into account both variations in Σ and in κ. Further details may be found in Balbus (1988).

III. Discussion

The clover-leaf appearance of fig. (2) is unmistakable and seems to be generic. Growth is most powerful for wavenumbers nearly along the x and y axes. The former are Jeans-like perturbations, and their relatively rapid growth has been discussed above; it is a consequence of the reverse shear flow. The lobes along the y axis correspond to wave crests initially pointing nearly radial outward, and I will refer to them in a slightly loaded way as "spurs". The maximum amplitude reached by the spurs changes its sign relative to that of the initial unit amplitude disturbance; initial peaks become troughs, and vice-versa. These disturbances are initially damped by reverse shear, but later come back with ferocity as divergence swing amplification and ultimately classical swing amplification become established as the fluid element moves through the arm.

I would like to propose that much of the seemingly chaotic appearance of spiral arms has a dynamical basis which can be understood as the interplay between the two types of behavior of the bimodal response just described. For example, the relative growth rates along the axes depend upon the underlying stability of the disk as measured by Q_T. The lower the value of Q_T, the more unstable are the Jeans-like disturbances compared with the spurs. At higher values of Q_T, the overall growth diminishes, but spurs grow as easily, or more so, than the Jeans solutions. The tendency of spur-like structures to appear in the outer regions of galactic disks may be due to this effect.

When the two directions of growth respond similarly, another interesting possibilty is present. Such conditions may well be present in the solar neighborhood of our own galaxy (Balbus 1988). The idea is that the intersection of Jeans wave crests and spur wave crests would reach nonlinear development more rapidly than other regions, and would thus be likely sites for interesting things— like molecular cloud complexes, say—to occur. It is a straightforward exercise to show that the characteristic masses and spacings of the wave crests intersections give rise to very agreeable solar neighborhood values: $10^5 \, M_\odot$, and a kiloparsec or so for the spacing. There is some sensitivity to the assumed velocity dispersion of the gas; we have taken a one dimensional dispersion of $6 \, \mathrm{km \, s^{-1}}$ (Spitzer 1978).

On galactic scales, classical swing amplification may well be resposible for the sweeping beauty of the grand design spirals. But I believe that the underlying mechanism can be even richer and more subtle than has generally been appreciated, even by the experts! It is surprising that expansion actually *destabilizes* some perturbations. The addition of underlying expansion gives rise to reverse shear flow and divergence swing amplificaton which shapes and destabilizes very distinct kinds of disturbances. The emergence of galactic spurs and the formation of embedded cloud complexes may in fact be dynamically related phenomena which have their origins in the sort of linear proceeses described here. An independent test of this very local swing amplifier would clearly be desirable, and an N-body simulation with the requisite resolution is currently being developed.

References

Balbus, S.A. 1988, *Ap. J.*, **324**, in press.

Bonnor, W.B. 1957, *M.N.R.A.S.*, **117**, 111.

Fujimoto, M. 1968, *Proc. IAU Symp. 29*, p. 453.

Goldreich, P., and Lynden-Bell D. 1965, *M.N.R.A.S.*, **130**, 125.

Julian, W.H., and Toomre, A. 1966, *Ap. J.*, **146**, 810.

Lifschitz, E.M. 1946, *J. Phys. U.S.S.R.*, **10**, 116.

Lubow, S.H., Balbus, S.A., and Cowie, L.L. 1986, *Ap. J.*, **309**, 496.

Roberts, W.W. 1969, *Ap. J.*, **158**, 123.

Safronov, V.S. 1960, *Sov. Phys. Dokl.*, **5**, 13.

Shu, F.H., Yuan, C., and Lissauer, J.J. 1985, *Ap. J.*, **291**, 356.

Spitzer, L. 1978, *Physical Processes in the Interstellar Medium*, p. 278.

Toomre, A. 1964, *Ap. J.*, **139**, 1247.

Toomre, A. 1981, in *Structure and Evolution of Normal Galaxies*, ed. S.M. Fall and D. Lynden-Bell, p. 111.

RESONANCE EXCITATION: A POSSIBLE
INTERPRETATION OF THE 3-KPC ARM

Chi Yuan and Ye Cheng

Department of Physics, The City College of New York
New York, NY 10031

Abstract

A minor oval distortion or an uneven distribution of mass in the central region of the Galaxy may excite an outgoing spiral wave that closely resembles the observed 3-kpc arm. The wave is generated by a resonance excitation mechanism at the outer Lindblad resonance corresponding to the pattern frequency of the oval distortion or of the uneven distribution of the mass. An expansion velocity of 53 km/sec associated with the wave crest requires a perturbational field of 5% of the local mean gravitational field. The calculations are based on a non-linear analysis of gas motions in a disk system with the inclusion of self-gravitation and viscosity.

I. Introduction

While the topic of our paper might not be considered as part of the main subject of this conference, it does relate to the early observations of Frank Kerr. We hope this connection will justify its presentation here as a tribute to Frank Kerr for his pioneering work in radio astronomy and for his other scientific achievements.

The 3-kpc arm may best be described as a coherent armlike feature of gas, spanning a galactocentric sector of 90^0. It is situated about 3.75 kpc from the galactic center (or 3 kpc if the galactic center is 8.0 kpc from the sun). The most striking aspects of the 3-kpc arm are its high radial expansion velocity (~ 53 km/sec) and the large mass associated with it ($\sim 4 \cdot 10^7 M_\odot$). The 3-kpc arm was discovered by Rougoor and Oort (1960), piecing together the Dutch and Australian data of HI emission and absorption observations. Recent observations of CO (Bania 1980) indicate that the 3-kpc arm is also populated with molecular clouds.

A theory based on an expulsion model was favored for sometime to explain the fast expansion of the arm (van der Kruit 1971; Sanders and Prendergast 1974). The model requires frequent explosions at the galactic center for an energy source. The model also needs a replenishment mechanism to account for the mass loss in the gas component due to the expansion of the arm, the rate of which is estimated to be $1 M_\odot / year$ from the center. Both the energy and the mass problems are difficult to resolve in the expulsion model.

In 1984, Yuan showed that a minor oval distortion or a barlike structure, rotating at a typical angular speed in the central region of the Galaxy, can excite an outgoing acoustic wave which bears a close resemblance to the observed 3-kpc arm. The idea that a barred structure may be responsible for the 3-kpc arm was suggested first by Kerr (1968), although the physical mechanism invoked is different from that in the present approach. The wave model has definite advantages over the expulsion one. It does not require frequent explosions at the center, therefore no energy problem exists. Nor does it invoke any replenishment mechanism to maintain the gas supply since the average radial flux integrated along a circle around the center is essentially zero, therefore no mass loss problem exists. Furthermore, observations of a central bar structure are not uncommon among disk galaxies. Thus, it is not unreasonable to assume that we have such a distortion in the center of the Galaxy.

However, Yuan's theory is not free from shortcomings. First, it is a local theory. Strictly speaking, it is only valid near the Lindblad resonance. Second, it does not include the effect of self-gravity of the gas and the problem is formulated in such a way that self-gravity cannot be easily included. Third, it deals exclusively with the bisymmetric disturbance and hence has no ability to explain the 135 km/sec arm on the other side of the Galaxy. In this contribution, we briefly sketch our theory and report our findings, and leave the detailed treatment to be published elsewhere (Yuan and Cheng 1987).

II. Excitation and Propagation of the Waves

The spiral density waves are generated by the periodic perturbational potential at the Lindblad resonances, which are located at radii satisfying the following relation

$$m\,[\,\Omega(r) - \Omega_p\,] = \pm n\,\kappa(r) \tag{1}$$

where Ω_p is the pattern frequency of the disturbance rotating about the galactic center, $\Omega(r)$ and $\kappa(r)$ are respectively the angular frequency of the galactic rotation and the epicyclic frequency of the particle oscillation in the radial direction. The $+$ sign corresponds to the inner Lindblad resonance and the $-$ sign, the outer Lindblad resonance. The terms "inner" and "outer" are in reference to the co-rotation which occurs at

$$\Omega(r) - \Omega_p = 0 . \tag{2}$$

In equation (1), m and n are positive integers. In the following discussions, we focus our attention only on the outer Lindblad resonance with $n = 1$. When $m = 2$, we have a two-arm spiral pattern, which is the case for an oval distortion; when $m = 1$, we have a one-arm spiral pattern, which is the case for an uneven distribution at the center. Equation (1) implies that at those radii, the particle will oscillate at an epicyclic frequency somehow in phase with the frequency it sees the periodic potential.

The external perturbational fields in general vary on the scale of the size of the galactic disk. They thus excite only the long waves at the Lindblad resonances. Since only trailing waves propagate away from the Lindblad resonances, carrying angular momentum from the disturbance to the disk, the waves excited at the Lindblad resonances are necessarily long trailing waves. At the outer Lindblad resonance, the long trailing waves, according to the density wave theory, will propagate towards the galactic center with a speed equal to the group velocity derived from the dispersion relation (see Goldreich and Tremaine 1978 for general discussions). These waves, however, cannot penetrate very far before reflected as short trailing waves at the Q-barrier. The

Q -barrier is the location dividing the wavy region from the evanescent region, which is located at

$$Q^{-2} + m^2(\Omega - \Omega_p)^2 / \kappa^2 - 1 = 0 . \qquad (3)$$

Q is the quantity introduced by Toomre (1963) to set the stability criterion for a self-gravitating disk, defined as

$$Q \equiv \kappa a / \pi G \sigma_0 \qquad (4)$$

where a in this case is the random speed of the gas clouds and σ_0 is the surface density of the gas component. When the effect of self-gravity is negligible, $Q = \infty$, the Q -barrier coincides with the Lindblad resonance. When the effect of self-gavity increases, the Q -barrier moves towards the co-rotation, or towards the center in the present case. The distance between the outer Lindblad resonance and the Q -barrier is very small for the 3-kpc arm problem. The long waves are hardly noticeable. Only the reflected short trailing waves are visible, seemingly emitted from the outer Lindblad resonance, propagating outward.

The viscosity in the interstellar gas plays a decisive role in confining the waves to a region not far from the outer Lindblad resonance. Using a crude estimate that the kinematic viscosity ν is equal to the product of the random speed of gas clouds, a, and the collision mean free path among them, l, or $\nu \approx al \approx 10 \cdot 0.1 \ km \cdot kpc / sec = 1 \ km \cdot kpc / sec$, we reach the conclusion that these short trailing waves are effectively attenuated within 1.5 kpc of the resonance. Therefore, the spiral density wave for the 3-kpc arm does not interfere with the grand design spiral structure of the Galaxy outside.

The angular momentum removed from the central region by the waves is estimated to be only a small fraction of the total. The time scale for depleting a substantial fraction of the total angular momentum in the central region will be of the order of 10^{11} years (Yuan 1984).

III. The Calculation

In order to reach an expansion velocity of 53 km/sec, it is necessary that we use the non-linear theory. For this, we adopt the mathematical formulation developed by Shu, Yuan and Lissauer (1985), in which a singular integro-differential equation for the Lagrangian displacement of a fluid particle is derived. To that formulation, we add the pressure contribution and the Newtonian viscosity contribution. Following the same heuristic arguments, we replace the singular integro-differential equation by a non-linear ordinary differential equation. This differential equation yields the same linearized equation and the same non-linear amplitude relation (or equivalently the statement of conservation of angular momentum flux) of the integro-differential equation. Yet it does not reproduce the same non-linear dispersion relation as claimed by Shu et al. (1985).

To integrate this second order, highly non-linear, inhomogeneous differential equation, not only do we need to specify the boundary conditions which are outgoing waves at the one side and algebraic decay at the other side, but also the precise numerical values of those conditions.

The reason for the requirement of high precision is that the equation admits solutions of exponential growth and decay. Any small error involved in the boundary conditions will lead to an exponentially growing result which destroys the validity of the entire calculation. To obtain accurate boundary values, we must solve the linearized equation first. The solution of the linear equation is necessarily coincident with the non-linear equation in the evanescent region where the amplitude of the waves diminishes and also at the far side of the wavy region where the amplitude of the waves approaches zero due to viscous attenuation. Because of the exponentially growing component in the solution, we have to integrate the equation from both ends and to match the integration somewhere in between. The detailed discussions of the numerical work will be presented elsewhere (Yuan and Cheng 1987).

We have considered two cases, $m = 1$ and $m = 2$. For $m = 2$, we take $\Omega_p = 105 \ km \ / \ sec \cdot kpc$ and a perturbational field equal to 5% of the mean field. The resulting two-arm spiral pattern with radial streaming velocity marked by arrows is plotted in Figure 1 and a density profile along the ray from the galactic center to the sun is plotted in Figure 2.

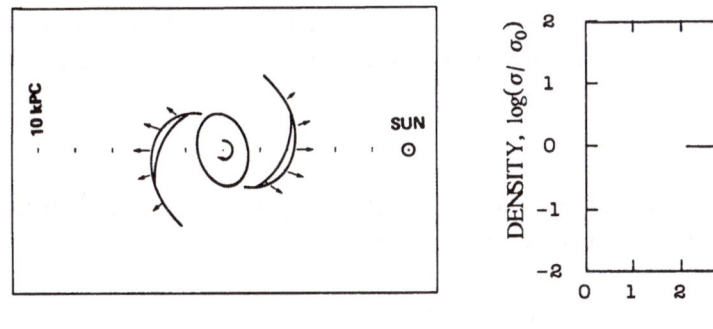
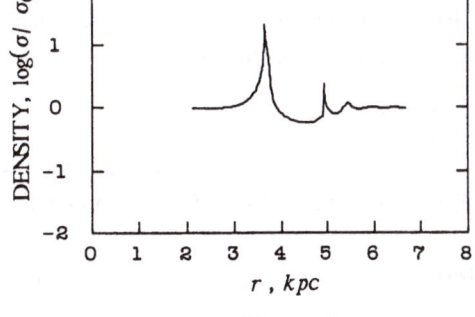

Figure 1

Figure 2

For $m = 1$, we take $\Omega_p = 152 \ km \ / \ sec \cdot kpc$ and use the same perturbational field. The corresponding results are plotted in Figures 3 and 4.

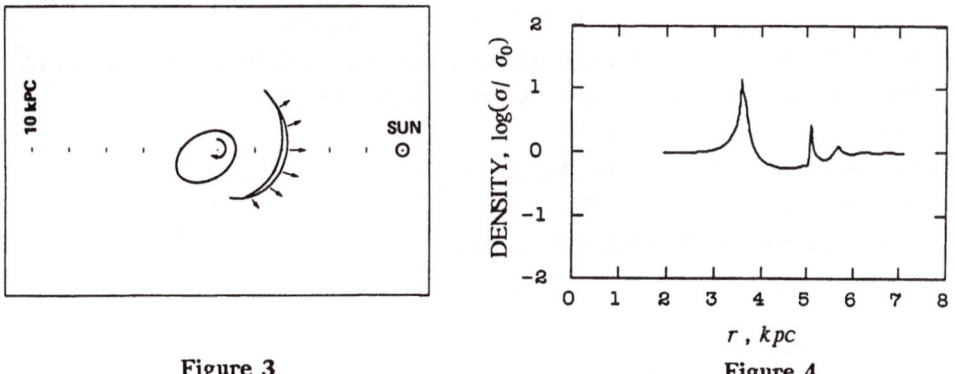

Figure 3

Figure 4

The oval distortion responsible for the perturbation is estimated to have semi-major to semi-minor axis ratio equal to 1:0.95. It is sketched schematically in Figure 1. A schematic representation of an uneven distribution of mass is plotted in Figure 3.

The results we obtained are in excellent agreement with observations as far as the 3-kpc arm is concerned. We have not made any effort to match our results to the 135 km/sec arm. It is conceivable that another uneven distribution of mass rotating at a higher pattern frequency, in the galactic plane or tilted to the plane, may give rise to the 135 km/sec arm.

IV. Conclusion

The present study confirms and strengthens the early results of Yuan (1984) that the 3-kpc arm may be a part of the waves generated by a minor oval distortion in the central region of the Galaxy. The strength of the perturbational field is reduced from 10% of the mean field in the previous calculation to 5% in the present one. We are no longer confined to the case of an oval distortion in the center. Another uneven distribution of the mass in the center, either in the plane or tilted with respect to the plane, can produce a one-arm spiral wave This may provide grounds to interpret the 135 km/sec arm as a separate wave feature caused by another periodic disturbance. Inclusion of the effect of self-gravity modifies the required perturbational field strength, while not qualitatively changes our previous results.

Acknowledgment

This study is supported in part by NASA-Ames University Consortium NCA2-186. We wish to thank Dr. P. Cassen for his interest in this research.

References

[1] Bania, T. M. 1980, *Ap. J.*, **242**, 95.

[2] Kerr, F. J. 1968. In *radio Astronomy and the Galactic System*, *IAU Symp. No.31*, p. 239, ed. H. van Woerden

[3] Goldreich, P., and Tremaine, S. 1978a, *Icarus*, **34**, 227.

[4] ———. 1978b, *Icarus*, **34**, 240.

[5] Rougoor, G. W., and Oort, J. H. 1960, *Proc. Nat. Acad. Sci.*, **46**, 1.

[6] Sanders, R. H., and Prendergast, K. H. 1974, *Ap. J.*, **188**, 489.

[7] Shu, F. H., Dones, L., Lissauer, J. J., Yuan, C., and Cuzzi, J. N. 1985, *Ap. J.*, **299**, 542.

[8] Shu, F. H., Yuan, C., and Lissauer, J. J. 1985, *Ap. J.*, **291**, 356.

[9] Toomre, A. 1963, *Ap. J.*, **138**, 385.

[10] van der Kruit, P. C. 1971, *Astr. Ap.*, **13**, 405.

[11] Yuan, C., 1984, *Ap. J.*, **281**, 600.

[12] Yuan, C., and Cheng, Y. 1987, in preparation.

IV. THE INTERSTELLAR MEDIUM AND THE NEAR OUTER GALAXY

HI AND THE DIFFUSE INTERSTELLAR MEDIUM

CARL HEILES

Astronomy Department, University of California, Berkeley

1. A 'NEO-ESTABLISHMENT' VIEW OF THE INTERSTELLAR MEDIUM

Why the title of this heading? Shri Kulkarni and I have written a long review article entitled 'H I and the Diffuse Interstellar Medium' for the new edition of the book, 'Galactic and Extragalactic Radio Astronomy' (Verschuur and Kellerman 1988). Excerpts of this article appear in two other books: 'Interstellar Processes' (Hollenbach and Thronson 1987) and 'Physical Processes in Interstellar Clouds' (Morfill and Scholer 1987). Now, Shri was my graduate student and received his degree not too many years ago, and I am still young enough to feel like a graduate student. Being erstwhile graduate students, we feel that the point of view we express in these articles is a carefully considered one. However, I find that we have deluded ourselves. There are those who say that some of our major points are just myths whose veracity has been built up over the years by repetitive statements of probable truths—veracity established by the 'Madison Avenue' approach. Some famous historical examples of this incorrect scientific procedure include the evolution of experimental determinations of the velocity of light, the rotation rate of the planet Mercury, and the Hubble constant.

'Establishment' old timers have heard 'truths' repeated so often that they accept them as really true. Have Shri and I really come so far along the path to the establishment? Maybe—but some of our views conflict with what is *really* the establishment. So I apply the prefix 'neo', which is accurate in that some of our views were definitely not establishment a short time ago, and maybe not even now.

As I mentioned above, Shri and I have written a comprehensive review article which appears in three separate books. Rather than present yet another excerpt here, I am going to briefly discuss only some of those points that depart from the real establishment. or that I consider important. To keep things simple and clear, I am going to shy away from giving references here. There are many more points to consider, and many more details; the interested reader should consult the abovementioned Kulkarni/Heiles reviews directly.

Until recently, the neutral diffuse ISM was thought to consist of just three major components: a cold, dense, and largely neutral H I ($T \sim 80$ K; cold medium or CM); a warm H I (Shri and I believe $T \lesssim 8000$ K; see discussion below), either surrounding the cold clouds in an envelope or pervading much of the space as an 'intercloud medium' (warm neutral medium or WNM); and a hot, highly ionized medium ($T \sim 10^6$ K; the hot ionized medium or HIM).

Recently, diffuse $H\alpha$ emission line data (Reynolds 1984) indicate that a nontrivial fraction of the interstellar volume is filled by a fourth phase: a highly ionized, warm phase ($T \sim 8000$ K; warm ionized medium or WIM). The WIM is important: it contains substantial mass, and requires a substantial energy input to keep it ionized. However, it is hard to observe: it has low volume density, and therefore low emission measure (~ 2 cm^{-6} pc toward the Galactic poles), so its optical emission lines are very weak. This means we don't know in detail how it is distributed. The WIM is responsible for most of the pulsar dispersion, and indeed this is the easiest way to detect it.

Conventional thinking imagines all four components to be in pressure equilibrium. The temperatures are approximately as defined above; thus the volume density of each component is reasonably well-defined, too. The total Galactic mass in each component, assuming that they are distributed everywhere as they are in the Solar vicinity, is about 3, 2, 2, and 0.02 $\times 10^9$ M$_\odot$ for the CM, WNM, WIM, and HIM, respectively.

IRAS, Space Telescope, and X-ray observers take note: The WIM is almost certainly associated with dust, as are the CM and the WNM. However, the WIM is so difficult to observe that there is no way to ascertain the dust, other than by direct measurement. Since the total column density of the WIM is roughly equal to that of the WNM, *i.e.* $\sim 10^{20}$ H nuclei cm^{-2} toward the Galactic poles on average, there may exist a nontrivial column density of dust, even in regions where the H I column density is small. The WIM is distributed in an unknown fashion; in the long-term future, we hope someone will map the WIM using IR fine-structure lines from heavy elements, which are important coolants for the WIM: the line intensities are proportional to the column density, not the emission measure, which is an important advantage.

Now here are the 'neo-establishment' and most important points I promised earlier.

First, the assumption of pressure equilibrium. This assumption is based mainly on a theoretical consideration: self-gravity is negligible for diffuse cloud structures, so pressure fluctuations must rapidly generate motions that will eliminate the fluctuations. However, it is not just gas pressure. There are also contributions from magnetic fields and cosmic rays. The energy densities of cosmic rays, magnetic fields, and interstellar gas are all comparable, so we cannot neglect the any of the components. What really matters is the pressure gradient, and we don't know over what length scales the magnetic and cosmic ray components change. In addition, the magnetic pressure is not isotropic. And finally, how do the magnetic and cosmic ray pressures couple to the gas, both for individual clouds and for the large-scale z-distribution?

Second, the temperature of the WNM. It is often quoted as ~ 1000 K, because this value was obtained in several *early* 21-cm observations. However, it is unlikely to be correct for the following reasons: one, observationally derived temperatures are, in fact, lower limits because of the possible presence of cooler gas; two, the maximum *measured* temperature is about 6000 K; three, the careful analysis of the Arecibo H I absorption survey implies a temperature of ~ 5000 K; and four, the reliable but very sparse UV absorption line measurements for gas near the Sun give ~ 8000 K. A temperature of 8000 K makes sense from the theoretical standpoint, because just above 8000 K the cooling curve jumps up sharply; consequently, a very large range of heating rates results in an equilibrium temperature of ~ 8000 K. In direct opposition to this appealing argument is the embarassing absence of a sufficiently powerful heating mechanism to put the WNM into this range (see 'third' below)!

However, the temperature of the WNM is uncertain and we desperately need more data. The best data are the UV absorption lines, and this is an excellent project for the Hubble Space Telescope. More 21-cm line measurements would be wonderful. And we dream of the day when we can directly map the IR ($\lambda \approx 30\mu$m) emission lines from Si and Fe, which should be important coolants for the WNM.

Third, the heating sources for the two neutral components are a long-standing problem. Cooling is so efficient that it is difficult to keep the CM as hot as its usual temperature of 80 K, and even more difficult to keep the WNM at its probable temperature of 8000 K. Cosmic rays are inadequate by a factor of ~ 40; soft X-rays cannot penetrate enough column density to heat more than the outer skin of a neutral gas structure. Grains are the probable answer, but one must push the efficiencies to the limit for conventional grain models (Draine 1978). The new IRAS-discovered 12-μm emitting component, which must involve very small grains, may be the answer.

Fourth, the widespread misconception concerning the ionization of the two neutral components. The question is important, because, among other reasons, the ionization is required for the interpretation of optical absorption line data. It is usual practice to assume that the only ionization in these components comes from heavy elements, mainly Carbon. This leads to an ionization fraction $x_e = n_e/n_{HI} \approx 4 \times 10^{-4}$. This is incorrect, for two reasons. First, much of the Carbon is depleted onto grains. Second, it neglects the contribution from cosmic ray ionization of H I. The latter contribution exceeds that from Carbon, even if Carbon were not depleted onto grains. We expect $x_e \approx 6 \times 10^{-4}$ and 0.03 for the CM and WNM, respectively.

Fifth, the statistics of the 'clouds' containing the CM. The notion of a 'standard cloud' became popular as a result of statistical studies of optical extinction from clouds: E(B–V) ~ 0.05 mag, $N_{HI} \sim 3 \times 10^{20}$ cm^{-2}, and ~ 6 clouds per kpc in the Galactic plane. However, 21-cm absorption surveys show a power-law distribution in column density, with the number of clouds per kpc $\sim 6(N_{HI}/10^{20}$ cm$^{-2})^{-0.8}$. With a power-law distribution, there is no well-defined average! The power law distribution appears to be well defined only within certain limits of N_{HI}, which allows one to define a median cloud: it has $N_{HI} \sim 0.6 \times 10^{20}$ cm^{-2}, T ~ 135 K, $n_{HI} \sim 22$ cm^{-3}, and a diameter of 0.9 pc. Note that the median N_{HI} is a full factor of *five* smaller than that of the 'standard cloud'! In fact, the associated E(B–V) is too small to detect optically.

This leads one to conclude that the properties of the 'standard cloud' were defined by the *technique*. In fact, we might have this same problem with the above-defined median H I cloud; it is defined by the limits of the power-law distribution, which are perhaps partly a result of the technique.

Finally, the questions of topology and connectivity. These include the small scale structure of the CM: are clouds spherical, filamentary, or planar? They range up to the largest scales, and the question is

'What kind of Swiss cheese do we have up there?' If supernovae are rare, then each explosion will blow a hole in a mainly quiescent medium. The WNM and WIM occupy most of the volume (the 'cheese'), with the HIM occupying an occasional very large supernova-blown bubble and the CM occupying a small fraction of the volume. If the CM is predominantly in spherical clouds, the CM blobs are individually distinct and not connected, as are the occasional very small bubbles in Swiss cheese; if they are large planar sheets or long filaments, the CM in one region might be connected to that in another.

But when supernova dominate, as they must at least in the interior of the Galaxy, theorists predict that the HIM occupies most of the volume. The CM, WIM, and WNM all huddle together, as they must to protect themselves from the devastating effects of the HIM, so they become the 'holes' in the cheese. However, observational data don't look like that, at least near the Sun, and they don't seem to look like that in external galaxies either.

I cannot overemphasize the importance of this disagreement between theoretical expectation and observation. It would seem that the basic questions of topology and connectivity of the various components of the interstellar medium are so fundamental that their qualitative aspects should be well-understood from both the theoretical and observational standpoints. However, they are not, and on one or perhaps both of these fronts we are missing some very fundamental points.

2. SOME NEW IRAS-BASED RESULTS.

Enough of review; sometimes I think I have done nothing during the past two years but deal with the Kulkarni/Heiles review papers. However, I have also been doing some new things. Here I consider just one: my students and I have been looking at a sample of 26 IRAS-defined clouds (Heiles, Reach, and Koo 1988). These are all blobby clouds about a degree in diameter, and isolated from other clouds. This makes them easy to study because it is easy to pick a nearby 'off' region for a comparison measurement. These clouds exhibit a remarkable diversity of kinematical properties, as defined by the 21-cm line. Furthermore, we have been able to derive accurate IR flux densities in all four bands. The IR colors tell something about the grain size distribution. As a result, we have been able to ascertain some tentative results concerning the effect of kinematics on grain size distribution, and also on molecular content.

First, let me just summarize the contents of the Heiles *et al.* (1988) paper. The easiest way is just to quote the abstract:

We have compared the IR and H I properties, and CO content, of a set of 26 isolated, $\sim$ degree-sized interstellar clouds. These comparisons offer some tentative conclusions concerning the effects of velocity on molecular content and grain size distribution, although these conclusions are rendered uncertain by the small sample. The departure of S_{100}/N_{HI}, where S_{100} is the 100μm surface brightness, from the value in purely atomic clouds is a measure of the H_2 content of clouds. Even clouds with low column density, $\lesssim 2.4 \times 10^{20}$ H-nuclei cm^{-2}, may contain more H_2 than H I, in contrast to results obtained from UV absorption line studies. The [H_2/H I] ratio of a cloud is large only for quiescent clouds. The dependence of the S_{60}/S_{100} ratio on velocity implies that fast shocks preferentially destroy large grains and/or produce small grains. The dependence of S_{12}/S_{100} on velocity possibly implies that very small grains (VSG's) are formed in shocks in the 10 to 20 km s^{-1} velocity range, although nearly *all* clouds appear to contain VSG's. Some members of our cloud sample emit more power in the IRAS 12μm band than in the 100μm band. Such clouds must have very large fractions of their total Carbon in the form of PAH's, if VSG's are exclusively PAH's. Finally, the absence of correlation of S_{12}/S_{100} with S_{60}/S_{100} implies that VSG's are not formed preferentially from the breakup of large grains.

Now, let me emphasize an aspect that may be more important and fundamental. Furthermore, it is certainly most appropriate for the present symposium, which is in honor of Frank Kerr and his pioneering and extensive work in the 21-cm line.

As I mentioned above, these clouds exhibit a remarkable diversity of kinematical properties. By this I mean that there is an overwhelming tendency for these clouds to have multiple velocity components with velocity separations typically of order 10 km/s and ranging up to 40 km/s. In short, the spatial structure at a particular velocity is correlated with that at another velocity.

This was a total surprise to me. My main technique for examining H I data has been to consider the two-dimensional angular distribution of H I at a particular velocity. This technique automatically makes it very difficult to see spatial correlations at different velocities! Now the reader might reply, 'Why not look at position-velocity diagrams?' Well, I have done that often enough during my lifetime (*e.g.* Heiles and

Habing 1974). In those diagrams there is some indication of correlation, but it is not obvious whether it is significant.

What we need is a statistical study of such correlations. To what extent are these correlations a property of isolated clouds, such as we have in this particular IRAS sample, and to what extent are they a *general* property of the interstellar H I? I know of no investigation of this question. If such correlations exist as a general property, then in our ∼ 30 years of 21-cm line research we have missed a fundamental property of the H I. Such a property would certainly have implications for all theories and models of the interstellar medium. I am looking forward to attacking this question.

In considering this question, I am reminded of a pet theme of mine: we find what we look for. This theme reverberates in all astronomical research, and indeed in all areas of human endeavor. Frank Kerr is aware of this theme and its importance: his current project, involving searching for galaxies that happen to be located behind the obscuring dust of the Milky Way (Kerr and Henning 1987), is based in part on understanding the importance of this theme. And, in considering this theme, we come full circle to the beginning of this talk: the establishment point of view always emphasizes the results obtained from the tendency of humans to follow this theme. The mark of a true pioneer is to depart from this theme: and the importance of making that departure is the most important message in this contribution.

It is a very great pleasure to thank Frank Kerr for his role in defining, and making stimulating and important contributions to, an area of research in which I have been priveleged to work during my entire tenure in astronomy. I am pleased and honored to have been selected to present this paper at this meeting.

This work was supported by a grant to me from the National Science Foundation. My travel to this meeting was supported by a grant from the University of California.

REFERENCES

Draine, B.T. 1978, *Ap. J. Suppl.*, **26**, 595.

Heiles, C. and Habing, H.J. 1974, *Astron. Ap. Suppl.*, **14**, 1.

Heiles, C., Reach, W. T., and Koo, B-C 1988, *Ap. J.*, submitted.

Hollenbach, D.J. and Thronson, H.A. Jr. 1987, 'Interstellar Processes', Astrophysics and Space Science Library 134, D. Reidel Publishing Co.

Kerr, F.J. and Henning, P.A. 1987, *Ap. J.*, **320**, L99.

Morfill, G.D. and Scholer, M. 1987, 'Physical Processes in Interstellar Clouds', NATO ASI Series C210, D. Reidel Publishing Co.

Reynolds, R.J. 1984, *Ap. J.*, **282**, 191.

Verschuur, G.L. and Kellerman, K. 1988, 'Galactic and Extragalactic Radio Astronomy', Springer-Verlag Publishing Co.

THE DISK-HALO CONNECTION AND THE NATURE OF THE INTERSTELLAR MEDIUM

Colin A. Norman
Johns Hopkins University and Space Telescope Science Institute

Satoru Ikeuchi
Tokyo Astronomical Observatory

ABSTRACT

We discuss briefly some new results on the nature of the interstellar medium that are specifically concerned with the disk halo interaction. Over the last five years or so it has become clear that the supernovae rate in our Galaxy is spatially clumped and the consequences of such clumping are superbubbles and supershells fed by tens or hundreds of supernovae per shell. These objects evolve and expand rapidly and soon break out of the disk of the Galaxy, feeding the halo with very significant mass, energy, and momentum. As cooling occurs, gas will rain down onto the disk of the Galaxy completing the cycle.

Here we sketch out the basic flow of physical quantities from disk to halo and vice versa. Some of the many implications are noted including aspects of dynamo theory, quasar absorption lines, the theory of galactic coronae, and the nature of the X-ray background. The essential difference here with the McKee-Ostriker (1977) theory is that the filling factor of the hot gas in the disk is significantly less than unity.

INTRODUCTION

Our understanding of the interstellar medium seems to be as follows. There are a number of phases in rough pressure balance distributed throughout the disk. There is a source of energy that keeps the higher temperature phases heated. It is generally accepted that this is ultimately due to supernovae. Hot gas can flow into the halo from the disk where it cools and subsequently falls back to the disk. The details of the structure of both the disk and halo interstellar media are complex and subject to considerable debate. One key issue that we discuss here is the filling factor of hot gas in the disk. Arguments are presented for a filling factor of much less than unity and that furthermore most of this is due to the clumped energy input of tightly bunched aggregations of supernovae that generate very large structures called supershells or superbubbles.

Supershells are obviously a major constituent of the Galaxy as discussed by Carl Heiles at this meeting (c.f. Kulkarni and Heiles, 1987). The large, ubiquitous holes seen across the face of M31 in the Westerbork survey seem to be quite similar (Brinks 1984). A bona fide local example of a superbubble is probably the Local Bubble seen in X-ray and UV observations (Cox and Reynolds, 1987). Typical energies and scales of superbubbles are 10^{54} ergs and 100-1000 pcs respectively. Starburst systems show extreme examples of such systems as evidenced by the prodigious outflows (1-100 $M_\odot$ yr^{-1}) observed by Heckman et al. (1987).

The coronal gas around our own Galaxy requires both energy and momentum for its heating, support, and possibly (collisional) ionisation. The relevant observations here include: the TENMA observations of a smooth 10^7K X-ray emitting ridge; IUE observations of highly ionised species in the galactic halo; and observations of an infalling neutral hydrogen component- the so-called high velocity clouds. Metal lines of, for example, CIV seen at large distances from the centres of the associated galaxies demonstrate that hot, highly ionised halos have significant extent, typically of order tens of kiloparsecs.

2. SUPERBUBBLE EVOLUTION AND STATISTICS

In the initial stages a superbubble will expand as a standard bubble solution with a continuous energy input with spherical geometry (Tomisaka *et al.* 1981) however, as the bubble propagates through a scale height or so of an exponential atmosphere it will accelerate down the rapidly decreasing pressure gradient and break out of the disk with an opening angle given (in radians) by the inverse of the Mach number at the point of break out. The resulting structure is a wide angle jet or chimney acting as a conduit for material to flow from the disk to the halo (Tomisaka and Ikeuchi 1986, MacLow and McRay 1987).

The number of superbubbles at any given time in the Galaxy has been estimated by Ikeuchi (1987) using the available data on the number of OB associations in the Galaxy and the average number of supernovae of type II per typical OB association. He finds a superbubble formation rate of order 10^{-4} per year and with an estimated lifetime of ten million years per superbubble one expects the steady state number of superbubbles in the Galaxy to be of order a thousand.

The filling factor of this hot superbubble generated gas is of order ten per cent of the total volume for the numbers given above. Of course this filling factor will increase for a larger and less clumped supernova rate and this could occur at an earlier phase of the evolution of the Galaxy or in galaxies of different types. We shall return to this question later.

Chimneys supply energy to the halo at a rate of order 10^{40}-10^{42} erg s^{-1}. The mass supply rate is of order a solar mass per year. The scale height of the hot component is of order a few kiloparsecs and after a cooling time of order 10^7 yr, gas in the halo can cool, condense and infall with a velocity of order of two hundred kilometres per second.

Generally a large scale mass circulation is set up with a commensurate energy and momentum input into the halo from superbubbles. As the gas rains back down onto the disk there will be a further exchange of mass energy and momentum—this time from the halo to the disk. Chimneys should be observed as independent entities from hard X-ray measurements in our own Galaxy and from absorption line studies, HI observations, etc. of external galaxies.

3. THE MULTIPHASE STRUCTURE OF THE DISK

Many of the details of our calculations will be presented elsewhere (Norman and Ikeuchi 1988). Here we will sketch qualitatively some of our basic results. A particularly important parameter to determine is the extent to which the hot component of gas is more or less all pervasive or merely occupies a small part of the volume of the overall disk interstellar medium.This property is described by the filling factor.

One way to look at this is to study the dependence of filling factor on the clumping of supernovae which is of course intimately related to the number and strength of the resulting superbubbles. It is convenient to include the analysis of the dependence of this parameter with mean ambient density. Normalising to an average power injected into the interstellar medium of order 10^{42} erg s^{-1} we find that the Galaxy has a filling factor of about 10% for the hot gas component. In fact, this is exactly the region in parameter space where the chimney model applies. The standard McKee–Ostriker model seems more applicable to systems with ambient densities and superbubble rates lower by about an order of magnitude each. A two-phase model (c.f. Field 1986) will be relevant to galaxies with significantly higher densities than would be found in later type galaxies. Clearly the chimney model has pleasing aspects of both the two and three phase models—it is essentially a two phase model with the third hot phase being the the hot chimneys with their current Galactic filling factor of about 10%.

For recent overviews of our knowledge of the interstellar medium in both halos and disks and flows from disk to halo one should consult the books by Bregman and Lockman (1986), Hollenbach and Thronson (1987) and the recent preprint by Corbelli and Salpeter (1987). We can also study the global mass and energy flow from the disk to the halo. One important physical point is to determine whether or not conditions are suitable for the superbubbles to burst out of the disk into the halo. This is a crucial part of the chimney picture. We find that for canonical Galactic parameters the chimney phase is associated with a mass flow rate of 0.3-3 $M_{\odot}$ yr^{-1} and a global power input of 10^{40}–10^{42} erg s^{-1}. These numbers which emerge naturally from the calculation are those conventially thought to apply to the Galaxy and give us additional confidence in the applicability of the chimney model to the Galaxy.

4. THE STRUCTURE OF THE HALO

Our halo model is one where most of the physical quantities circulate-mass, metallicity, and magnetic field for example. The circulation does not start directly from the disk as in other models. The mass flow, for example, is injected at a distance from the disk of approximately 1 kpc as the chimney structure widens appreciably and the material is subsequently injected into upper halo. In the upper halo, significant—more or less complete—mixing occurs. Cooling takes place and the mass can rain back down on the disk. Note that in this case the temperature increases with scale height in distinction to some other models. This overall mass flow can eventually be determined observationally by using the full range of measurements available including neutral hydrogen studies in both emission and absorption of the gas distribution and its kinematics, X-ray measurements in both the hard and soft energy bands, and quasar absorption line studies. Radio continuum studies of the thermal and non-thermal distribution and its spectral index variation are also potentially important indicators of the size and distribution of shock waves propagating into the halo.

It is important to emphasise that in this model the halo is in a crucial even prima donna role regulating the overall pressure balance, and mass and energy flow between the halo and the disk. In many ways it has the a similar physical role as the all pervading hot gas in the McKee Ostriker

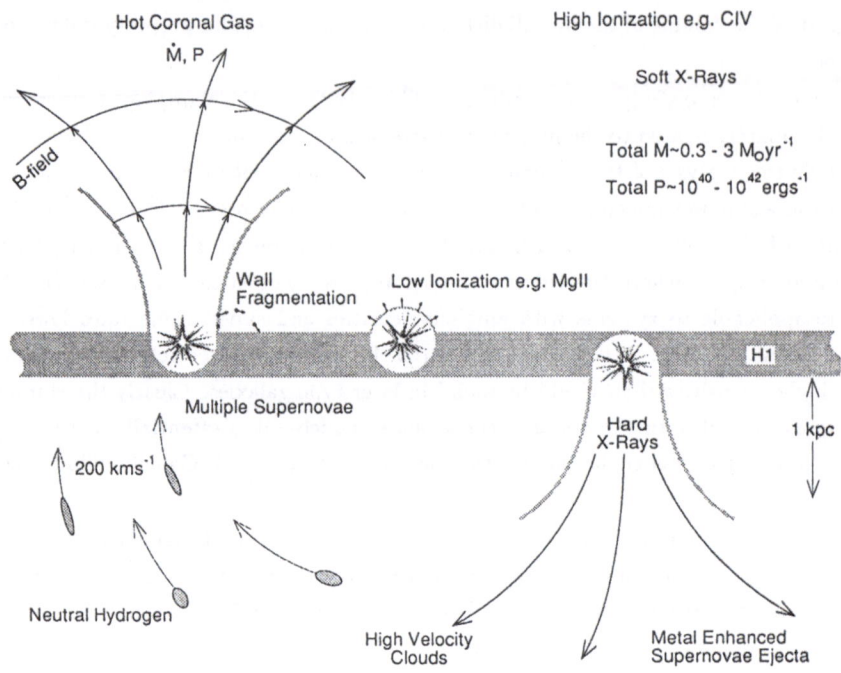

Figure 1 shows the global mass and energy flow from halo to disk and vice versa.

model. The connectivity of the hot gas is not, however in the disk, but via the chimneys and general halo gas (Figure 1).

5. FURTHER IMPLICATIONS AND SUMMARY

This model has some interesting implications for our understanding of quasar absorption lines, the dynamo mechanism in spiral galaxies, the nature of cosmic rays, the structure of edge on galaxies, and the nature of interstellar media in external galaxies.

The interpretation of quasar absorption lines depends on our understanding of the evolution of the intervening galactic interstellar media and hot gaseous coronae as a function of cosmic time. Halos can evolve quite significantly from, perhaps, enormous wind driven structures at early epochs to chimney driven coronae, and in some cases to cooling halos where the supernova rate drops below the critical value for the blowout condition to be satisfied. One could envision an early stage of extended high ionisation followed by a slow shrinking of the halo through the CIV stage to the coooler MgII phase. Many details are as yet unclear but a two pronged attack on both the interstellar medium and quasar will be increasingly productive (Figure 2).

In the chimney mode the galactic dynamo models may have to be significantly modified. Buoyancy may no longer be the dominant mechanism for vertical magnetic flux transport. Here the magnetic flux is transported into the halo by an essentially convective mode in conjunction with the hot gas ejected through chimneys formed from superbubbles. The whole problem of cosmic ray acceleration and propagation is also altered significantly in this picture.

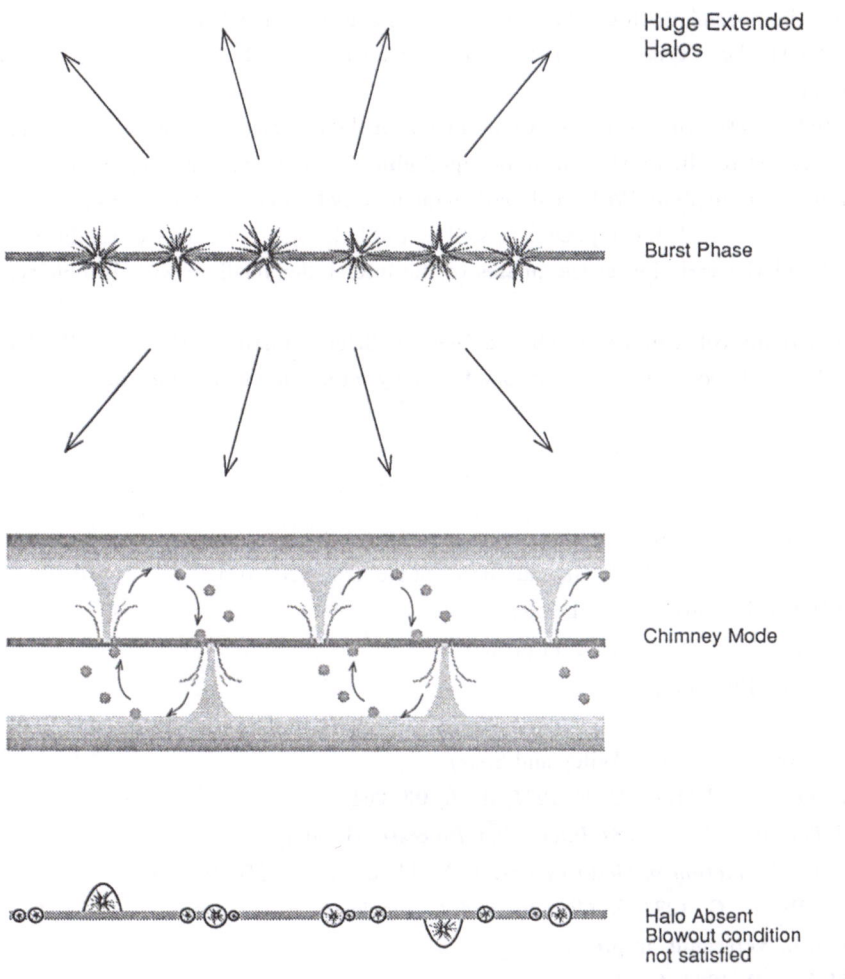

Huge Extended Halos

Burst Phase

Chimney Mode

Halo Absent
Blowout condition
not satisfied

HALO EVOLUTION

Figure 2 shows how galaxies may evolve from a massive starburst system in the early phase to a
relatively quiescent late type galaxy.

Chimneys should be observable in nearby galaxies. The problem is similar to that of finding
a Heiles-type supershell in say M31. One very interesting observation relating to both edge on
galaxies and the galactic dynamo problem is to determine whether or not the vertical component
of the galactic magnetic field changes sign across the disk or not. In other words it would be most
interesting to know whether the galactic dynamo is, in fact even or odd.

Some preliminary statements can be made on the nature of interstellar media in other galaxies.
Very crude estimates of the mean ambient gas density and the power and clumpiness of supernovae in
disk galaxies as a function of Hubble type can be used to infer the state of the phases of the interstellar
medium as a function of Hubble type. Using a fairly broad brush, the picture is one where galaxies

of type SO/Sa have a homogeneous three phase medium, the chimney model is relevant to galaxies of type Sb/Sc and for the later types, Sc/Irr, the predominant state of the interstellar medium is expected to be two phase.

We have presented our attempt to incorporate in models of the gaseous component of both halo and disk some of the recent results on the nature of superbubbles and their substantial influence on the physics of the interstellar medium.We have described qualitively how crucial the mass and energy flow through the halo is for the disk component as well as the halo gas. A number of implications have been discussed and the variation of the phases of the interstellar medium with Hubble-type briefly noted.

It is a pleasure to thank colleagues at the Space Telecope Science Institute, the Johns Hopkins University and the Tokyo Astronomical Observatory for many interesting discussions in the course of this work.

REFERENCES

Bregman, J. N. 1980 *Ap. J.,* **236**, 577.

Bregman, J. N. and Lockman, F. J. 1986, Gaseous Halos of Galaxies (NRAO)

Brinks, E. 1984, PhD Thesis, University of Leiden.

Corbelli, E. and Salpeter, E. E. 1987 *Ap. J.,* in press.

Cox, D. and Reynolds, R. 1987, preprint.

Field, G. B. 1986 in *Highlights of Modern Astrophysics: Concepts and Controversies,* eds. S. L. Shapiro and S. A. Teukolsky (John Wiley and Sons).

Heckman,T. M., Armus, L. and Miley, G. K. 1987, *A. J.,* **93**, 264.

Hollenbach, D. and Thronson, H. A. 1987 *Interstellar Processes* (Reidel).

Ikeuchi, S. 1987 in *Star Formation in Galaxies,* eds. T. X. Thuan and T. Montmerle.

Kulkarni, S. R. and Heiles, C. 1987 in *Galactic and Extragalactic Radio Astronomy,* eds. K. I. Kellerman and G. L. Verschuur, in press.

MacLow, M. and McRay, R. 1987 *Ap. J.,* in press.

McKee, C. F. and Ostriker, J. P. 1977, *Ap. J.,* **218**, 148.

Norman, C. A. and Ikeuchi, S. 1988, in preparation.

Tomisaka, K., Habe, A., and Ikeuchi, S. 1981, *Ap. Sp. Sci.,* **78**, 273.

Tomisaka, K., and Ikeuchi, S. 1987, *Publ. Astron. Soc. Japan,* **38**, 697.

GIANT HI CLOUDS IN THE GALAXY

Bruce G. Elmegreen
IBM Thomas J. Watson Research Center
P.O. Box 218, Yorktown Heights, N.Y. 10598 USA

A substantial fraction of the mass of all the interstellar material in the outer Galaxy is in the form of giant cloud complexes that extend for several hundred parsecs each, at a mean density of ~ 10 cm^{-3} and a mass of $10^7 M_\odot$ (McGee and Milton 1964; Kerr 1964; Burke, Turner and Tuve 1964). McGee and Milton found 29 such clouds in the outer Galaxy. The HI map in Henderson, Jackson and Kerr (1983; Figure 1) has approximately 12% of its emission inside ~ 8 closed contours where the column density exceeds 7×10^{20} cm^{-2}; the average mass inside each contour is $6 \times 10^7 M_\odot$. Grabelsky *et al.* (1987) find that the CO clouds in the outer galaxy cluster inside these giant HI clouds.

Observations of other galaxies show equally large clouds, usually in the spiral arms and throughout the spiral region. This is the case for HI in the galaxies M33, M101, M81, M31, M106 and IC342 (Wright, Warner and Baldwin 1972; Newton 1980a,b; Allen, Goss and van Woerden 1973; Allen and Goss 1979; Viallefond, Allen and Goss 1981; Viallefond, Goss and Allen 1982; Rots 1975; Emerson 1974; Unwin 1980a,b; Bajaja and Shane 1982; van Albada 1980). Equally large CO clouds are in M31, M101, and M51 (Blitz *et al.* 1981; Blitz 1985; Boulanger *et al.* 1981, 1984; Linke 1982; Stark 1985; Nakano *et al.* 1987; Rydbeck *et al.* 1986; Verter and Kutner 1987; Lo *et al.* 1987).

Giant atomic and molecular cloud complexes similar to those seen in other galaxies and beyond the solar circle in our Galaxy should also be present in the inner regions of our Galaxy. Approximately 30 such clouds are expected in the first quadrant if a large fraction of the bright HI is in this form. They should have the following properties: (1) an extent of several hundred parsecs parallel to the galactic plane, and 100 pc or more perpendicular to the plane; (2) a modest density enhancement over the surrounding medium, perhaps a factor of ~ 10, with the density exceeding the minimum value for self-gravitational binding in the tidal force field of the Galaxy, and, possibly, (3) a virialized velocity dispersion.

These expected physical properties correspond to the following observable properties: (1) a longitude and latitude extent of 6° to 10° in the nearby portion of the Sagittarius spiral arm and 1° to 2° in the far side, with intermediate sizes in the Norma-Scutum arm; (2) brightness temperatures for the clouds that may be a factor of ~ 10 larger than the brightness temperatures in the surrounding regions, but only at high galactic latitudes, because there the opacity is small and the emission from peculiar velocities at different distances in the galactic plane do not contribute, and (3) a total velocity extent of the emission, down to

~1/5 of the peak temperature (i.e., 20K), that may exceed 30 or 40 km s^{-1}. This total velocity extent comes from the virial-theorem Gaussian dispersion,

$$v_{VT} \simeq (G\rho)^{1/2}R \simeq 7 \text{ km s}^{-1} \ (n/10 \text{ cm}^{-3})^{1/2} \ (R/200 \text{ pc}),$$

and from a full width at 1/5 peak given by

$$\Delta v_{1/5} = (8 \ln 5)^{1/2}v_{VT} \simeq 50 \text{ km s}^{-1} \ (n/10 \text{ cm}^{-3})^{1/2} \ (R/200 \text{ pc}).$$

If such clouds exist, then the latitude-integrated (l, v) diagram of the first quadrant should be a composite of ~30 giant emission features, which, for all but the nearest clouds, will appear as long streaks parallel to the velocity axis and several degrees wide.

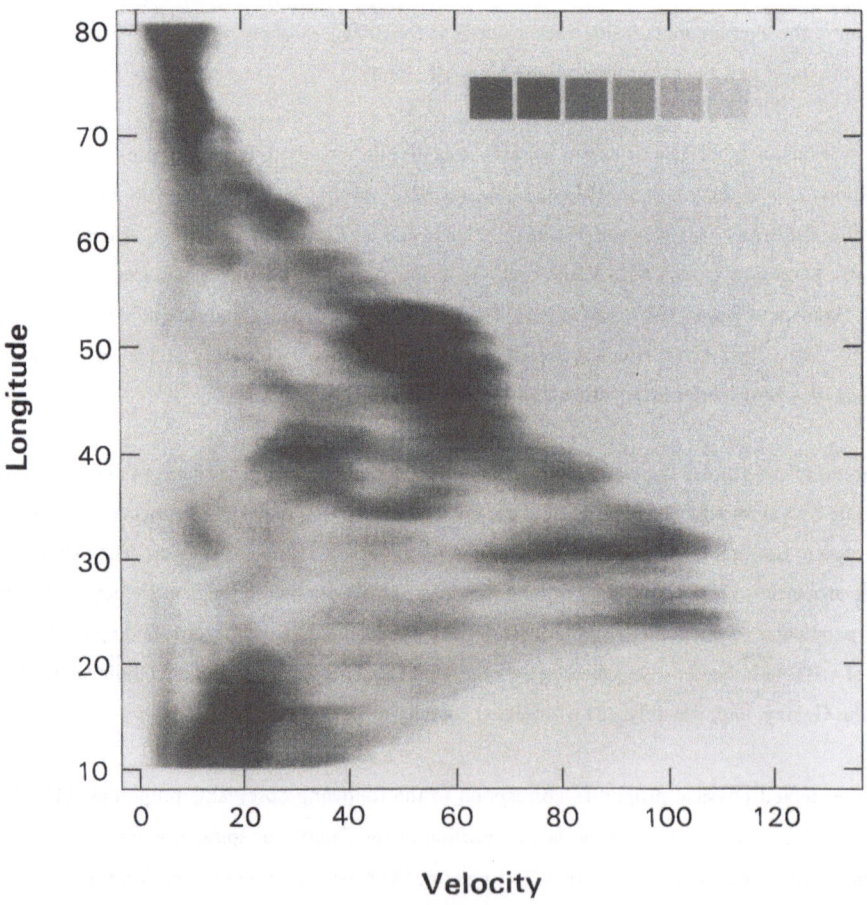

Figure 1 - Longitude-velocity diagram for HI emission from the Weaver-Williams survey. The emission has been clipped at 40K, integrated over latitude and multiplied by the near kinematic distance.

Figure 1 shows a longitude-velocity diagram of HI emission from the first quadrant (Elmegreen and Elmegreen 1987; hereafter EE87), using data from the Weaver and Williams (1973) survey. The plotted quantity is the brightness temperature clipped at 40K, integrated over the full latitude extent of the survey ($\pm 10°$), and multiplied by the near kinematic distance. The clipping removes the low level of emission that is expected to occur between, above and below the giant cloud complexes, and separates the expected cloud complexes from each other. A value of 40K was chosen; this includes ~60% of the total emission. The integration over latitude accentuates the expected difference between the latitude extent of the nearby clouds and the latitude extent of the distant emission at the same velocity. Integration also accentuates the difference between cloud and intercloud emission at near distances. Conventional (l, v) diagrams, made of emission entirely from the midplane, should not show giant complexes as well as integrated (l, v) diagrams because the emission from the low latitude intercloud regions is often optically thick and at about the same brightness temperature as the clouds. The clouds are most obvious away from the galactic plane, where the intercloud gas is optically thin and fainter than the cloud emission, and where more-distant gas is not confusing.

There is a problem with a straightforward latitude integration, however, because it gives a larger integral for nearby clouds that cover a large latitude extent than it does for similar, more distant clouds that cover a small latitude extent. The result is that distant clouds on an integrated (l, v) diagram tend to be fainter than nearby clouds. This problem can be offset by multiplying the integral by the near kinematic distance for each l and v. Then the result gives the same level of brightness for all intrinsically similar clouds that are located at their near kinematic distances. This unconventional procedure for presenting the data was designed to highlight the cloudy structure in the nearby part of the first quadrant, up to and including the terminal velocity distance. Many of the far-distance clouds are probably made less detectable by this presentation, and some clouds at the greatest distances may be lost altogether.

Figure 1 shows a considerable amount of structure that has not been present in previous versions of the (l, v) diagram. Many of the features with velocities less than ~40 km s^{-1} are roundish in the diagram, with velocity extents of ~ 20 km s^{-1} and longitude extents of ~5° to 10°. Many of the features at larger velocities are elongated parallel to the velocity axis with the same total velocity extent but smaller longitude extents. This is to be expected if each feature results from a similar cloud, with only the distance varying as the velocity increases.

Some of the features are expected to be blended emission from several clouds. The features at $(l, v) =$ (24,102), (30.5,94), (50.5,53) and (55.5, 38) are located at the tangent points to spiral arms (cf. EE87). The features at (19.5,42), (23,54), (24,102), (31,47) and (31.5,13) probably contain contributions from both near and far kinematic distances, because both near and far CO clouds are within the same (l,b, v) coordinate intervals. These potentially blended clouds will not be included in the following discussion of mass, molecular fraction and gravitational binding.

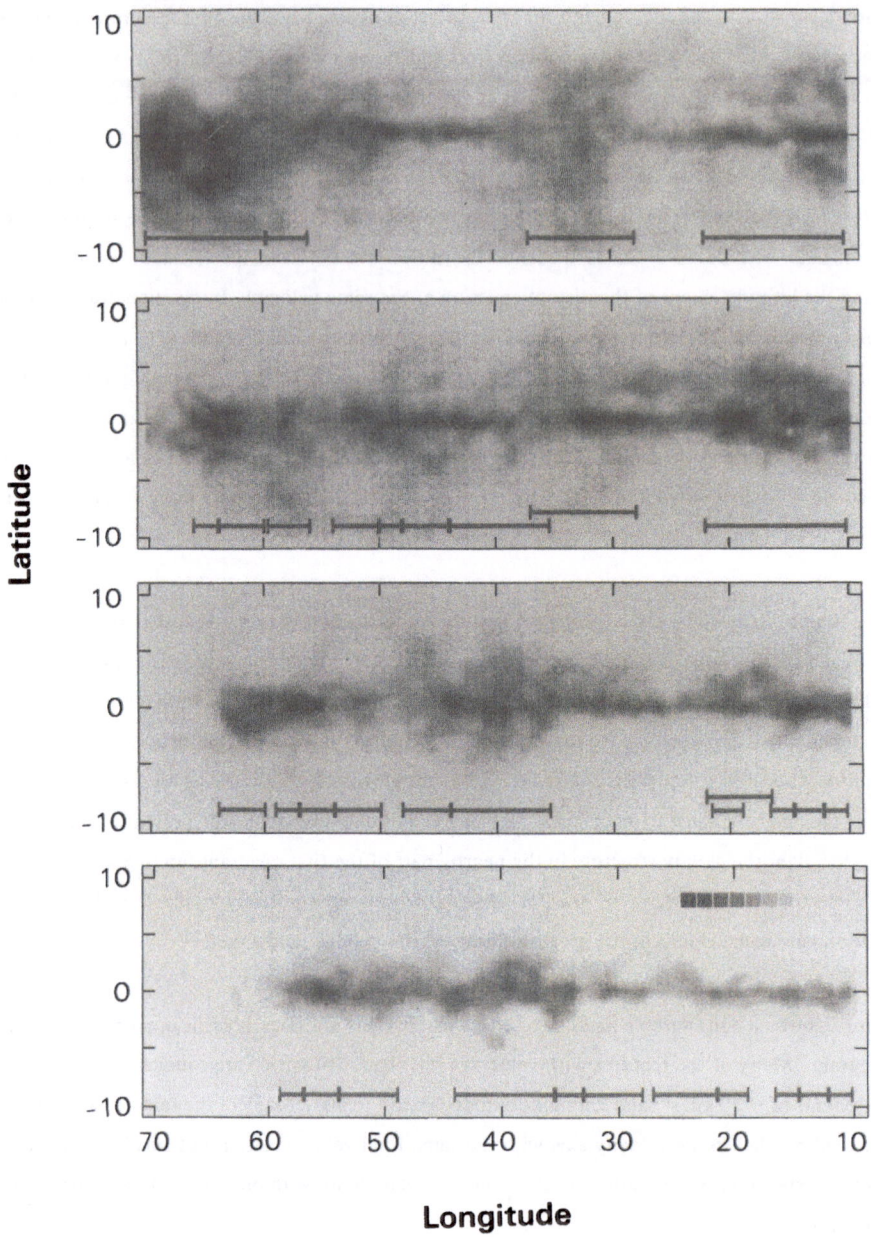

Figure 2 - Longitude-latitude diagrams for HI emission from the Weaver and Williams (1973) survey. The four panels are for different velocities, 10 km s^{-1} (top), 20 km s^{-1}, 30 km s^{-1} and 40 km s^{-1} (bottom).

Figure 2 shows longitude-latitude maps of the unclipped emission for $v = 10\pm1$ km s^{-1} (top), 20 ± 1 km s^{-1}, 30 ± 1 km s^{-1} and 40 ± 1 km s^{-1} (bottom). Other velocities are displayed in EE87. Because most of the emission features on the clipped and integrated (l, v) diagram also have relatively large latitude extents on (l,b) maps, each unblended feature is probably a distinct cloud, and not the result of spurious velocity crowding.

Table 1 lists for each of the 25 unblended clouds the total mass, radius, one-dimensional Gaussian velocity dispersion, average density, distance from the galactic center, ratio of the density to the critical density for gravitational self-binding in the tidal force field of the Galaxy, ratio of the virial-theorem velocity dispersion for the observed mass and radius to the observed velocity dispersion, and molecular fraction (from EE87). The molecular fraction is defined to be the ratio of the mass of the bright CO emission associated with each HI cloud, as catalogued by Dame *et al.*(1986) and Myers *et al.* (1986), divided by the sum of this molecular mass and the atomic mass. The atomic mass derivation assumes that the complexes contain numerous unresolved clumps with an excitation temperature of 120K (see EE87). The tabulated quantities are listed in order of increasing distance from the galactic center.

TABLE 1: PROPERTIES OF UNBLENDED CLOUDS

(l,v) ($^\circ$, km s^{-1})	Mass ($\times 10^6 M_\odot$)	Radius (pc)	σ (km s^{-1})	n (cm^{-3})	D (kpc)	$\dfrac{n}{n_{crit}}$	$\dfrac{\Delta v_{VT}}{\sigma}$	$\dfrac{M(H_2)}{M(H_2) + M(H)}$
26 105	4.0	140	8.1	10.8	4.4	2.2	0.69	...[1]
13.5 47	5.8	146	4.5	13.9	4.9	3.0	1.46	0.68
10.5 33	2.4	113	4.7	11.9	5.4	2.8	1.00	...
34 95	3.1	116	7.1	14.8	5.6	3.5	0.76	...
13.5 32	3.3	119	5.1	14.4	5.8	3.5	1.08	...
30 78	10.7	217	7.0	7.7	5.8	1.9	1.05	0.47
15 28	5.2	150	4.2	11.4	5.9	2.8	1.47	0.61
27 64	1.3	97	6.0	10.4	6.1	2.6	0.63	0.25
37.5 81	21.7	362	6.1	3.4	6.3	0.87	1.33	0.46
41 62	24.0	285	5.4	7.6	7.0	2.2	1.76	0.21
36 58	16.5	304	5.1	4.3	7.1	1.2	1.51	0.30
40 64	33.2	420	8.1	3.3	7.2	0.93	1.14	0.19
34 50	3.4	108	7.5	19.9	7.6	5.9	0.77	0.33
47.5 57	42.6	430	5.8	4.0	7.8	1.2	1.78	0.11
12 14	9.2	161	5.6	16.4	7.9	5.1	1.41	0.16
39.5 33	5.6	249	6.6	2.7	8.2	0.85	0.75	0.047
57.5 35	1.8	109	5.4	10.7	8.6	3.5	0.80	0.086
46 28	0.9	118	5.7	4.2	8.8	1.4	0.51	0.017
52 24	0.6	84	5.1	7.1	9.0	2.5	0.53	0.070
62.5 26	1.8	135	4.5	5.3	9.1	1.8	0.83	0.045
48.5 20	13.0	236	3.4	7.3	9.2	2.6	2.27	0.03
58 18	0.8	106	4.2	4.9	9.3	1.8	0.68	...
64.5 20	0.7	88	4.2	7.5	9.3	2.7	0.68	...
12 6	0.1	53	1.1	3.7	9.7	1.4	1.11	...

[1] No clear correlation with CO in Dame *et al.* (1986).

The generally large values of n/n_{crit} (> 1), and the closeness of the ratio $\Delta v_{VT}/\sigma$ to 1 (the average value for all clouds equals 1.08 ± 0.44) implies that most of the clouds are gravitationally bound. They are distinct entities, like giant molecular clouds, although their mass is generally larger and their density less.

The masses, densities and velocity dispersions of the complexes do not vary much with galactocentric distance, but the molecular fraction decreases with increasing distance. This decrease may explain the origin of the molecular ring. The molecular ring results from a general lack of all gas in the inner regions of the Galaxy, and from a decrease in the molecular fraction of gas outside approximately 5 kpc. The present observations suggest that the decrease outside 5 kpc results in part from a decrease in the molecular fraction per cloud. The basic cloud size does not change with galactocentric distance: the largest clouds in the inner Galaxy have about the same total mass as the largest clouds in the outer part of the Galaxy. The decrease in the molecular fraction could result in part from a decrease in the ambient pressure of the interstellar medium, because a low pressure corresponds to a low density and column density for a given cloud mass, and this corresponds to less shielding of uv radiation by molecules and dust, and to a smaller molecular fraction. The interstellar pressure should decrease along with the mean radiation field in an exponential fashion, as the radiation field does for other galaxies. A factor of 1.5 decrease in the pressure and radiation field from 5 kpc to 10 kpc can explain the observed variation in the molecular fraction from ~70% to 10% (EE87). The clouds in the outer region of the Galaxy, such as those in the Henderson, Jackson and Kerr (1983) survey, should be less molecular than similar clouds in the inner region.

The largest HI clouds appear to be confined to the main spiral arms, as are the largest CO clouds. The Sagittarius arm is particularly clear on Figure 1. It consists of clouds in a sequence starting at the near side with $(l, v) = (12,14)$, and continuing onto a cluster of clouds containing (31,47N), (39.5,33) and (34,50), and then to the spiral arm tangent point at (50.5,53), with a possible spur at (55.5, 38), and onto an arc of clouds on the far side at (47.5,57), (41,62), (40,64), (36,58) and (31,47F). The Norma-Scutum arm appears to extend along a sequence beginning with a cluster of clouds at (10.5,33), (13.5,32) and (15,28), and continuing onto another cluster including (19.5,42) and (23,54), and onto the clouds (27,64) and (30,78) until the spiral arm tangent occurs at (30.5,94), with a possible weak spur at (34,95). Then the arm continues back down in longitude on the far side of the tangent point, including (26,105), (24,102F) and possibly (23,54F). A next inner arm seems to include clouds at (13.5,47) and (24,102N), but this is unclear.

The observations can be summarized as follows: (1) Giant atomic cloud complexes have been found in the inner part of the Milky Way. They are similar to those observed in the outer Milky Way and in other galaxies. Each contains approximately $10^7 M_\odot$. (2) The giant clouds in the inner Galaxy have approximately the same mass as the giant clouds in the outer Galaxy, but the molecular fraction per cloud is larger in the inner Galaxy. (3) Each cloud complex appears to be gravitationally bound. (4) Most of the HI clouds are associated with giant CO clouds. (5) The HI clouds lie in the main spiral arms.

REFERENCES

Allen, R.J., Goss, W.M., and van Woerden, H. 1973, *Astron.Astrophys.*, **29**, 447.
Allen, R.J., and Goss, W.M. 1979, *Astron.Astrophys.Suppl.*, **36**, 135.
Bajaja, E., and Shane, W.W. 1982, *Astron.Astrophys.Suppl.*, **49**, 745.
Blitz, L. 1985, *Astrophys.J.*, **296**, 481.
Blitz, L., Israel, F.P., Neugebauer, G., Gatley, I., Lee, T.J., and Beattie, D.H. 1981, *Astrophys.J.*, **249**, 76.
Boulanger, F., Stark, A.A., and Combes, F. 1981, *Astron.Astrophys.*, **93**, L1.
Boulanger, F., Bystedt, J., Casoli, F., and Combes, F. 1984, *Astron.Astrophys.*, **140**, L5.
Burke, B.F., Turner, K.C., and Tuve, M.A. 1964, in IAU Symposium No.20, ed. F.J. Kerr and A.W. Rodgers, Australian Academy of Sciences, p. 131.
Dame, T.M., Elmegreen, B.G., Cohen, R.S., and Thaddeus, P. 1986, *Astrophys.J.*, **305**, 892.
Elmegreen, B.G., and Elmegreen, D.M. 1987, *Astrophys.J.*, **320**, in press.
Emerson, D.T. 1974, *Monthly Not.Roy.Astron.Soc.*, **169**, 607.
Grabelsky, D.A., Cohen, R.S., May, J., Bronfman, L., and Thaddeus, P. 1987, *Astrophys.J.*, **315**, 122.
Henderson, A.P., Jackson, P.D., and Kerr, F.J. 1982, *Astrophys.J.*, **263**, 116.
Kerr, F.J. 1964, in IAU Symposium No.20, ed. F.J. Kerr and A.W. Rodgers, Australian Academy of Sciences, p. 81.
Linke, R.A. 1982, in *Extragalactic Molecules*, ed. L. Blitz and M. Kutner (Green Bank: NRAO Publications Office), p. 87.
Lo, K.Y., Ball, R., Masson, C.R., Phillips, T.G., Scott, S., and Woody, D.P. 1987, *Astrophys.J.(Letters)*, **317**, L63.
McGee, R.X., and Milton, J.A. 1964, *Austral.J.Phys.*, **17**, 128.
Myers, P.C., Dame, T.M., Thaddeus, P., Cohen, R.S., Silverberg, R.F., Dwek, E., and Hauser, M.G. 1986, *Astrophys.J.*, **301**, 398.
Nakano, M., Ichikawa, T., Tanaka, Y.D., Nakai, N., and Sofue, Y. 1987, *Pub. Astr. Soc. Japan,* in press.
Newton, K. 1980a, *Monthly Not.Roy.Astron.Soc.*, **190**, 689.
Newton, K. 1980b, *Monthly Not.Roy.Astron.Soc.*, **191**, 615.
Rots, A.H. 1975, *Astron.Astrophys.*, **45**, 43.
Rydbeck, G., Hjalmarson, Ä., Johansson, L.E.B., and Rydbeck, O.E.H. 1986, in *Star Forming Regions* , IAU Symposium No. 115, ed. M. Peimbert and J. Jugaku (Dordrecht: Reidel), p. 535.
Stark, A.A. 1985, in *The Milky Way Galaxy*, IAU Symp. No. 106, ed. H. van Woerden, R.J. Allen, and W.B. Burton (Dordrecht: Reidel), p. 445.
Unwin, S.C. 1980a, *Monthly Not.Roy.Astron.Soc.*, **190**, 551.
Unwin, S.C. 1980b, *Monthly Not.Roy.Astron.Soc.*, **192**, 243.
van Albada, G.D. 1980, *Astron.Astrophys.*, **90**, 123.
Verter, F., and Kutner, M.L. 1987, in preparation.
Viallefond, F., Allen, R.J., and Goss, W.M. 1981, *Astron.Astrophys.*, **104**, 127.
Viallefond, F., Goss, W.M., and Allen, R.J. 1982, *Astron.Astrophys.*, **115**, 373.
Weaver, H., and Williams, D.R.W. 1973, *Astron.Astrophys.Suppl.*, **8**, 1.
Wright, M.C.H., Warner, P.J., and Baldwin, J.E. 1972, *Monthly Not.Roy.Astron.Soc.*, **155**, 337.

THE SOUTHERN EXTENSION OF THE TAURUS
MOLECULAR CLOUDS

Loris Magnani
E.O. Hulburt Center for Space Research
Code 4130MA, Naval Research Laboratory
Washington, D.C. 20375

In the direction of the Galactic anticenter, the nearest molecular complex to the Sun is the Taurus-Auriga-Perseus system of dark clouds. The distance to this complex is traditionally quoted to be 140 ± 25 pc for the Taurus-Auriga clouds (Elias 1978) and 300 - 380 pc for the Perseus clouds (Eklöf 1958). Recently, the entire region has been mapped in the CO(J=1-0) transition by Ungerechts and Thaddeus (1987; hereafter UT). Their CO survey covers the region extending from $l = 155°- 190°$ and $b = -10° - -30°$, with cloud 1 (also known as L1453, 1454, 1457, and 1458 in the Lynds (1962) catalog) situated furthest from the Galactic plane at $b \sim -35°$.

The Taurus dark clouds are thought to extend from $l = 160°- 180°$ and from $b = -10°$ to $-25°$, but as the existence of UT1 implies, there may be other components south of $b = -25°$. In fact, the IRAS 100 μm data clearly shows emission extending south of the traditional Taurus dark cloud region to Galactic latitude $b \sim -45°$. Most of the 100 μm emission from this southmost region is similar in morphology and emissivity to the emission from the IRAS 100 μm cirrus clouds (Low et al. 1984). These objects are for the most part diffuse, atomic, dust clouds; a small fraction of which are associated with local molecular clouds (Weiland et al. 1986). A portion of the infrared emission from this region may thus be molecular and associated with the Taurus-Auriga molecular clouds.

Eight of the high-latitude molecular clouds in the Magnani, Blitz, and Mundy (1985; hereafter MBM) catalog are located in this region and five are mapped or partially mapped by the UT survey (MBM 11 - 13, 14, and 17; MBM 11 - 13 are also known as L1453, 1454, 1457, and 1458). The largest of the MBM clouds in this region is MBM16, which subtends more than five square degrees and is located at $l = 172°$ and $b = -38°$.

The morphology of this region as revealed by the IRAS data and the average distance of the high-latitude clouds (~100 pc; MBM) indicate that the MBM clouds may be related to the Taurus-Auriga cloud complex. Star counts conducted by Magnani and de Vries (1986) in the direction of MBM16 give a distance of 100 ± 50 pc for this object, further supporting its association with the nearby Taurus dark clouds.

The most reliable method for determining the distance to a local molecular cloud is to map the interstellar absorption lines toward stars of differing distances in the direction of the cloud. Such a program was carried out for some of the MBM clouds and is reported by Hobbs, Blitz, and Magnani (1986) and by Hobbs et al. 1987).

Optical spectra of the NaI D lines were obtained for seven lines of sight in the direction of MBM16 and ten lines of sight in the direction of MBM12. Absorption lines due to the interstellar cloud were seen in both instances and upper and lower limits to the distances of the two objects were obtained.

The distance to MBM12 obtained in this manner is 70 - 145 pc and, for MBM16, 60 - 95 pc. With these distance determinations and the IRAS 100μm morphology, the association of MBM12 and 16 with the Taurus-Auriga molecular clouds is more secure.

The average velocity of the CO(J=1-0) lines in MBM16 is $\sim$ 7 km s^{-1} with respect to the LSR, a value similar to that found in most of the Taurus-Auriga molecular clouds (UT). For MBM12, the average CO velocity is $\sim$ -1 km s^{-1} (UT), but the cloud is comprised of several clumps whose velocity ranges from -7 to +4 km s^{-1}. This object shows a very fragmented and complicated velocity structure which is perhaps indicative of energetic events which may have disrupted the cloud (Hobbs, Blitz, and Magnani 1986). Although MBM12 does not share the average velocity typical of most other Taurus clouds, there are sufficient clouds within the complex with average velocities other than 7 km s^{-1} that we do not question the association of this object with the complex.

The complicated velocity structure of MBM12 is typical of the high-latitude molecular clouds and leads to an interesting phenomenon: the virial mass of most of these clouds is an order of magnitude or more greater than the mass estimated from the CO data by conventional methods. This large discrepancy in the mass estimate implies that the high-latitude clouds are not gravitationally bound. This characteristic was first noted by Blitz, Magnani, and Mundy (1984) and MBM. Subsequent to that work, various studies have noted the discrepancy between the kinetic energy and the gravitational potential energy of some, typically small, molecular clouds: Murphy and Myers (1985), Keto and Myers (1986), Maddalena (1986), Heiles and Stevens (1986), Stark and Bania (1986), de Vries, Heithausen, and Thaddeus (1987), and UT. In particular, the UT survey lists 18 small clouds with $M_{vir}/M_{co} > 10$. All but 7 of these objects are located at a distance of 140 pc or less and they are associated with the Taurus-Auriga complex. Without exception, all these gravitationally unbound objects at the distance of Taurus are located at the periphery of the complex. MBM16 also shows a large discrepancy between its virial and CO mass estimates, and it also lies in the outer regions of the Taurus-Auriga dark clouds.

The complicated velocity structure in the small clouds at the edge of the Taurus complex may be a relic of the velocity structure of the atomic gas from which the molecular complex originally formed. Alternatively, the velocity structure may be a result of supernovae or stellar winds which have lashed the edges of the molecular cloud. If MBM12 and 16 are at the nearer distances (70 and 60 pc, respectively), they may be within or at the edge of the hot bubble of gas in which the Sun is located (Perry, Johnston, and Crawford 1982; Frisch and York 1983; and Paresce 1984). These clouds would thus provide an opportunity for studying the transition region between the cold and hot phases of the interstellar medium.

I take this opportunity to thank and acknowledge Dr. Frank Kerr who was a member of my thesis committee and patiently read and commented on several drafts of the thesis.

Blitz, L., Magnani, L., and Mundy, L. 1984, Ap. J. Letters, 282, L9.

Eklof, O. 1958, Uppsala Obs. Medd., No. 119, in Ark. Astr., 2, No. 21 (Stockholm: Almqvist & Wiksell), p. 213.

Elias, J.H. 1978, Ap. J., 224, 857.

Frisch, P.C. and York, D.G. 1983, Ap. J. Letters, 271, L59.

Heiles, C. and Stevens, M. 1986, Ap. J., 301, 331.

Hobbs, L.M., Blitz, L., and Magnani, L. 1986, Ap. J. Letters, 306, L109.

Hobbs, L.M., Blitz, L., Penprase, B., Magnani, L., and Welty, D.E. 1987, Ap. J., submitted.

Keto, E.R. and Myers, P.C. 1986, Ap. J., 304, 466.

Low, F.J. et al. 1984, Ap. J. Letters, 278, L19.

Lynds, B.T. 1962, Ap. J. Suppl., 7, 1.

Maddalena, R.J. 1986, Ph.D. Thesis, Columbia University.

Magnani, L., Blitz, L., and Mundy, L. 1985, Ap. J., 295, 402 (MBM).

Murphy, D.C. and Myers, P.C. 1985, Ap. J., 298, 818.

Paresce, F. 1984, A. J., 89, 1022.

Perry, C.L., Johnston, L., and Crawford, D.L. 1982, A. J., 87, 1751.

Stark, A.A. and Bania, T. 1986, Ap. J. Letters, 306, L17.

Ungerechts, H. and Thaddeus, P. 1987, Ap. J. Suppl., 63, 645 (UT).

de Vries, H.W., Heithausen, A., and Thaddeus, P. 1987, Ap. J., in press.

Weiland, J.L., Blitz, L., Dwek, E., Hauser, M.G., Magnani, L, and Rickard, L.-J 1986, Ap. J. Letters, 306, L101.

KINEMATICS OF 21 CM SELF ABSORPTION
TOWARDS THE TAURUS MOLECULAR COMPLEX

W.L.H. Shuter (UBC) and R.L. Dickman (FCRAO)

The relationship between the cold absorbing atomic hydrogen observed at 21 cm in interstellar clouds and the molecular gas inferred from observations of ^{13}CO (McCutcheon, Shuter, and Booth 1978) is complex, and can be quite variable. To investigate this subject further we mapped a 10^{0} x 7.5^{0} grid (1950 center: $\alpha = 04^{h}\ 30^{m}$, $\delta = +27^{0}$; $l = 172^{0}$, $b = -14^{0}$) towards the Taurus Molecular Complex (TMC) at Arecibo Observatory in the 21 cm line of atomic hydrogen. Details of the observing procedures and the analysis are given elsewhere (Shuter, Dickman, and Klatt 1987). The same grid had previously been mapped in ^{13}CO at Five College Radio Astronomy Observatory (Kleiner and Dickman 1984; 1985). In this paper we place the main emphasis on comparing the large-scale kinematics of the 21 cm absorption and ^{13}CO.

The column density distribution of 21 cm self absorption is shown in Fig. 1.

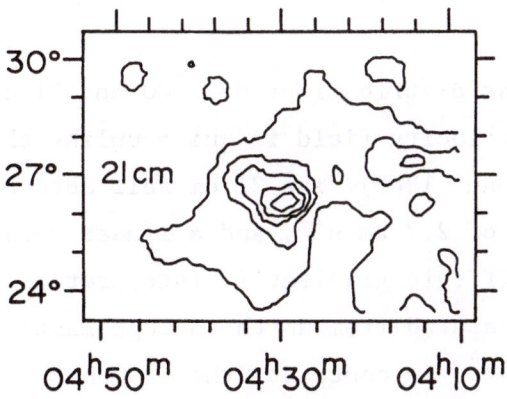

Fig. 1 - Column density of 21 cm self absorption towards TMC. Contours are equally spaced, but have arbitrary units.

The distribution of absorbing hydrogen is similar in general outline but differs in detail and covers a wider area than that of ^{13}CO (Kleiner and Dickman 1984); moreover, both distributions are concentrated about the same central point. The overall similarity leads us to believe that the 21 cm absorption is associated with the molecular gas in TMC, which is 140 pc from the Sun, rather than being an unrelated feature along the line-of-sight to TMC.

The velocity field of the 21 cm absorption is shown in Fig. 2.

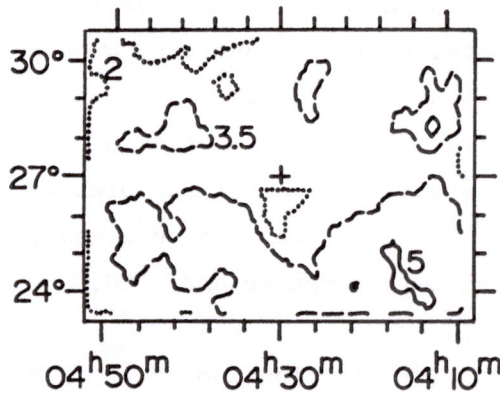

Fig. 2 - The velocity field of 21 cm self absorption towards TMC. Units are km s^{-}.

Despite the similar distributions of ^{13}CO and 21 cm self absorption, this velocity field is quite unlike that of the ^{13}CO (Kleiner and Dickman 1985). The 21 cm self absorption has a systemic velocity of 2.7 km s^{-1}, and a linear velocity gradient of 85 km s^{-1} kpc^{-1}. If this gradient is interpreted as a rotation, the position angle (measured from North through East) of the rotation axis is towards 297°. In contrast, the ^{13}CO has a systemic velocity of ~ 250 km s^{-1} kpc^{-1} with an inferred rotation axis towards 25°.

A complete first order Taylor series representation of the line-of-sight component of the three-dimensional velocity field in the Solar Neighborhood (Goulet and Shuter 1984) can be used to predict the systemic velocity and velocity gradient of ambient gas at the location of TMC, based on global fits to 21 cm absorption data, ^{13}CO data, and other tracers of the velocity field.

In this representation, the line-of-sight velocity, v_1, with respect to the LSR, for a tracer with galactic coordinates l,b at distance d is given by

$$
\begin{aligned}
v_1 = \quad & K_1 \\
& + (K_2 + U_0) \cos b \cos l + (K_3 + V_0) \cos b \sin l + (K_4 + W_0) \sin b \\
& + K_5 \, d \cos^2 b \cos^2 l + K_6 \, d \cos^2 b \sin^2 l + K_7 \, d \sin^2 b \\
& + K_8 \, d \cos^2 b \sin 2l + K_9 \, d \sin 2b \sin l + K_{10} \, d \sin 2b \cos l
\end{aligned}
$$

In the above expression U_0, V_0, W_0 are the three velocity components of the Standard Solar Motion defining the LSR, and we have adopted standard values of 10.2, 15.3, 7.9 km s^{-1} respectively. The coefficient K_1 is the so-called "K term". The terms with coefficients K_2 - K_4 represent the reflex of the Solar Motion for the specific tracer used, K_5 - K_7 represent gradient terms, and K_8 - K_{10} are shear terms. The coefficient K_8 corresponds to the Oort constant, **A**.

Values of the coefficients K_1 - K_{10} used in deriving the systemic velocities v_1, and their gradients, are listed in Table 1.

Using these data, one can predict the gradient and position angle expected for each velocity field tracer at $l_0 = 172°$, $b_0 = -14°$, d = 140 pc, by calculating v_1 at $l_0 = \pm 1°$ and $b_0 = \pm 1°$. Table 2 lists the predicted velocity field parameters for each kinematic tracer, along with those observed.

TABLE 1

TAYLOR SERIES COEFFICIENTS FROM GLOBAL FITS TO THE VELOCITY FIELD IN THE SOLAR NEIGHBORHOOD

	Standard Galactic Rotation	21 cm(abs)	^{13}CO	21 cm(em)	BV stars d < 200 pc	Units
K_1	0.0	...	5.6	...	0.7	km s^{-1}
K_2	-10.2	-11.0	-11.4	-10.1	-9.1	km s^{-1}
K_3	-15.3	-16.3	-19.8	-16.2	-18.2	"
K_4	- 7.9	- 9.5	-10.6	-11.7	- 7.0	"
K_5	0.0	8.1	...	4.2	- 2	km s^{-1} kpc^{-1}
K_6	0.0	- 0.2	...	- 0.1	30	"
K_7	0.0	-11.5	...	20.3	24	"
K_8	12.9	(15.0)	(15.0)	(15.0)	12.7	km s^{-1} kpc^{-1}
K_9	0.0	6.9	...	3.3	-29	"
K_{10}	0.0	3.1	...	2.3	15	"

In this table, the coefficients K_1 - K_{10} for hydrogen in absorption, 21 cm(abs), and in emission, 21 cm(em), and those for nearby BV stars are taken from Goulet and Shuter (1984). Coefficients for ^{13}CO were derived from the analysis by Frogel and Stothers (1977).

Table 2 shows that while the systemic velocities of the absorbing hydrogen and ^{13}CO towards TMC differ significantly - the ^{13}CO is red-shifted by 3.7 km s^{-1} with respect to the hydrogen - they are in reasonable accord with the predictions from global fits. However, the gradients, particularly that of ^{13}CO, are quite discordant. If interpreted as rotations, the rotation axis of ^{13}CO in the Taurus Complex is retrograde with respect to differential galactic rotation (Kleiner and Dickman 1985) and parallel to the magnetic field direction, while that of the absorbing hydrogen is

TABLE 2

VELOCITIES AND GRADIENTS
(For l = 172^0, b = -14^0, d= 140 pc)

	Systemic Velocity	Magnitude of gradient	Position Angle (of inferred rotation axis)
	km s^{-1}	km s^{-1} kpc^{-1}	deg.
Data			
TMC 21 cm absorption	2.7	85	297
TMC ^{13}CO	6.4	250	25
Global Fit Predictions			
Standard Galactic Rotation	-0.5	24	227
21 cm absorption	1.6	42	232
^{13}CO	6.3	64	241
21 cm emission	1.0	54	270
BV stars	-0.2	49.	285

For purposes of comparison, the magnetic field at this location
lies at position angle 25^0 (or 205^0) (Hemeon-Heyer 1986), the
North Galactic pole is at 42.8^0, and the North pole of Gould's belt
at ~ 60^0 (Taylor, Dickman, and Scoville 1987).

perpendicular. Interestingly, the hydrogen <u>emission</u> towards TMC,
which we have not yet studied in detail, appears to have a systemic
velocity of ~ 6.5 km s^{-1}, similar to that of the ^{13}CO. This value
matches that of Gould's belt (Sodroski, Kerr, and Sinha 1985),
while that of the absorption is just somewhat greater than the
value expected for ambient gas.

This kinematic picture is complex, and difficult to account for. One view, which we are presently studying, is that the molecular gas in TMC, traced by ^{13}CO, is expanding away from the center of Gould's belt, a position that is still not well defined (Taylor, Dickman, and Scoville 1987). In doing so, it compresses and heats the atomic hydrogen gas ahead of it, which acquires the same systemic velocity. It also leaves a swirling wake in the ambient hydrogen on the side nearest the Sun, that we see as 21 cm absorption. We hope to be able to confirm this picture by studying the 21 cm emission towards TMC. If our present view is correct, we would expect the systemic velocity and its gradient to match that of the ^{13}CO quite closely; we might also perhaps see a larger density concentration in the North-West compared to that in the South-East, produced by the larger initial relative velocity which results from the rotation of the molecular complex.

In a related kinematic issue, we have discovered the presence of velocity waves in the absorbing hydrogen (Shuter, Dickman, and Klatt 1987). These have a projected wavelength of ~ 16 pc, an amplitude of ~ 1.5 km s^{-1}, and appear to be propagating parallel to the direction of the magnetic field. In a preliminary analysis of the column density field of 21 cm absorption, we also find waves of comparable wavelength that are ~ 100% amplitude modulated. In previous work (Kleiner and Dickman 1984; 1985) waves with a similar wavelength were found in CO column density, but not in velocity. We intend to analyse our 21 cm emission data on TMC both for the presence of column density and velocity waves, and to determine the systemic velocity and the velocity gradient. We hope that, when this work has been completed, we will have gained a much clearer insight into the large-scale kinematics of the Taurus Complex.

W.L.H.S. acknowledges a research grant from NSERC. R.L.D.'s research was supported by NSF Grant AST-85-12903 to Five College

Radio Astronomy Observatory, and in part by the 1986 Ernest F. Fullam award of Dudley Observatory. We thank M. Hemeon-Heyer, F. Kerr, S. Kleiner, and R. White for helpful information.

The Arecibo Observatory is part of the National Astronomy and Ionosphere Center, which is operated by Cornell University under contract with the National Science Foundation.

REFERENCES

Frogel, J. A., and Stothers, R. 1977, A. J., **82**, 890.
Goulet, T., and Shuter, W. L. H. 1984, in "Local Interstellar Medium," ed. Y. Kondo, F. C. Bruhweiler, and B. D. Savage (NASA Conf. Pub., No. 2345), p. 319.
Hemeon-Heyer, M. C. 1986, Ph. D. thesis, University of Massachusetts.
Kleiner, S. C., and Dickman, R. L. 1984, Ap. J., **286**, 255.
Kleiner, S. C., and Dickman, R. L. 1985, Ap. J., **295**, 466.
McCutcheon, W. H., Shuter, W. L. H., and Booth, R. S. 1978 M. N. R. A. S., **185**, 755.
Shuter, W. L. H., Dickman, R. L., and Klatt, C. 1987, Ap. J. (Letters), in press.
Sodroski, T. J., Kerr, F. J., and Sinha, R. P. 1985, in "The Milky Way Galaxy," ed. H. van Woerden, R. J. Allen, and W. Butler Burton (Reidel: Dordrecht), p. 345.
Taylor, D. K., Dickman, R. L., and Scoville, N. Z. 1987, Ap. J., **315**, 104.

THE VARIABLE HII REGIONS IN CEPHEUS A

V.A. Hughes
Astronomy Group, Department of Physics, Queen's University
Kingston Ontario K7L 3N6 Canada

Cepheus A has become a fascinating and intriguing star formation region. The name CepA was originally assigned by Sargent (1977, 1979) to a condensation in the much larger Cep OB3 molecular cloud observed in CO. Early observations at λ6cm using the Westerbork Synthesis Radio Telescope (Hughes and Wouterloot, 1982) showed the presence of two continuum sources, namely CepA East and CepA West. The west component correlated with an optical source, which does not as yet fit any normal classfication (Hartigan and Lada 1985). The east component is completely obscured optically and forms the subject of the present paper.

Evidence for star formation comes from the detection of radio HII regions (Beichmann, Becklin and Wynn-Williams 1979) of bipolar outflows in CO (Rodriguez et al 1980), and of OH and H_2O masers (Lada et al. 1981). In addition, infrared sources have been detected, the region is a source of molecular lines including H^2, HCO^+, NH_3, and there are some nearby Herbig-Haro objects (for a list of references see Hughes and Wouterloot, 1984).

High angular resolution observations made in 1981 and 1982, using the VLA[1] at λ20cm and λ6cm in both the "A" and "B" configurations (Hughes and Wouterloot 1984; Hughes 1985) showed the presence of two lines of compact radio sources contained inside a region ∿ 50" in size, corresponding to a linear dimension of < 0.2 pc. There were about 14 sources, of which the central sources were not resolved at 0".3 (200 au), though the outer sources were more diffuse. Values of spectral index, α, were determined, (where we define α by $S_\lambda \propto \lambda^{-\alpha}$ where S_λ is the flux density at wavelength λ,) and showed that the diffuse components had $\alpha \sim -0.1$, as for optically thin HII regions, but the more compact sources had $\alpha > -0.1$, consistent with their being HII regions, but containing optically thick regions. When the position of OH and H_2O masers as determined by Cohen, Rowland and Blair (1984) were plotted, it was seen that H_2O was situated at the edge of the central compact sources, while the OH was well outside, indicating the presence of a cocoon. Maps of the regions at λ20cm and at λ6cm, with angular resolutions of 1" and 0".3, respectively, are shown in Figures 1 and 2. When the received flux density was used to determine the excitation parameters of the HII regions, it was found that each appeared to be excited by a B3 star. Thus, from all the above factors, it appeared reasonable to associate the central compact regions with stars, and since the age of the HII regions could not be more than 1,000 yrs, with the pre-main-sequence stage of star formation, and the outer more diffuse regions with stars in a later stage where the HII regions has expanded.

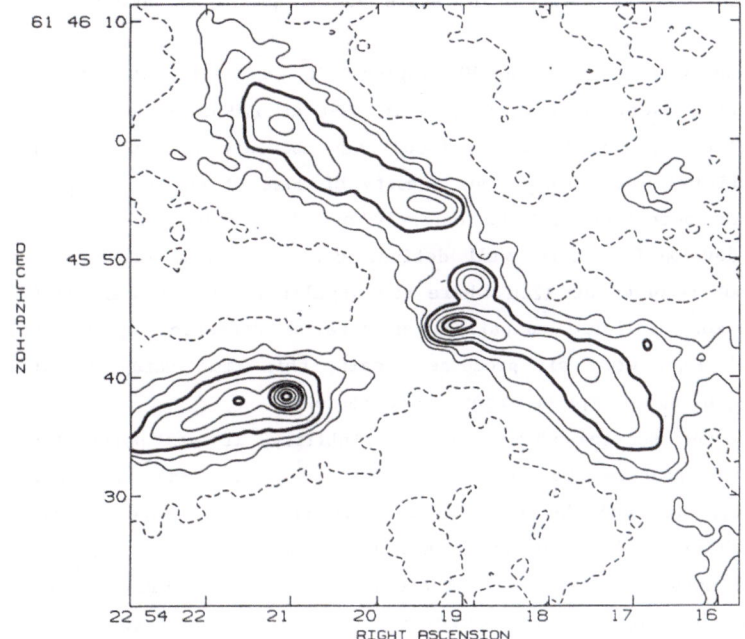

Figure 1. The 1982L map of Cep A East obtained using the VLA in the "A" configuration. Angular resolution is ∿ 1."5. Contour levels are at −0.1, 0.1, 0.2, 0.4, 0.8, 1.6, 2.4, 3.2, and 4.0 mJy/beam.

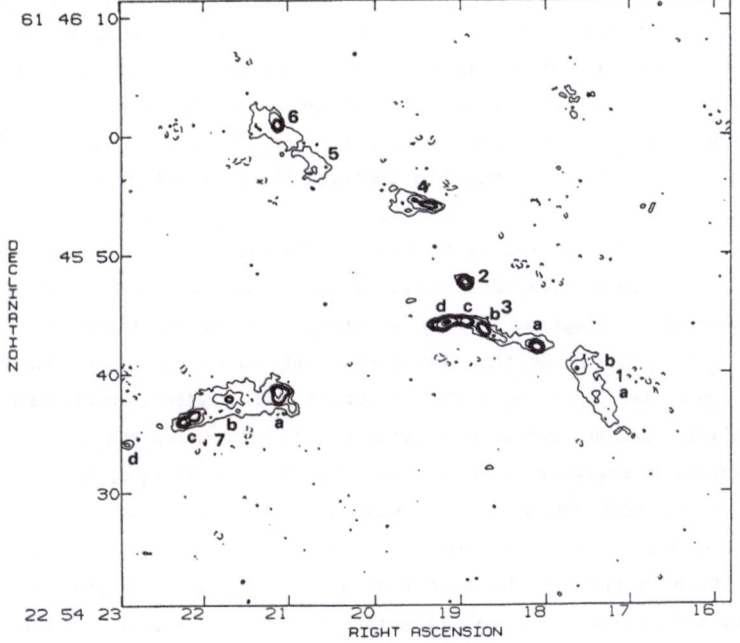

Figure 2. The 1982C map of Cep A East obtained using the VLA in the "A" configuration. Angular resolution is ∿ 0."5. Contour levels are as in Figure 1.

On this basis, there appeared to be about 14 B3 stars all at about the same stage of evolution.

Because of the apparent young age of the HII regions, it was decided to check on possible variability, and new observations were made in 1986 at $\lambda 20cm$, $\lambda 6cm$ and $\lambda 2cm$, using the VLA in the "A" configuration. It was found that not only were there variations by a factor of 3 times in the total flux density of some of the regions, some showing increases and some decreases, but an entirely new HII region had appeared. Moreover, source 2 had developed a more well defined core, and some of the more compact components were not resolved at $\lambda 2cm$ where the angular resolution was 0''.1 (70 au). In order to obtain more reliable comparison of data at different epochs, the data for 1981, 1982, and 1986 were all processed again, this time using "natural weighting" of the antennas, so as to improve signal/noise ratio.

Though some of the outer sources showed some variability, it was particularly pronounced for the central sources. To illustrate this the $\lambda 6cm$ maps containing sources 2, 3(a), 3(b), 3(c), 3(d) and the new source 8 are shown in Figures 3 and 4 for 1982 and 1986. In addition, the 1986 high resolution map at $\lambda 2cm$ is shown in Figure 5. The principal changes from 1982 to 1986 are the increase in flux density in source 2, chiefly brought about by the development of a more pronounced core, which is clearly seen in the high resolution $\lambda 2cm$ map of Figure 5, and the appearance of the very compact new source 8, as seen in Figure 4. In addition, the flux density of source 3(a) has apparently decreased and its size increased, source 3(c) appears to have a very compact core, and source 3(d) is now resolved into two compact components, separated by 450 au, as clearly seen in Figure 5. It would appear that if any sources are elongated, it is the result of having two unresolved sources present. Based on the properties of these central compact HII regions, namely their compact size, the presence of a more compact core, and the presence of the OH and H_2O maser emission, there is a high probability that they are individually produced by stars in their early evolutionary stages.

There is evidence for very high optical extinction in the direction of CepA, for which some measure can be obtained from IRAS data. If data from the IRAS sources are used to plot a colour-colour diagram, namely a plot of Y vs X, where $Y = \log(S_{25}/S_{12})$ and $X = \log(S_{60}/S_{25})$, (S_λ is the flux density at wavelength $\lambda\mu m$,) then it is seen that those associated with optically observed HII regions, reflection nebulae, and planetary nebulae, occupy separate regions of the diagrams; the 100 μm flux can be used to distinguish extra-galactic sources (Hughes and MacLeod, in preparation). Of interest is the fact that, in this way, the IRAS source in CepA is identified as an HII region, but it has the highest value of Y for any HII region in the IRAS catalog. We attribute this to the fact that in this direction there is $\sim$ 3^m of extinction at $\lambda 12\mu m$, which relates to 75^m of extinction in the optical assuming that extinction $\propto \lambda^{-1}$. Such a value for extinction was determined independantly by Lenzen, Hodapp, and Solf (1984). Since the resolution of the IRAS data is about $2'$, the data refer to the integrated infrared flux from all the sources in the region,

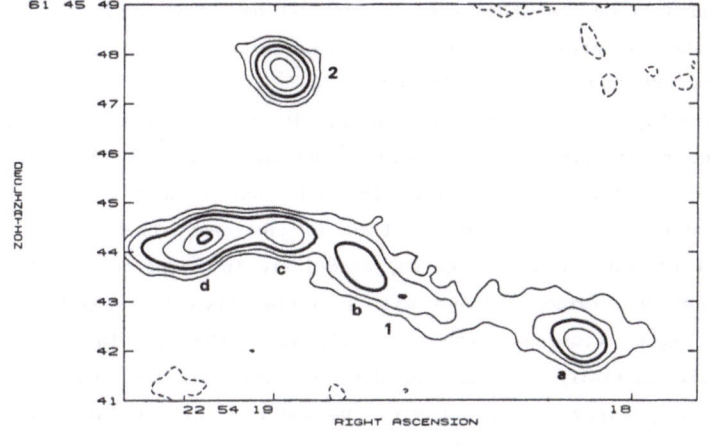

Figure 3. Subset of Figure 2, showing the central compact components in 1982. Contour values as for Figure 2.

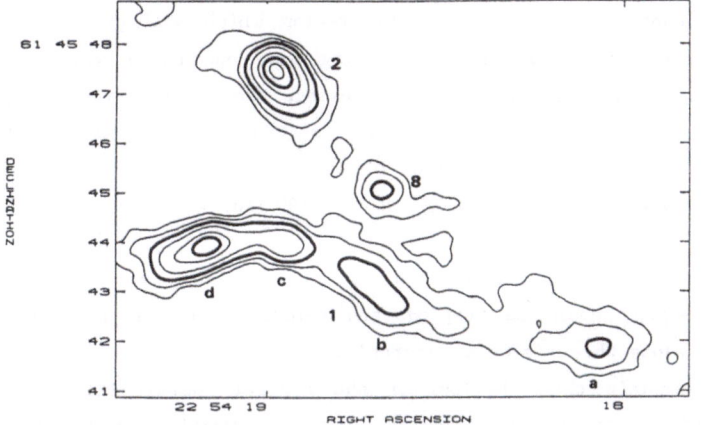

Figure 4. The same area of CepA as for Figure 3, but obtained in 1986.

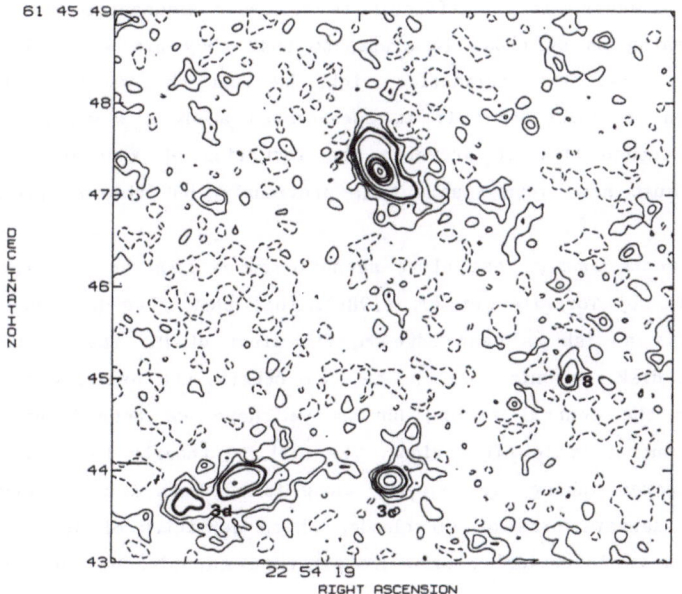

Figure 5. The 1986U map of Cep A East, using the VLA in the "A" configuration. Contour levels are at −0.1, 0.1, 0.2, 0.4, 0.8, 1.6, 2.4, 3.2, and 4.0 mJy/beam. Angular resolution is 0."15.

and thus the derived extinction is the overall extinction. Individual exciting sources, or stars, could show a greater local extinction, since if they are in an accretion stage, the density around the star would vary as R^{-2}.

Following these arguments, a model to explain at least the central compact sources is one based on the early work by Larson (1972). He showed that when stars of about $10M_\odot$ form, the total duration for accretion is such that a star could evolve onto the main-sequence while still accreting. If this is the case, then it is expected that the stars will be very highly obscured, especially in a region like CepA. Only when the star has become luminous enough will radiation pressure stop the accretion. Since radiation escaping from near the star will be in the near-infrared, it is expected that this will initially act on the dust grains, expelling them to a distance such that radiation pressure is balanced by the momentum of the infalling material. At this radius a cocoon will form. The star will ultimately become luminous enough in the ultra-violet that it can produce an HII region which will grow in size at a rate determined by the density of the surrounding gas, but there will be an extended corona close to the star. As the star progresses, the corona will collapse and the HII region expand until it finally evaporates most of the cocoon, and the HII region becomes larger and diffuse. The position of the OH masers is at a distance from the center of the associated HII region of about 1700 au, as is expected for the cocoon.

Initial calculations show that the cocoon could become relatively very dense, so that some of the stellar energy could be trapped, and released into the surrounding medium as a bipolar outflow along the axis of an accretion disk.

Such a scenario could explain the behaviour of the central compact HII regions, including the appearance of source 8, and a logical extension would account for the outermost diffuse regions. However, it is difficult to devise a mechanism that can highly synchronize the turning on of about 14 stars, so that they produce their HII regions at about the same time. One explanation might be that the whole cloud is undergoing compression, such as would occur if it encountered a shock wave, and the density has become great enough that it prevents the formation of appreciable HII regions, though the stars must have formed some time previously, but this at present is highly speculative.

One of the chief arguments against the above scenario comes from the interpretation of observations of NH_3 by Torrelles et al. (1986). They show that NH_3 forms a line, and suggest that it is a disk as seen edge-on, the lines of HII regions being seen along the edge of the disk. In this case, as in some others, the exciting source or sources are situated in the central region, but as yet have not been identified with any of the compact central radio HII regions. The disk is thought to be rotating, as evidenced by an apparent change in velocity along it of 2 km sec^{-1}. However, NH_3 normally forms in the denser cores of CO clouds, where densities of $10^5 - 10^6$ cm^{-3} may exist, rather than in individual clouds. Its presence may also be the result of local chemistry, and in any case, the probability that the line of NH_3 represents a disk seen edge on is about 1:20.

It could be argued that the line of HII regions is in fact a line of blobs, excited by some external source, such as seen in Orion by Garay et al. (1987), and though the spectral index of the Orion sources may be similar to the outer diffuse HII regions of CepA, the central regions of CepA have steeper spectral indices and the presence of OH and H_2O masers suggest an internal source of excitation.

As mentioned above, attempts to detect an exciting source for CepA have been unsuccessful. Most of the evidence comes from infrared polarization observations, which place it somewhere between sources 3(a) and 4, but these results are clearly affected by large scattering; since the total infrared luminosity is $2.8 \times 10^4\ L_\odot$, one or two early type stars are required. The fact that each of the HII regions could be produced by a B3 star, of luminosity $2 \times 10^3\ L_\odot$, led to the original suggestion that 14 B3 stars were present. To account for CepA in terms of one or two early type stars would require a particular geometry, such that, for instance, they are situated inside a cavity. But it still appears that the central compact radio HII regions are internally excited, and there is as yet little evidence for other sources of excitation.

CepA East is thus unique in a number of ways, in particular in its form and for the fact that it has the highest value of $\log(S_{25}/S_{12})$ for any of the IRAS sources. But if it is a region which contains B3 stars in the pre-main-sequence stage, and since this stage can only last for a short time, then there cannot be many regions like it. On the other hand, it may be the first observational evidence for the formation of massive stars.

Most of this work has been carried out under an Operating Grant from the Natural Sciences and Engineering Research Council of Canada.

REFERENCES

Beichman, L. A., Becklin, E. E., and Wynn-Williams, C. G. 1979, Ap.J.Lett., 232, L47.
Cohen, R. J., Rowland, P. R., and Blair, M. M. 1984, M.N.R.A.S., 210, 425.
Garay, G., Moran, J. M., and Reid, M. J. 1987, Ap.J., 314, 535.
Hartigan, P., and Lada, C. J. 1985, Ap.J.Suppl., 59, 383.
Hughes, V. A., and Wouterloot, J. G. A. 1982, Astr.Ap., 106, 171.
Hughes, V. A., and Wouterloot, J. G. A., 1984, Ap.J., 276, 204.
Hughes, V. A. 1985, Ap.J., 298, 830.
Lada, C. J., Blitz, L., Reid, M. J., and Moran, J. M.1981, Ap.J., 234, 769.
Lenzen, R., Hodapp, K.-W., and Solf, J. 1984, Ast.Ap., 137, 202.
Larson, R. B. 1972, M.N.R.A.S., 157, 121.
Rodriguez, L. F., Moran, J. M., Ho, P. T. P., and Gotlieb, E. W. 1980, Ap.J., 235, 845.
Sargent, A. I. 1977, Ap.J., 218, 736.
Sargent, A. I. 1979, Ap.J., 233, 163.
Torrelles, J. M., Ho, P. T. P., Rodriguez, L. F., and Canto, J. 1986, Ap.J., 305, 721.

[1]The VLA is operated by the National Radio Astronomy Observatory, which in turn is managed by Associated Universities, Inc. under contract with the National Science Foundation.

THE NEARBY MOLECULAR CLOUDS: A COMPLETE SURVEY

F. X. Desert
NAS-NRC Resident Research Associate
Goddard Space Flight Center
Greenbelt, MD 20771

D. Bazell
Applied Research Corporation
Landover, MD

F. Boulanger
IPAC
Pasadena, CA 91125

The infrared brightness of the sky, once the zodiacal light is removed, has been shown to be highly correlated with the total gaseous (HI and H_2) content of the interstellar medium. We have used this correlation to deduce a map of the molecular gas by subtracting the HI component from the infrared brightness. The method is limited to part of the sky where the heating of the dust is relatively uniform and allows the determination of the nearby molecular gas at high galactic latitude ($|b| \gtrsim 10°$) and heated or dustier than average HI clouds.

The 100 μm IRAS map at half a degree sampling and with the zodiacal light removed using the method by Boulanger and Perault (1986, preprint) has been used for the whole sky. The HI survey by the Berkeley group (Heiles and Habing 1974, Weaver and Williams 1973) and the Parkes group (Kerr et al. 1974) for the northern and southern skies respectively have been used. Having computed the correlation between the infrared and HI data for each small solid angle in the sky (about 10 deg^2), we have deduced the residual for each pixel. The large positive connected residuals are likely to constitute clouds of particular properties in the local IS medium: either molecular or infrared peculiar (high dust-to-gas ratio) clouds. About half the CO complexes at b>25° listed by Magnani, Blitz and Mundy (1985) coincide with clouds in our survey. An all-sky map and a catalog of 578 clouds are generated and statistically analyzed.

V. STAR FORMATION AND MOLECULAR AND HIGH-VELOCITY CLOUDS

EXTINCTION AND METAL ABUNDANCES IN THE OUTER GALAXY

Michel Fich
Physics Department, University of Waterloo
Waterloo, Ontario, Canada N2L 3G1

Abstract

Several surveys for HII regions in the outer Galaxy have failed to reveal the large number of "new" (i.e. previously unknown) HII regions expected. Previous catalogs of HII regions were primarily derived from optical surveys and it was thought that extinction probably hid many more HII regions. One possible explanation for this lack of HII regions is that the extinction throughout the outer Galaxy may be substantially less than previously thought, as low as 0.4 mag/kpc on average. This is consistent with the extinction being dependent on the dust/gas ratio, the dust/gas ratio depending on the metallicity, and the metallicity continuing to decrease with galactocentric distance at the same rate as the metal abundance gradient observed near the Sun.

Introduction

A complete sample of all of the HII regions in a large fraction of the outer Galaxy would serve many purposes. Such a sample could be used to study the properties of HII regions, to compare these properties within and outside of spiral arms, and to look for galactic radial gradients in the properties of the HII regions. These properties could be related to star formation rates, the initial mass function, the metal abundance, and the properties of the interstellar medium. In an attempt to produce a complete sample of HII regions in the outer Galaxy several surveys have been carried out at radio (both centimeter and millimeter) wavelengths and in the far infrared with IRAS.

The surprising result of these surveys has been that very few "new" (i.e. previously unknown) HII regions have been found. It was expected that many HII regions would be hidden by extinction but this has not been found to be the case. This paper will briefly describe the most important (for these purposes) of the radio surveys, show the implications of the survey results on the galactic extinction, and discuss a possible explanation involving the metal abundance gradient in the outer Galaxy.

The Radio Continuum Survey

There are several issues to be addressed when using radio continuum survey data. One must choose a wavelength, telescope beamsize, and a sensitivity limit that will enable one to identify the objects being searched for. The survey must cover enough area in the sky to produce a statistically significant number of objects. The survey by Kallas and Reich (1980) (hereafter KR) is a particularly useful one for studying HII regions in the outer Galaxy.

The KR survey was carried out at 21cm in the continuum with the 100m Effelsburg telescope. The survey covers the galactic plane from $l = 93°$ to $163°$ and $b = -4°$ to $+4°$ with a beamsize of 10 arcminutes and a point source sensitivity limit of 0.1 Jy. All objects (236 of them) with flux densities greater than 0.3 Jy and smaller than 30 arcminutes were catalogued. Not catalogued but included in the KR survey images are 30 objects that are larger than 30 arcminutes.

Known HII regions in the outer Galaxy are typically a few arcminutes or larger in diameter. The Perseus spiral arm, an area in which many known optical HII regions are found, is prominent in this part of the Galaxy spanning distances from the Sun of between 2.5 and 5 kpc. Additionally this wavelength is where most HII regions, all except the very youngest and densest, emit most strongly. An ionization bounded HII region surrounding an O9.5 star at a distance of 5 kpc from the Sun will be observed to have a flux density 0.3 Jy at the Sun at 21 cm. An earlier (more massive) star or a closer star will produce an even brighter HII region. All of these characteristics together would lead one to expect that the KR survey should detect all HII regions excited by O stars (i.e. more massive than $20M_\odot$) within the area they surveyed out to beyond the Perseus arm. HII regions excited by stars earlier than O8 will be detected to the edge of the Galaxy.

There remains the problem of identifying which radio sources are the HII regions, as there are other objects, primarily extragalactic, that emit at this wavelength. The most certain way to identify a radio source as an HII region is to measure recombination lines but this can take a prohibitively long time. The quickest way to be reasonably certain that an object is an HII region is to measure its emission at a different wavelength (e.g. at 6 cm), determining the spectral index, and by measuring the polarization of the radio emission. HII regions should not be significantly weaker at 6 cm than at 21 cm and they should show no polarized flux.

Measurements at other wavelengths, primarily at 6 cm, were used to eliminate a number of objects from the list of potential HII regions. In addition, all of the objects not resolved in the KR survey were observed at the VLA at 6 cm. Further observations at 2cm and 20cm were also carried out at the VLA. A few radio recombination line observations were made of some of the stronger KR sources.

The results of all of these studies can be summarized as follows: Of the 266 KR objects 126 have been definitely identified (52 HII regions, 59 extragalactic sources, and 15 supernova remnants); Of the remaining 140 objects most (115) have steep spectral indices and are not HII regions; Of the remaining 25 objects 10 are smaller than 0.3 arcseconds, very small for HII regions with 5 kpc of the Sun. Some, and probably most, of the remaining flat spectrum sources will be extragalactic objects. Using the above numbers it is possible to conclude that at least 78 percent of the HII regions have been identified. Using extragalactic radio source counts to estimate what fraction of the remaining objects are extragalactic the conclusion is that probably 89 percent of the HII regions have been identified.

This is where the surprising result arises. Only one of the HII regions detected in this survey was "new" (i.e. not previously identified from optical surveys). It would be natural to expect that extinction in the Galactic plane would have hidden a far larger fraction of all of the HII regions from our (optical) sight.

The average extinction in the Galaxy is usually thought to be of the order of 1.5 mag/kpc and other values in the range of 1.2 to 1.8 have been used by some authors. The average surface brightness of an HII region corresponds to an emission measure of a few thousand. At 5 kpc distance 1.5 mag/kpc corresponds to 7.5 magnitudes of extinction, or a factor of 1000 in brightness. Thus an HII region with an emission measure of 10^4 (surface brightness somewhat higher than average) would only appear as bright as an HII with no extinction and an emission measure of 10. This would not be visible on any of the classical optical surveys for HII regions. (The Palomar Observatory Sky Survey limit is approximately an emission measure of 35.)

Even the brighter HII regions would in general be hidden by extinction at the distance of 5 kpc. It has been assumed that the HII regions that are seen at such large distances are seen through "holes" or "windows" in the extinction, places where the semi-random placement of absorbing clouds has "conspired" to leave an opening through the interstellar medium. In retrospect there seem to be (too) many such holes. HII regions are seen at large distances in many different directions in the outer Galaxy.

The extinction can be directly measured in most HII regions. In some HII regions the extinction towards the exciting stars is measured as a part of the spectrophotometric technique of determining the distance to the stars. The extinction to HII regions with measured Hα and Hβ fluxes, either through spectroscopy or narrow-band imaging, can be determined from the line ratio. There are problems with both of these methods. For example, the extinction measured from the Hα to Hβ ratio usually gives a larger extinction than the stellar photometry. This is because the presence of dust inside the HII region will permit the Hα radiation to escape from a greater physical distance within the HII region, enhancing the brightness of the observed Hα line in comparison with the Hβ line. Both of these methods have been used to measure the extinction to a large number of optically visible HII regions in the outer Galaxy. The results are shown in Figure 1.

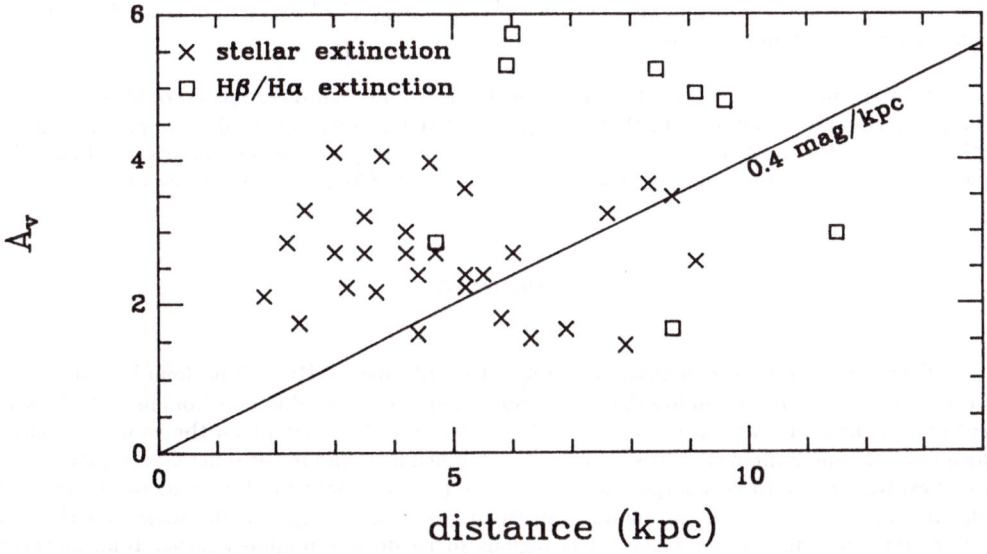

Figure 1. The extinction to HII regions measured from both the exciting stars (UBV) and from the Hα to Hβ ratio of the ionized gas.

There have been some other studies that have produced similar results. Lyngå(1979) used open clusters to measure the extinction and concluded that the extinction on a large scale was closer to 0.7 mag/kpc and decreased by 0.1 mag/kpc^2. Forbes (1985) in a study of objects between galactic longitudes of 40° and 90° noted that the extinction seemed to be independent of distance. Wramdemark (1976) and Sarg and Wramdemark (1977) found the same thing in fields at $l = 102°$, 120°, and 190°.

A Possible Explanation?

The abundance of heavier elements is observed to decrease at larger distances from the Galactic center. The amount of the decrease depends on the element but for the most abundant elements (C, N, and O) is generally on the order of 0.1 dex kpc^{-1}. Over a distance of 10 kpc this corresponds to a decrease in abundance of a factor of 10. The typical HII region (and the Perseus spiral arm) in the field observed above is several kpc further from the Galactic Center then the Sun. The abundances of heavy elements there are presumably lower by a factor of 2 to 3. If the dust/gas ratio depends linearly on the abundances then the ratio is presumably also lower by a factor of 2 to 3 towards these HII regions.

Furthermore, the measurements of extinction that give the results of 1.5 mag/kpc were made near the Sun. The Sun is in a spiral arm, or at least in an interstellar medium feature with spiral arm gas density properties (this feature is often referred to as the Orion arm, or Orion Bridge). Thus this value of the extinction is appropriate for inside spiral arms. Between spiral arms the gas density is at least a factor of 3 lower than in the interiors of the arms.

Thus if extinction is dependent on the amount of dust one would expect that the extinction between spiral arms in the outer Galaxy should be 5 to 10 times less than the local value; In other words the outer Galaxy extinction may be less than 0.3 mag/kpc. This low a value would completely account for the results of the radio survey. Extinction towards the HII regions in the outer Galaxy would be primarily due to material either very close to the Sun (in the "local" spiral arm) or close to the HII region (in the Perseus spiral arm). The extinction would be essentially independent of distance, as is observed.

Further support for this explanation is available from recent observations of M31. Walterbos and Schwering (1987) found that there is a strong radial decrease in the dust to gas ratio in that galaxy. They found that this decrease is at least as steep as the abundance gradient. The extinction is apparently much lower in the outer parts of M31 than in the inner area.

Discussion

There are alternate explanations for the lack of "new" HII regions found in the recent surveys. There is some possibility that HII regions are intrinsically different from inner Galaxy HII regions and that this difference causes the observed result. For example, if the volume density of the interstellar medium in the outer Galaxy is very much less than in the inner Galaxy, HII regions would evolve quickly to very large sizes and have very low surface brightnesses, and therefore be difficult to observe. Also, a low volume density of gas, and perhaps smaller molecular clouds as well, could cause most outer Galaxy HII regions to be density-bounded rather than ionization bounded. This means that some fraction of the ionizing photons emitted by the exciting star do not end up in the HII region but escape into the general interstellar medium. It seems unlikely that either of these effects can be important enough to account for the observations. However more work needs to be done to show this convincingly.

Future work planned to improve this result focuses in two areas: further searches for HII regions, especially using the infrared, and in making extinction measurements to all of the known HII regions in the outer Galaxy.

References

Fich, M. 1986, *Astron. J.*, **92**, 787
Forbes, D. 1985, *Astron. J.*, **90**, 301.
Kallas, E. and Reich, W. 1980, *Astron. Astrophys. Suppl.*, **42**, 227
Lyngå, G. 1979, IAU Symposium # 84 *The Large Scale Characteristics of the Galaxy*, ed W. B. Burton (Dordrecht:Reidel), 87
Sarg, K. and Wramdemark, S. 1977, *Astron. Astrophys. Suppl.*, **27**, 403.
Walterbos, R. A. M. and Schwering, P. B. W. 1987, *Astron. Astrophys.*, **180**, 27.
Wramdemark, S. 1976, *Astron. Astrophys. Suppl.*, **26**, 31.

IRAS RESULTS ON OUTER GALAXY STAR FORMATION TOWARDS GALACTIC LONGITUDE l=125°

S. Terebey
Astronomy Department 105-24
California Institute of Technology
Pasadena, CA 91125

M. Fich
Physics Department
University of Waterloo
Waterloo, Ontario N2L3G1, Canada

Abstract

We have systematically studied an infrared defined (60 µm) sample of IRAS sources in order to investigate star formation in the outer Galaxy. Five percent of the sample are point sources with IRAS spectra that suggest the emission is from a dust shell surrounding a mature star. Ninety-five percent have spectra where flux density strictly rises with wavelength. The sources are extended and we show that Point Source Catalog fluxes seriously underestimate total fluxes. We have reliably assigned CO kinematic distances to two thirds of the sources. Most of the infrared luminosities correspond to B spectral types. We detect 6 cm continuum emission from all sources inferred to have spectral type B1 or earlier. The combined IRAS/CO/6 cm data show these sources are young, moderately massive stars that are embedded in interstellar clouds. The young embedded sources define a distinct band in an IRAS color-color diagram. Normal IRAS galaxies fall in the same band, consistent with the interpretation that their infrared emission is due to star formation.

Introduction

The IRAS survey provides a very complete and sensitive look at star formation in our Galaxy. Source confusion aside, IRAS provides a complete flux limited sample that spans the entire Galaxy of young stars that are still embedded in molecular clouds. This stellar census is a terrific resource for studying the global properties of current star formation. However, to make full use of the IRAS data for star formation studies requires, in addition, methods to identify the young stars and to find their distances.

We focus on the outer Galaxy where infrared source confusion is small and where we can assign kinematic distances without the distance ambiguity inherent inside the solar circle. The outer Galaxy is of particular interest since we can study star formation in a region of very low molecular to atomic gas content, a very different physical condition than is characteristic in the inner Galaxy molecular ring.

Our approach has been to define a uniform infrared sample from the 60 μm Sky Brightness images and study two areas of the outer galactic plane in detail, areas near galactic longitudes l = 125° and l = 215°. In this paper, we describe results for a subsample of about 40 sources near l = 125°. The first step is to identify and understand the sources we see with IRAS. To this end, we use CO measurements to find distances and derive infrared luminosities. We use VLA radio continuum measurements as an independent means to infer the stellar type of the massive stars in our sample.

Data

Our subfield subtends 8°x8° centered on the galactic position l = 125°, b = 2°. In this direction the galactic midplane lies near b = 2° at distances more than several kpc away from the sun. We initially identified "hot spots", i.e. small or unresolved sources on the IRAS Sky Brightness Images and later on the higher resolution Coadded images. The sources cluster toward the galactic plane implying they are mostly galactic objects. The lumpy diffuse galactic background can obscure sources at low flux levels; we chose a limiting flux density of 25 Jy at 60 μm to ensure a complete sample. Inspecting the data shows the IRAS sources divide fairly clearly into two categories by their spectra, point source apparently stellar sources which are strongest at 12 μm and mostly invisible by 60 μm, as well as small but extended sources which are visible at 12 μm but strongest at 100 μm. By selecting our sample at 60 μm, we both discriminate against stellar sources and gain over using 100 μm by having higher spatial resolution and lower diffuse galactic background to contend with.

We measured fluxes from the Coadded data. Representative points from the perimeter of a 10' diameter circle defined the background level. The fluxes were calculated over a 5' diameter aperture. The IRAS survey beam is complicated but basically rectanglar with 5' resolution along the major axis for all four bands. We used a modified procedure for sources that were more extended than 5' or had multiple overlapping components. The main uncertainty in flux comes from the background level, especially at 100 μm. Our subsample contains a total of 39 sources.

Our VLA data is 6 cm radio continuum with 12" spatial resolution and typical rms noise level of 0.2 mJy. The CO data is from the NRAO 12 m telescope with velocity resolution 0.65 km/s and typical rms sensitivity of 0.2 K. We have coarse spatial maps for all sources. There are characteristically 1 or 2 distinct velocity features towards each source. In most cases, one component strongly peaks at the central infrared position, allowing us to identify the IRAS source with a galactic molecular cloud.

Analysis

The IRAS sources typically appear unresolved along the 5' major axis but extended along the minor axis (roughly 0.5', 0.5', 1', 2' for 12, 25, 60, 100 μm) in all four IRAS bands. We find 12 sources or 31% of our sample have multiple components. Since the sources are extended, Point Source Catalog (PSC) fluxes will underestimate total fluxes. In Figure 1 we plot the ratio of PSC flux to Coadd flux for several wavelengths. The actual value of the ratio will depend on the spatial profile of the source and the effective

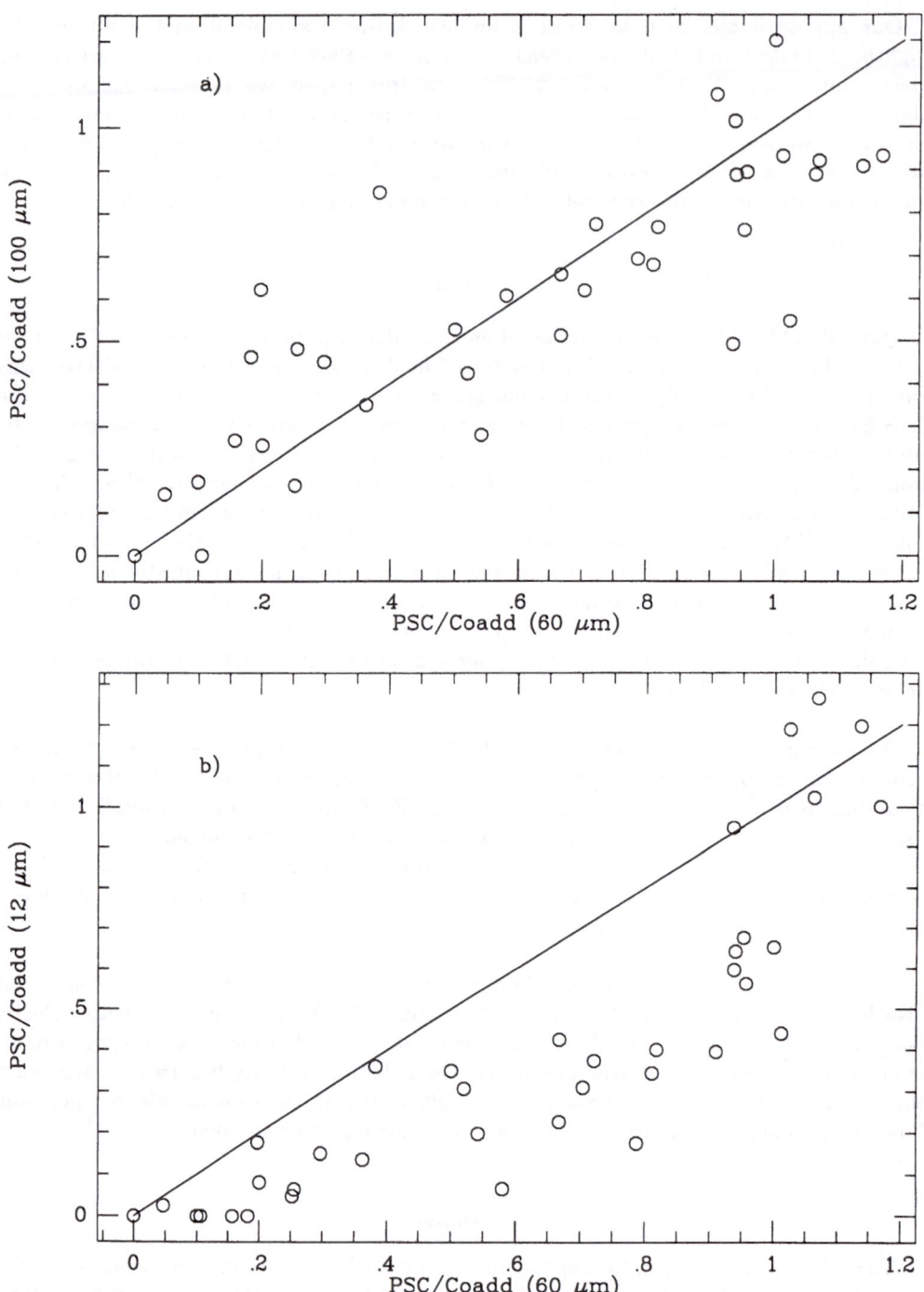

Figure 1. The ratio of PSC flux to Coadd flux for several wavelengths. The ratio depends on source spatial structure and IRAS beam. Unresolved sources have a ratio of one. Fig. 1a: one third of the sources have a ratio less than one half at 60 or 100 μm. Fig. 1b: the PSC flux severely underestimates the total flux at 12 μm. Two thirds of the sources have a ratio less than one half.

IRAS beam for a point source. An unresolved source will have a ratio equal to one, as is seen for some of the sources. Very extended sources would have low ratios. The trend seen in Fig. 1a for the 100 μm ratios to follow the 60 μm ratios quite reasonably suggests that sources which are extended at 60 μm are also extended at 100 μm. In detail, we see that PSC fluxes are less than 1/2 the Coadd fluxes for 1/3 of the 60 and 100 μm sources. A similar behavior is seen in Fig. 1b which plots 12 μm versus the 60 μm data. However, it is clear that the higher resolution 12 μm point source beam severely underestimates the total flux; a full 2/3 of the 12 μm sources have PSC fluxes less than 1/2 the Coadd values. This amply demonstrates that the common practice of using PSC fluxes for young galactic sources is not adequate, particularly in IRAS color-color diagrams where the variation of IRAS resolution with wavelength will introduce systematic biases. A number of authors have used positions in PSC color-color plots to discriminate between classes of objects. We emphasize that although objects might be distinguishable by this means, because of the resolution effects these PSC colors should not be directly applied in physical models of the emitting regions.

The spectra of 37 out of 39 IRAS sources show flux density strictly increasing with wavelength consistent with other known examples of young stars embedded in molecular clouds. Table 1 gives the median flux density for each band, normalized to 100 μm. The remaining 2 sources are quite different in character. They appear to be IRAS point sources, have spectra that peak at 25 μm, and IRAS colors like that of a single temperature black body. No CO emission was detected toward either source. Much less dust surrounds these sources implying they represent an older population of stars.

Using our CO data we can reliably assign a velocity/distance to 70% of the sample. These sources have a median kinematic distance of 4.5 kpc and median IRAS infrared luminosity corresponding to a B2 star. The IRAS sources that have multiple infrared components share the same molecular velocities showing they are not chance coincidences along the line of sight. The CO data, combined with the rising IRAS spectra imply that the IRAS sources are mostly young and moderate mass stars, most of which are still associated with their parent molecular clouds.

Before using infrared luminosities to determine stellar types, it is prudent to independently estimate the bolometric luminosities of our sources. Radio continuum measurements provide a way, since nearly all of these sources are optically invisible. We observed most of the sources at 6 cm and detected all that had infrared luminosities corresponding to a spectral type of B1 or earlier. The one additional source we detected is S185 which turns out to be too close and cover too large a solid angle to reliably measure its total infrared flux above the background.

Figure 2 is a color-color diagram where we plot the ratio of flux densities $F(12)/F(25)$ versus $F(60)/F(100)$. For comparison we plot PSC data for some planetary nebulae (Pottasch et al. 1984) and compact HII regions (Chini et al. 1986) and indicate a band where normal IRAS galaxies are found (Helou 1987). The HII region data should be viewed with some caution given the problems of using PSC fluxes for young galactic sources. Notice our two "old" stellar sources have colors near that of a blackbody, consistent with the suggestion we are seeing a dust shell around an older star. Our embedded IRAS sources show a distinctive trend with color that differs with the

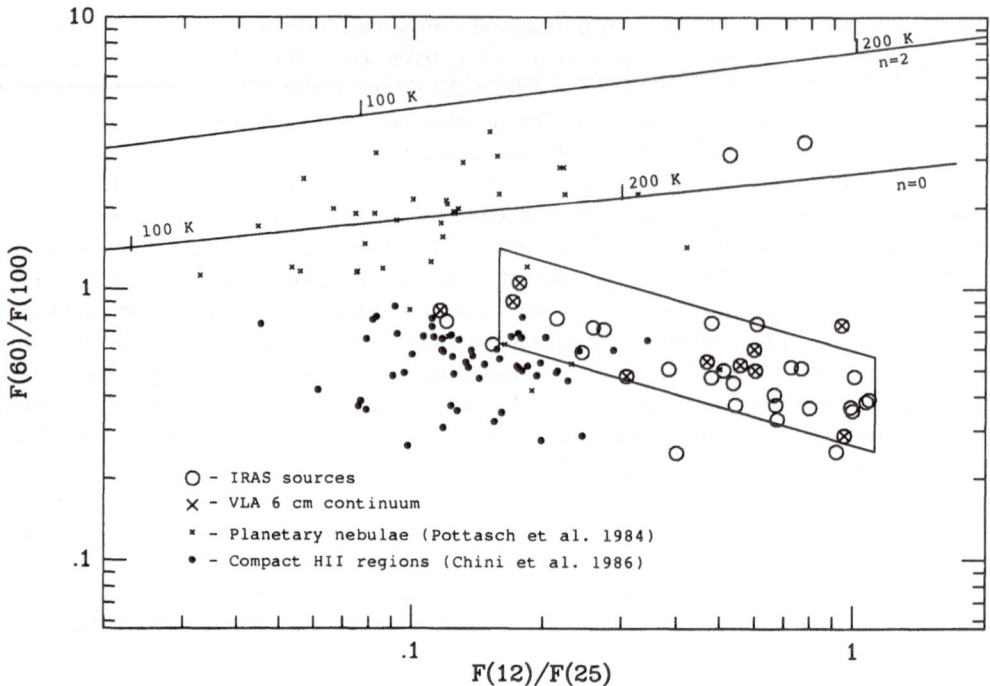

Figure 2. An IRAS color-color diagram. Plotted is the ratio of 12 to 25 μm flux density (Jy) versus the 60 to 100 μm ratio. The solid lines show the Planck curve for dust emissivities of λ^{-n} for n=0,2. The outlined region indicates the colors of normal IRAS galaxies (Helou 1987). Our IRAS sources have the same colors as galaxies. There is no correlation between color and detection of an HII region at 6 cm. Also plotted for comparison are planetary nebulae (Pottasch *et al.* 1984) and compact HII regions (Chini *et al.* 1986).

planetary nebulae and ultracompact HII regions. In fact, the IRAS galaxies show exactly the same infrared colors as the young embedded B stars. This is quite reasonable if the infrared emission from IRAS galaxies is due to star formation.

The specific trend with color has been interpreted as due to a mixture of "active" and cirrus components (e.g. Helou 1987). In the "active" component, an average dust grain sees a higher interstellar radiation field for those sources that have low $F(12)/F(25)$ but high $F(60)/F(100)$ colors. A galaxy high in the activity sequence might have younger stars or a higher star formation content that could lead to the higher radiation field. This physical mechanism can also be applied to the young embedded sources in our Galaxy. In this case, we might expect to see color correlate with infrared luminosity since the mean stellar radiation field scales with luminosity. However the data shows no correlation with infrared luminosity--or with the 6 cm detection of an HII region. This does not rule out the mechanism since evolutionary effects can potentially destroy a color-luminosity correlation. As the stars age, they modify or wander away from their dusty environments. This means the dust will be further from the star, in a lower stellar radiation field; in addition the infrared luminosity will become a less reliable measure of the bolometric luminosity. If there is a way to parametrize the evolutionary effects, say in terms of stellar age, then there might be an observable correlation of color versus evolutionary state. At present it is unclear how well the model describes the young galactic sources.

Table 1. Median IRAS flux density for embedded sources, normalized to 100 μm.

12 μm	25 μm	60 μm	100 μm
0.04	0.08	0.50	1.0

References

Chini, R., Kreysa, E., Mezger, P.G., and Gemund, H.-P. 1986, *Astro. Ap.*, **154**, L8.
Helou, G. 1987, *Ap.J. Letters*, **311**, L33.
Pottasch, S.R., Baud, B., Beintema, D., Emerson, J., Habing, H.J., Harris, S., Houck, J., Jennings, R., and Marsden, P. 1984, *Astr. Ap.* **138**, 10.

THE OUTER GALAXY AS A STAR FORMATION LABORATORY

Marc L. Kutner
Physics Department, Rensselaer Polytechnic Institute
Troy, NY

INTRODUCTION

In astronomy, we are fond of saying that we cannot perform experiments in the traditional sense of physics experiments. That is, we cannot take a sample of interstellar material, alter its environment, and see how it responds. We can, however, try to find similar objects in different environments, and see how the evolution of those objects differs. The outer galaxy provides us with a means of performing such experiments. By comparing star formation in the inner and outer galaxy, we can see how that process is affected by the environmental differences between those locations.

Mead et al. (1987, hereafter Paper I) have summarized important environmental factors that are different in the inner and outer galaxy. Briefly, they are: (1) The outer · galaxy should have fewer cloud–cloud collisions. This is because the surface density of molecular material is less in the outer galaxy, while the scale height is greater, so the volume density is lower. (2) The time between spiral arm passages is longer in the outer galaxy. This is because the arms are probably less tightly wound (as in other galaxies) as one gets farther out, and the angular rotation speed falls off with distance from the galactic center, R. (3) The cosmic ray flux may be lower in the outer galaxy (Bloemen et al. 1984, Bhat et al. 1985). This would change the energy balance within clouds.

Based on a study of 31 outer galaxy molecular clouds by Mead (1987a, hereafter Paper II), Mead and Kutner (1987, hereafter Paper III) have analyzed the properties of outer galaxy clouds and compared them with inner galaxy clouds. Some of these results are summarized by Mead (1987b, this volume). Outer galaxy clouds have sizes and masses that make them comparable to inner galaxy GMCs, except that there are no outer galaxy clouds that correspond to the most massive inner galaxy GMCs. The envelopes of the outer galaxy clouds are cooler than those of inner galaxy clouds. This could result from a lower cosmic ray flux in the outer galaxy.

With these differences in cloud environment and cloud properties, it is of interest to see if outer galaxy molecular clouds are forming massive stars with the same efficiency as inner galaxy GMCs. To this end we have chosen a set of 17 clouds from Paper II, and carried out a series of infrared and radio continuum observations. The infrared observations give the total luminosity of embedded stars, and the radio continuum obervations can be used to infer the rate of at which hydrogen–ionizing photons are being given off. Both of these are tracers of star formation.

OBSERVATIONS and RESULTS

The radio contiuum observations were made with the Very Large Array (VLA). Observations were carried out at two wavelengths, 6 cm and 20 cm. The 6 cm observations were in the D–array (smallest), and the 20 cm observations were in the C–array. In this way, both observations were sensitive to structures of approximately the same size, and could be used to obtain reliable spectral indices. The spectral indices were for the purpose of distinguishing between thermal and non–thermal emission. The resolution for the observations was 10" at 6 cm. More details of the observations, along with a tabulation of the individual sources, appears Paper IV.

Since we are looking at the radio continuum, we have no velocity check to make a secure identification of a radio source with a particular outer galaxy cloud. We can only require that a source be thermal, and that it fall within the cloud boundaries, or sufficiently close to a cloud edge that the morphology looks like that for inner galaxy clouds with HII regions on the cloud edge. Using these criteria, 14 HII regions are found to be associated with the 17 clouds in this sample. The 6 cm fluxes range from 1 mJy to 3.7 Jy, with most in the 1 to 10 mJy range.

We can use the fluxes and kinematic distances to deduce N_L, the number of ionizing photons per second radiated by the exciting star (Matsakis et al. 1976; Rubin 1968). These can then be converted into spectral types for the exciting star, following the calculations of Panagia (1973) and Avedisova (1981). Most of the exciting stars in our sample fall in the B0 to B1 range (with B1 corresponding to our sensitivity limit in most cases), with two O stars also being present.

Far infrared data fluxes were obtained from IRAS data, using four different data products, the Point and Small Extended Source catalogues, survey co–adds, produced for fields centered on these clouds of interest. In addition, individual Hours–confirmed images (HCONs) were inspected when extended emission was present. More details of the observations, along with a tabulation of the individual sources, appears in Paper IV. In addition, we have observed six sources from the Kuiper Airborne Observatory. These observations are described in Paper I.

An infrared source was detected for each of the clouds in the study. The range of infrared luminosities is 47 to $3.9 \times 10^5 \, L_\odot$. For clouds with radio continuum sources, the infrared luminosity is generally in good agreement with that inferred from the spectral type of the star required to produce the ionization. The advantage of the IRAS observations is that they provide more sensitivity to lower mass stars than B1. The average luminosity of the clouds is $4 \times 10^4 \, L_\odot$.

DISCUSSION

We can use these results to make a preliminary comparison of massive star formation efficiency in the inner and outer galaxies. We have found of the order of one HII region per molecular cloud studied. This number is comparable to (but probably slightly less than) what is deduced for the inner galaxy.

To be more quantitative, we can look at two indicators of the efficiency of massive star formation. One is the number of ionizing photons per second per cloud mass, and the other is the ratio of far infrared luminosity to CO luminosity. For the first quantity, we find 5×10^{48} ionizing photons per second per 10^5 $M_\odot$ of material for the outer galaxy. For the inner galaxy, the corresonding number is $\sim 2 \times 10^{49}$, or a factor of four higher. For the second quantity, $L_{IR}/L_{CO} = 3.4\ L_\odot K^{-1} km^{-1} s\ pc^{-2}$. The average value for the inner galaxy is ~ 12. Again, the inner galaxy number is a factor of ~ 4 larger.

We would therefore reach the preliminary conclusion that star formation in the outer galaxy is slightly less efficient than in the inner galaxy. However, this is not a great difference. For example, if cloud—cloud collisions are responsible for massive star formation, then we would expect the efficiency to fall off by a much larger factor in the outer galaxy.

REFERENCES

Avedisova, V. S., 1979, *Sov. Astron.*, **23**, 5.
Bhat, C. L., Issa, M. R., Houston, B. P., Mayer, C. J., and Wolfendale, A. W., 1985, *Nature,* **314**, 511.
Bloeman, J. B. G. M., et al., 1984, *Astr. and Ap.*, **135**, 12.
Matsakis, et al., 1976, *A. J.*, **81**, 172.
Mead, K. N., 1987a, submitted to *Ap. J.* (Paper II).
Mead, K. N., 1987b, *Proc. Kerr. Symp. Outer Galaxy* (this volume).
Mead, K. N., and Kutner, M. L., 1987, submitted to *Ap. J.* (Paper III).
Mead, K. N., Kutner, M. L., and Evans, N. J. II, 1988, submitted to *Ap. J.* (Paper IV).
Mead, K. N., Kutner, M. L., Evans, N. J. II, Harvey, P. M., Wilking, B. A., 1987, *Ap. J.*, **312**, 321 (Paper I).
Panagia, N., 1973, *Astr. and Ap.*, **78**, 929.
Wright, A. E., and Barlow, M. J., 1975, *M. N. R. A. S.*, **170**, 41.

POINTS TO PONDER ABOUT THE MOLECULAR OUTER GALAXY

Kathryn N. Mead[1]
E. O. Hulburt Center for Space Research
Naval Research Laboratory
Code 4138ME, Washington, DC 20375-5000

[1]National Research Council / Naval Research Laboratory Cooperative
Research Associate

ABSTRACT

After a brief introduction reviewing our results on molecular
clouds and star formation in the outer galaxy, we pose some questions
about the implications for understanding molecular clouds, star
formation and galactic structure. Molecular clouds and HII regions are
found as distant as 18kpc from the galactic center. Evidence for a
molecular arm coincident with the HI arm at ~13kpc is discussed.
Questions are raised about arm/interarm contrasts, and the cloud size
distribution and initial mass function in the outer galaxy.

INTRODUCTION

Though the existence of atomic hydrogen to distances of over 20kpc
from the galactic center has been well known for some time, molecular H
and associated star forming activity has only begun to be appreciated.
Observation and analysis of star formation regions out to 18 kpc
(Brand; Tereby; Fich all in this volume) reveals a new and contrasting
(compared to inner galaxy and local star forming regions) environment
for molecular clouds and star formation. In the next paper (Kutner),
some of these environmental differences will be discussed.

In this paper I will raise several questions concerning outer galaxy
molecular clouds and star formation. Many of these questions are not
new, the intent is merely to see how certain programs of study can lead
to an understanding not only of molecular clouds and star formation but,
one hopes, of galactic structure as well.

Some background on our work is in order since it began and proceeded
somewhat differently than other studies (such as Brand; Tereby; Fich

this volume). The starting point and basis for our work is an understanding of the relationship between molecular clouds and star formation. Therefore, in order to find molecular clouds to study, a CO survey of the outer galaxy was done. Data were taken with the NRAO 11m telescope over the longitude range $45< \ell <225$ excluding ℓ's between $160°$ and $200°$ and a latitude range which followed the HI warp. The region was quite undersampled (see Kutner and Mead 1981, Mead 1982, Kutner 1983). Twenty-eight clouds were mapped at one beamwidth spacing in CO(J=1-0). Subsets of these were (at least partially) mapped in ^{13}CO(J=1-0) and CO(J=2-1) (Mead 1987; Mead and Kutner 1987). We also have VLA and IRAS data on 17 clouds (Kutner, this volume; Mead, Kutner and Evans 1987).

The placement of the clouds in an overhead view of the galaxy is shown in Figure 1. (All the distances are derived from a flat rotation curve with $R_\odot=8.5$kpc and $v_\odot=220$ km/s.) Note that the clouds lie in a band at R=13kpc. For comparison, a plot of HI surface density is shown in Figure 2 (taken from Henderson, Jackson and Kerr 1982). (Figure 2 is

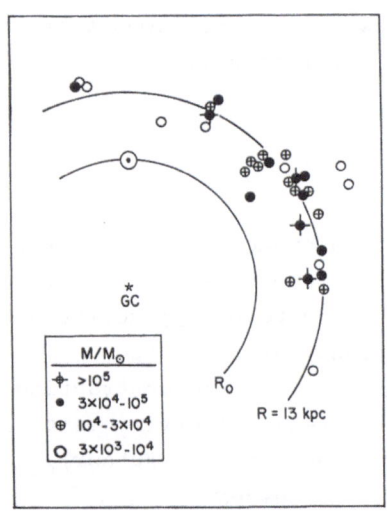

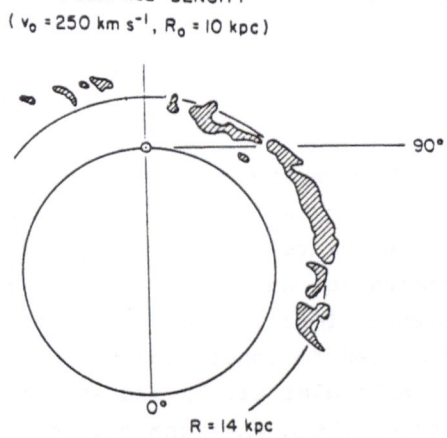

Figure 1. Clouds are shown in an overhead view of the galaxy. Kinematic distances were determined by using a flat rotation curve with R_o=8.5kpc and v_o=220km/s.

Figure 2. HI surface density from Henderson, Jackson, and Kerr (1982). (The "old" galactic rotation constants were used in drawing this figure.) Notice the similarity between the arm-like distribution of the molecular clouds and the "outer arm" traced by the HI distribution.

drawn based on the "old" galactic rotation constants $v_\odot$=250km/s and $R_\odot$=10kpc, as noted in the figure.) An outward-looking view of the galaxy is shown in Figure 3 where the clouds positions are plotted as a function of ℓ and z. Notice that the clouds do not lie in the plane, but follow the HI warp.

A histogram of the masses of the clouds is shown in Figure 4. Though a fully sampled CO survey of the outer galaxy would probably reveal more low mass clouds, this should be a good representation of the high mass end of the distribution. When one thinks of the mass distribution of well known clouds, one thinks of a distribution which is up to a factor of 10 more massive than this one. Fewer cloud-cloud collisions are expected in the outer galaxy, and if clouds grow by coalescing upon collision (e. g. Kwan and Valdes 1987), then this mass distribution is consistent with a lower cloud collision frequency. Finally, the size distribution is shown in Figure 5. The size, r, is defined as half of the geometric mean of the major and minor axes: $r=0.5(ab)^{0.5}$. The sizes of these outer galaxy clouds are typical GMC sizes.

FUTURE WORK

The presence of molecular clouds and associated star formation in the outer galaxy not only raises new questions about those subjects but affords us a new perspective on the study of old questions about galactic structure. In this section I will briefly discuss a few questions with the intent that they be addressed in the future and that they stimulate the reader to raise additional questions. Though these thoughts are numbered for clarity, there is inevitably overlap between them.

1. How far out does star formation go? Does star formation (low mass or high mass) cease interior to the molecular edge of the galaxy? Where is the molecular edge of the galaxy relative to the atomic edge? How does the Milky Way compare with other galaxies?

The most distant HII regions we have found are 13kpc from the galactic center and we have observed CO as distant as 18kpc. Brand's (1986 and this volume) most distant HII region is at 18kpc. Fich (this volume)

and Tereby (this volume) have also studied outer galaxy star formation
via radio continuum and IRAS observations, respectively. HI is found
out to R=30kpc (e. g. Burton, this volume). Based on this information,
stars can form as long as there is molecular material, but molecules
don't exist as far out as H does. Of course the present studies of
molecular clouds and star formation were not aimed at the questions
posed here. One can envision a survey for molecular material at large R
and accompanying searches for HII regions and infrared emission.

2. What is the initial mass function in the outer galaxy?

Differences in the galactic environment in the outer galaxy may effect
star formation in such a way as to produce a different mix of stellar
masses than in the inner galaxy. Some environmental differences are the
following. a) Spiral arm passages are less frequent (they are less
tightly wound). b) The metallicity is lower in the outer galaxy. This
not only affects the stars but, perhaps more importantly, the molecular
clouds from which the stars are formed by altering the chemistry and the
cooling in the clouds. c) The scale height of both molecular and atomic
material is higher. d) The ratio of atomic to molecular mass is higher.
e)The cosmic ray flux is lower; this also affects the chemistry and
heating of the clouds.

3. What is the cloud size distribution in the outer galaxy?

Some of the environmental factors in #2 may affect cloud size. If
clouds grow via collisions (Kwan and Valdes 1987) then outer galaxy
clouds, as a group, should be smaller than inner galaxy clouds. This
may affect the IMF. The cloud size distribution and the IMF may vary
with distance above the plane, z.

4. (a) What is the nature and extent of the molecular arm?

Though there is clearly an arm in HI, the case for a molecular arm is
less strong. Though our clouds trace an arm-like feature, our clouds
were not chosen at random. A fully sampled CO survey would be a more
satisfying approach to studying the molecular component of the arm.

   (b) Do arm/interarm variations dominate inner/outer galaxy
   variations?

It may be that being in an arm is what is important in the cloud/star
life cycle. Though we find outer galaxy clouds to be somewhat smaller
and cooler, it may be that, because they are in an arm, much of the
chemistry and physics in the clouds and the star formation process
depends more on crowded nature and gravitational influence of an arm
passage than on galactocentric distance.

5. How do arms and interarms differ?

In other galaxies arms are traced from optical photographs. How should
we do it in the Milky Way? In practice, authors usually contrast the
quantity which they study: cloud sizes, massive stars, CO line
intensity, for example. In the outer galaxy, where it's easier to tell
an arm from an interarm (look at a photograph of another galaxy) this
question should be easier to address. Additionally, the outer galaxy
spares one the headaches associated with distance ambiguities. Even
with rotation curve uncertainties, a CO survey, for example, would show
where (in ℓ and v) the molecular material is and how the arm and
interarm properties of the CO differ.

6. What is the nature of outer galaxy clouds and what does this tell us
about the clouds themselves and about the effect of environment on the
clouds?

Properties such as density, temperature, size, and mass have already
been compared. But the study of isotope ratios and abundance variations
in the outer galaxy remains intriguing and unexplored. Though difficult,
such studies have always held the lure of disclosing information about
the nature of galactic evolution. Now that we have clouds with larger
R's than have been studied previouly we should continue the quest because
the potential for understanding is worth the difficulty.

REFERENCES

Brand, J.   1986, Ph. D. Thesis, Leiden University.
Henderson, A. P., Jackson, P. D., and Kerr, F. D. 1982, Ap. J., 263,
   116.
Kutner, M. L. 1983, in Surveys of the Southern Galaxy, ed. W. B. Burton

Kutner, M. L. 1983, in <u>Surveys of the Southern Galaxy</u>, ed. W. B. Burton and F. P. Israel (Dordrecht: Reidel) p. 143.

Kutner, M. L., and Mead, K. N. 1981, <u>Ap. J., Letters</u>, 249, L15.

Kwan, J., and Valdes, F. 1987, <u>Ap. J.</u>, 315 , 92.

Mead, K. N. 1982, <u>BAAS</u>, 14, 617.

Mead, K. N. 1987, <u>Ap. J.</u>, in press.

Mead, K. N., and Kutner, M. L. 1987, <u>Ap. J.</u>, in press.

Mead, K. N., Kutner, M. L., and Evans, N. J., 1987, in preparation.

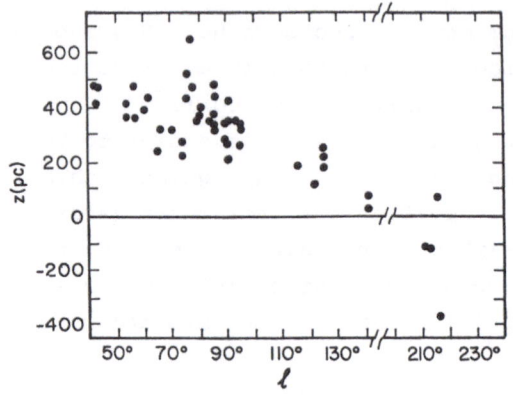

<u>Figure 3.</u> Locations of molecular clouds with respect to the plane are shown in this figure. Distances were determined as in figure one. Notice that the molecular material, like the atomic hydrogen, does not lie in the plane.

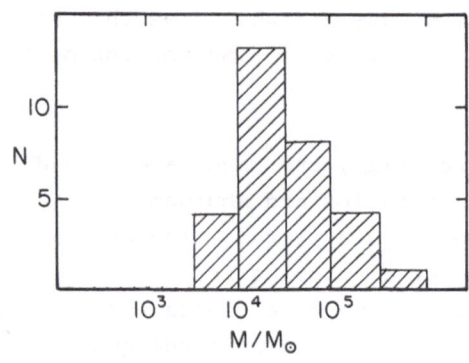

<u>Figure 4.</u> Cloud masses derived from ^{12}CO integrated intensity (Mead and Kutner 1987) are shown. This less-massive distribution (than in the inner galaxy) is consistent with clouds' growing via collisions and fewer collisions in the outer galaxy.

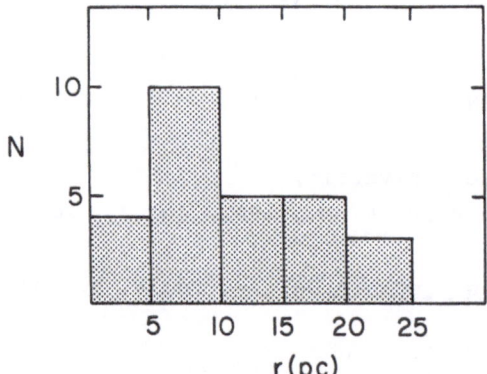

<u>Figure 5.</u> Cloud radii were estimated by calculating half the geometric mean of the long and short axes of the clouds: $r=0.5(ab)^{0.5}$. This figure shows that outer galaxy clouds have sizes typical of giant molecular clouds.

STAR FORMATION IN THE OUTER GALAXY

J.G.A. Wouterloot, J. Brand, C. Henkel
Max-Planck-Institut für Radioastronomie, Bonn

In the outer Galaxy (R>14 kpc; R_0=10 kpc) conditions for star formation are different from those in the solar neighbourhood (see Mead, Kutner, this volume). We have studied the star formation process by searching with the 100m telescope for H_2O and OH maser emission at the positions of IRAS sources within 30 arcmin from 29 HII regions with R>14 kpc selected from the catalogues of Blitz et al. (1982) and Brand et al. (1986) (Wouterloot et al., 1987). The IRAS sources had colours of compact HII regions (see Wouterloot and Walmsley, 1986) and we assume that they are embedded in the giant molecular clouds associated with the HII regions. In addition we observed 5 HII regions at 6 cm with the VLA.

We observed 77 sources and found 15 H_2O masers in the area 14<R<22.3 kpc, one of which was known (near S128 with R=14.1 kpc; Ho et al., 1981). The detection rate of H_2O emission for sources with L>$10^4 L_0$ is the same (about 35%) as in other parts of the Galaxy (solar neighbourhood and Perseus arm; see Wouterloot and Walmsley, 1986). An OH maser was detected near S128. Radio emission was found associated with all of the HII regions observed at 6 cm and with all but one of the luminous (L>6000 L_0) IRAS sources in the VLA fields. The excitation parameters of the regions studied are consistent with excitation by early B type stars.

Comparing the surface density of HII regions with that of H_2, we find that the rate of star formation per unit mass of H_2 at large R is similar to that in the solar neighbourhood. This means that once H_2 clouds are formed, they start forming stars independent of their location in the Galaxy and that external triggering mechanisms such as density waves or cloud collisions are not important. The small number of HII regions in the outer Galaxy is mainly due to the low H_2 surface density. Probable causes for the latter are the closeness of corotation, the decreasing volume density and/or the increasing scaleheight of HI.

REFERENCES

Blitz, L., Fich, M., Stark, A.A.: 1982, Astrophys. J. Suppl. **49**, 183
Brand, J.: 1986, Ph.D. Thesis, Leiden University
Brand, J., Blitz, L., Wouterloot, J.G.A.: 1986, Astron. Astrophys. Suppl. **65**, 537
Ho, P.T.P., Haschick, H.D., Israel, F.P.: 1981, Astrophys. J. **245**, 526
Wouterloot, J.G.A., Brand, J., Henkel, C.: 1987, Astron. Astrophys., in press
Wouterloot, J.G.A., Walmsley, C.M.: 1986, Astron. Astrophys. **168**, 237

THE HIGH-VELOCITY CLOUDS: WHY WERE THEY EVER A MYSTERY?

Gerrit L. Verschuur
8625 Greenbelt Rd #204
Greenbelt MD 20770

A mystery?

The existence of the high-velocity clouds (HVCs) appears to be the longest standing problem in radio astronomy, having been with us for a quarter of a century. Why was it ever a problem in need of a solution?

When we examine the foundations of our intellectual constructs we often find that the edifice depends very heavily on the order in which observations were made, and, more specifically, the sequence in which certain interpretations of the data were presented. If we look closer we may discover that, had the order of discovery been otherwise, very different theories would have been proposed. The high-velocity cloud mystery has suffered from this syndrome.

The first great paper on spiral structure of the outer galaxy was written by van de Hulst, Muller and Oort (1954). Their observations concentrated on the region between -10° and +10° latitude, a bias which is still with us in modern surveys. They used a scanning local oscillator and one receiver channel switched over 648 kHz, or 140 km/s. In retrospect, high-velocity gas, even in the plane, would have been invisible to them.

In the beginning, the study of galactic hydrogen was built upon important assumptions. First that the gas was planar and would be distributed symmetrically with respect to the mid-plane of the galaxy. Thus these observers saw fit to note an exception at $l^I=50°$ where "...the maximum intensity is at positive latitude: +0.°5 for the main peak and second peak, but +1.°5 for the third peak, so that the outermost spiral arm seems tilted from the plane of the other ones." They also noted the rather large latitude extent of the gas in some directions. For example, at $l^I=80°$ it covered 11° between half-intensity points which implied a 1 kpc thickness.

The second fundamental assumption, which has subtly hampered our understanding of the nature of both the outer galaxy and the high-velocity cloud phenomenon, was that of circular motion, although van de Hulst, Muller and Oort found small deviations of the order of a few km/s. They admitted that their discussions of the data were "...based on the hypothesis that the systematic motions in the Galactic System are everywhere circular and that, for a given distance from the centre, the circular velocities are the same at all position angles in the galactic plane." Their observations toward the galactic center and anti-center, "...give some support to this hypothesis."

Had their receivers been more sensitive, and had they switched over a larger velocity range, they might not have drawn this conclusion. They used circular motion as a working hypothesis and admitted that "...here and there deviations from this schematic picture occur."

They also drew the first spiral structure map and there we can see an "outer arm" which, with hindsight, does not fit the ordered spiral pattern, probably because non-circular motions are present. The authors discussed the outer arm, as well as a distant arm, but it is difficult to interpret their text to identify which is which. In any event, they wrote that, "An unexpected, striking feature of the "distant arm" is its great thickness perpendicular to the galactic plane. From a few tentative tracings across the arm ... we derive an average thickness of 800 ps between points where the density has become half the maximum density." The distant arm lay slightly north of the plane of symmetry of the more central part of the Galactic System.

In regard to the spiral structure, van de Hulst (1953) had said that, "It may be necessary ...to emphasize that these results are the beginning and not the end of an era."

This was an amusing reminder to the incautious that the presentation of the first map of spiral structure did not mean that the issue of galactic structure was closed, a caution to which the next quarter century has born ample witness. He also stressed that not all the arms were in the same plane and then touched on a very important point which remained overlooked until Burton (1971) resurrected it. The emission profiles showed clear peaks and, "They must indicate an irregular distribution of the densities or of the velocities of the interstellar gas. It seems most likely that most of the maxima and minima are due to a patchy density distribution, although it is already clear that some minor details cannot be explained in that manner."

He did not say how minor, nor which details these were. To this day 21-cm observers generally assume that spectral structure signifies density variation. Velocity-crowding, as it is now called, may nevertheless be very important in determining both large- and small-scale HI profile structure.

Westerhout (1957) was the next to discuss the gas in the Outer Galaxy. His data were also collected between b=±10°. Deviations from circular motion, which he called "group motion", were noted. The assumption of circular motion failed at some longitudes where positive velocities of the order of a few km/s were seen. He made the important point that, "Such group motion can be detected only in the neighborhood of the sun."

At great distance we have nothing with which to compare the velocities, so if non-circular motions are present they cannot be recognized. Westerhout presented data on the warping of the plane and found the gas to be non-planar by as much as 400 pc at l^{II}=110°, but admitted that he did not incorporate this into his further calculations.

Thus we reach into the ocean of uncertainty, treading on stepping stones which take us far from the secure base that was our conceptual home. The stepping stones read; flat galaxy, circular motion, ordered spiral arms, and density fluctuations. There is a hint that the galactic gas flairs and tilts at the outskirts. So we move forward, constructing new steps as we go.

The High-Velocity gas

In 1961 Oort and his collaborators began a search for evidence of infalling gas, hydrogen at high latitudes expected to show negative velocities. They immediately found evidence of excess emission at negative velocities in certain parts of the sky and this gas became to be called the "high-velocity clouds" discovery of the HVCs was not serendipitous, an important point seldom mentioned.

In 1963 the report of the discovery of the HV gas was published by Muller, Oort and Raimond (1963) in which this matter was described as being in the form of "clouds". Thus a subtle but important bias was immediately introduced.

Then Kerr (1965) pointed out that, "The main group of clouds is found directly opposite the Magellanic Clouds and might be associated with the Magellanic countertide."

A flurry of Dutch activity occurred in 1966 when numerous papers on the large-scale distribution of galactic hydrogen were published. Lindblad (1966) noted that the picture of galactic structure current at the time <u>depended almost entirely</u> on observations made in the Netherlands between 1953 and 1957. His study of the outer parts of the galaxy again used surveys which reached to latitude 10°. Examination of his data show that at l=135° gas could be seen to -100 km/s yet this gave no cause for alarm as regards the existence of non-circular motions. Compare this to Burton and te Lintel Hekkert's (1985) data which reveal emission out to -200 km/s close to this position. What would the astronomer's of the 50s and 60s have made of that?

The 1966 series of papers continued to reinforce the idea that non-circular motions were a local, not a galactic-scale, phenomenon.

The first directed study of high-velocity gas, by Hulsbosch and Raimond (1966), continued to refer to this gas in terms of a cloud-like nature. They said that a systematic search for "...high-velocity <u>objects</u> at latitudes |b|>20°..." had been made and that, "...it seems justified to conclude that high negative velocities predominate over positive velocities at high latitudes in the part of the sky visible from the Netherlands." However, by now Kerr had found positive velocity gas at high latitudes in the southern skies. This important discovery was essentially ignored in northern discussions. Hulsbosch and Raimond also made the delightful understatement that, "Most of the objects are so weak that it will take many more observations to obtain details about their sizes and internal velocity distribution." (Two decades later Westerbork measurements, by Schwarz and Oort (1981) revealed that the "objects" were hardly weak. The concentrations are very small, however, which is expected for distant HI features resembling local gas.)

Oort (1966) interpreted the data and stated that, "Characteristics of the high-velocity clouds above ±15° latitude are that <u>all of them</u> have negative radial velocities..." and "...they are concentrated in only slightly more than one quadrant of longitude, and...show a highly asymmetrical distribution in latitude." These statements, while no longer true, continue to exert a subtle influence on the thinking about the phenomenon.

Oort's paper listed models which included supernova shells, condensations in a hot corona, fragments ejected by the galactic nucleus, objects outside the galaxy, ejecta from the disk which are the principle constituents of the corona, or intergalactic gas accreted by the Galaxy. He was to come down heavily in favor of the notion that the HVCs were at z-heights of the order of 500 pc. This notion exerted a powerful influence on future work. He did not consider the possibility that the galaxy in its outer regions might be a very confused beast, manifesting serious non-circular motions as well as large non-planar excursions. The outskirts of many galaxies are now known to show such disorder and the galactic environment reveals at least one great tentacle of gas stretching through space; i.e., the Magellanic Stream.

In the subsequent years we often heard the voice of Frank Kerr reminding us that the observed radial velocities represented only one component of space motion (Kerr, 1967), a caution ignored by the Dutch school. Then Kerr and Sullivan (1969) and Verschuur (1969) considered models which placed the HVCs in intergalactic space. The referee told us that the evidence that the HVCs were local was so overwhelming that it was no longer necessary to consider alternatives. An appeal worked in our favor. Still the local school held sway.

Verschuur (1972,1973) and Davies (1972) then showed that HV phenomenon in the northern skies might represent distant extensions of galactic structure. Hulsbosch and Oort (1973) soon criticized these models and stated that, "... no reasonable rotation of the warped disk could produce the velocities observed at b≳45° and in the Anti-center region." Who said nature is always reasonable? Oort was always concerned about the clouds at the south pole which showed negative velocities. The influence of his original thinking about the nature of the negative velocities is still evident in a recent review by van Woerden, Schwarz and Hulsbosch (1985) who again suggest that the dominance of negative velocities indicate an inflow of gas into the Galaxy. Yet we know that the Magellanic Stream at the south pole is extragalactic and happens to show large negative velocities. The Kerr reminder strikes again!

Concerning the infall class of models, these theories have been favored because they Are amenable to manipulation. Fountain models are elegant but that does not make them relevant! Concerning Bregman's (1980) work, for example, van Woerden et al (1985) politely said that while it gave a, "...fair representation of HVCs, important features remain unexplained."

What was the problem?

The high-velocity clouds became a problem because of a set of fundamental assumptions about the galaxy, compounded by expectations about infalling matter, followed by a set of conclusions related to the nature of the clouds that did not withstand the test of time. These expectations obscured the nature of the phenomenon

What has happened to our stepping stones? The Galaxy is not flat – it always warps up to the latitude limit of any given survey (!) – the assumption of circular rotation in the outer regions is demonstrably inadequate, spiral arms in the outer regions of many galaxies are far from ordered, and velocity-crowding effects may be very important.

The original statements about the nature of the HVCs have also been proven false. The HV gas consists of more than a set of clouds. The Northern Stream is a vast structure extending from $l=0°$ to $160°$, $b=0°$ to $+60°$ (Giovanelli, 1980) revealing galactic rotation patterns. Negative velocities dominate for $30°< l< 210°$ while positive velocities dominate over $210°< l<330°$. The clouds are found over most of the sky, and apart from velocity patterns, are not asymmetrically located in latitude. Finally, high-resolution emission profiles now reveal that the clouds are anything but weak. Bright peaks of about 20 K (Schwarz and Oort, 1981) have been seen and a high-resolution profile of the HV gas looks very much like lower resolution profiles of local, non-planar, gas.

Table 1 compares the density and size estimates made in 1966 with the values found from the Westerbork data. At a distance of 500 pc HVC Al, for example, either has pockets of gas of diameter 0.5 pc with densities of 200 to 720 cm^{-3} at the 500 pc distance, or 10 pc diameter structures with densities of 10 to 36 cm^{-3} if it were at 10 kpc distance (the Verschuur/Davies class of models). If the HVCs were local they would be extraordinary objects, while if located of the order of 10 kpc away they are quite normal. Only their radial velocity components confuse us a little.

Several arguments now reveal that the HVCs are at distances of the order of 10 kpc (and greater for the very-high-velocity clouds, e.g., Cohen, 1982). The models of Verschuur and Davies have already been alluded to. In addition, Verschuur (1985) noted that the angular size of the typical HVC in the southern skies, taken from Hulsbosch's (1983) data, is a function of velocity. This suggests that the smallest, highest velocity features are indeed the more distant. Verschuur (1987) argues that the distance to the northern clouds can be derived from observed cloud parameters. When one consideres modern data, such as the results shown by Burton at this meeting, it is far easier to recognize that the major high-velocity cloud complexes exist in the Outer Galaxy and represent extensions of galactic (spiral) structure into near-galactic space.

Conclusions

The Dutch observers thus went looking for infalling matter and found the HVCs. From the start, the phenomenon was forced into a mold that did not fit, a mold that it continues to resist, 25 years after the discovery. The HVCs became, and remained, a mystery because authority insisted, and most workers accepted, that the Outer Galaxy was essentially ordered and showed no significant non-circular motions. In addition, the caution that one component of space motion does not a model make was ignored. Then, when the Magellanic Stream was discovered, the step of considering that the Galaxy might look like other interacting galaxies, with flailing streamers of material reaching into intergalactic space, was shunned. It is time to abandon the concept that the Milky Way is a perfect spiral galaxy.

References
Bregman, J. N. 1980, Astrophys. J. **236**, 577
Burton, W. B. 1971, Astron. Astrophys. **10**, 76
Burton, W. B. and te Lintel Hekkert. P. 1986, Astron. Astrophys. Suppl. **65**, 427
Cohen, R. J., 1982, Mon. Not. Roy. Astr. Soc. **200**, 391
Davies, R. D., 1972, Nature **237**, 88
Giovanelli, R. 1980, Astron. J. **85**, 1155
Hulsbosch, A. N. M. 1983 In *The Milky Way Galaxy.* (ed. H. van Woerden, R. J. Allen, and W. B. Burton) IAU Symposium **106**, page 409. (Reidel-Dordrecht)
Hulsbosch, A. N. M. and Raimond, E. 1966 BAN **18**, 413
Hulsbosch, A. N. M. and Oort, J. H., 1973, Astron. Astrophys. **22**, 153
Kerr, F. J. 1965 NATO *International Summer Course on Observational Aspects of Galactic Structure.* ed Mavridis. Athens.
Kerr, F. J. 1967. In *Radio Astronomy and the Galactic System.* (ed. H. van Woerden) IAU Symp. **31**, p 297. (Reidel-Dordrecht)
Kerr, F, J. and Sullivan, W. T. 1969, Astrophys. J. **158**, 115
Lindblad, Per Olof 1966. BAN Suppl. **1**, 77
Muller, C. A., Oort, J. H. and Raimond, E. 1963. C. R. Acd. Sci. Paris. **257**, 1661
Oort, J. H. 1966 BAN **18**, 421
Schwarz, U. J. and Oort, J. H. 1981, Astron. Astropphys. **101**. 105

Van de Hulst, H. C., Muller, C. A.. and Oort, J. H. 1954. BAN **12**, 117.

Van de Hulst, H. C. 1953 Obs. **73**, 129.

Van de Hulst, H. C., Muller, C. A.. and Oort, J. H. 1954. BAN **12**, 117.

Van de Hulst, H. C. 1953 Obs. **73**, 129.

Van Woerden, H. Schwarz, U. J. and Hulsbosch, A. N. M. 1985, In *The Milky Way Galaxy.* (ed. H. van Woerden, R. J. Allen, and W. B. Burton) IAU Symposium **106**, page 387. (Reidel-Dordrecht)

Verschuur, G. L. 1969, Astrophys. J. **156**, 771

Verschuur, G. L. 1972 *Report of Spiral Structure Meeting.* Tucson. Arizona, ed. B. J. Bok, C. Cordwell, and R. Humphreys.

Verschuur, G. L. 1973, Astron. Astropphys. **22**, 139

Verschuur, G. L. 1985, Bull. A. A. S. June 1985.

Verschuur, G. L. 1987. In preparation

Westerhout, G. 1957. BAN **13**, 201.

VI. THE MAGELLANIC CLOUDS AND
THE OUTER REGIONS OF OTHER GALAXIES

STRUCTURE, ROTATION AND MASS OF THE MAGELLANIC CLOUDS

G. de Vaucouleurs, University of Texas
Read by Frank Bash, University of Texas

1. INTRODUCTION

Frank Kerr, Jim Hindman, and Brian Robinson first detected and mapped the 21-cm line emission from the Magellanic Clouds in 1952-53 while I was at Mount Stromlo recognizing spiral structure and rotation in them. It was natural that we should join forces for the first combined radio and optical analysis of these systems. It is a pleasure to review this early work and later developments. I regret that circumstances do not permit me to be present in person today, but my good friend and colleague Frank Bash has generously offered to read this paper on my behalf. I am sure that he will deliver it much better than I could have done myself.

To begin with, it is appropriate to recall that the distances of the Clouds were already well determined in the mid-1950's. In 1955 a review of three distance indicators (novae, cepheids, RR Lyr) gave me a mean modulus of 18.6 (deV. PASP 67, 350, 1955). We know today that the Small Cloud is slightly more distant than the Large Cloud, by some 10 to 15% (0.2-0.3 mag), but the mean distance needs hardly to be revised. A recent, unpublished review of the six primary indicators currently available (including long-period variables, AB supergiants and eclipsing binaries) gives *apparent* moduli in the V and B bands $\mu_V = 18.7 \pm 0.1$, $\mu_B = 18.8 \pm 0.1$ for the Large Cloud and 18.9 (V), 19.0 (B) for the Small Cloud.

The variable extinction, both foreground and internal, is more difficult to evaluate. A recent review of a score of modern determinations in each Cloud suggests average reddenings of about 0.1 for each Cloud. Most authors agree that the true extinction-corrected modulus of the LMC is in the 18.2-18.6 range and that of the SMC is about 0.3 greater. For definiteness, let's say, 18.4 and 18.7, corresponding to distances of 48 and 55 kpc for the centroids of the LMC and SMC. At these distances one degree of arc corresponds to linear distances of 0.84 and 0.96 kpc.

2. THE MAGELLANIC CLOUDS AS DISK SYSTEMS

The Magellanic Clouds were first recognized as disk systems, and more precisely as late-type barred spirals, in the 1950's. This was established by several and mutually consistent lines of evidence: (i) the discovery of an extensive spiral structure emerging from the axial bar (Shapley 1950; deV. 1954, 1955a), most evident in the LMC (Figure 1), but detectable also in the SMC

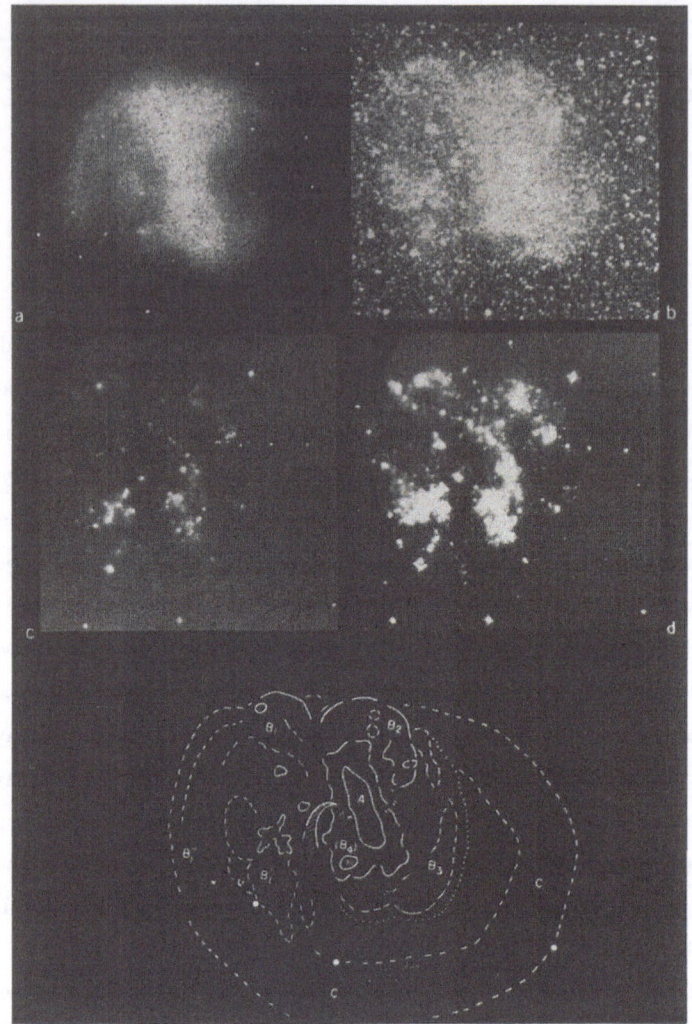

Figure 1. Four views of the main body of the Large Magellanic Cloud: visual, as drawn by J. Herschel (1836-38) (upper left); small-scale direct photograph (Mt Stromlo, 1953) (upper right); ultraviolet electronographs at $\lambda 1050 - 1550$, 1 min exposure (lower left) and $\lambda 1230 - 1550$, 10 min exposure (lower right) by C. Young from the Moon (1972).

when allowance is made for its greater inclination and tidal distortion (Shapley's "wing"); (ii) the interpretation of the observed velocity field as due to general rotation (deV. 1954, 1955a; Kerr and deV. 1955, 1956) rather than translation; (iii) the close agreement between the photometric major axis (deV. 1957) and the kinematic line of nodes (Kerr and deV. 1956; deV. 1960; Feast, Thackeray and Wesselink 1955, 1961; Feast 1964) both in position angle $170° \pm 5°$; (iv) the small value of the z velocity dispersion (≈ 10 km/s) of HII regions and OB stars (deV. 1955b; Feast 1961, 1964) in agreement with the corresponding values in our Galaxy and other spirals (Heidmann, Heidmann and deV. 1971; Shostak and van der Kruit 1982); (v) the similarity of the LMC – SMC pair with

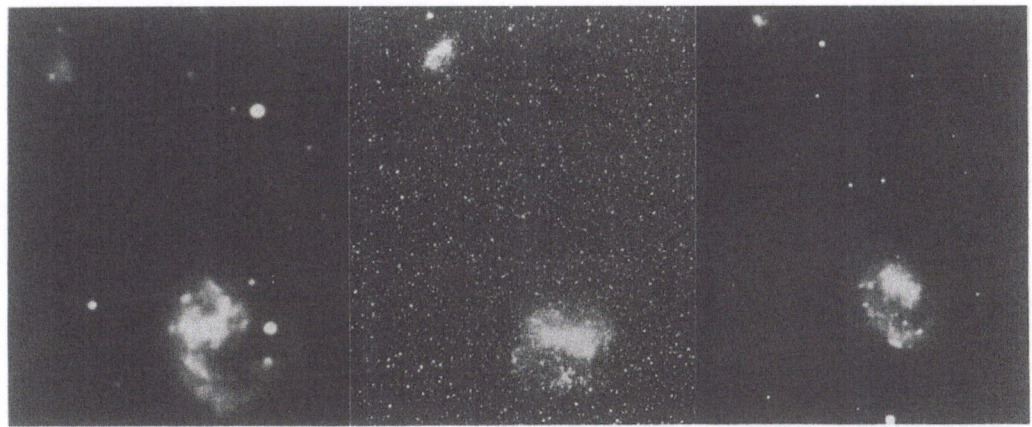

Figure 2. Comparison of three pairs of asymmetric, late-type barred spirals of the Magellanic type: NGC 4618-4625 (left), LMC- SMC (center), NGC 4027-4027A (right). McDonald and Mt Stromlo Observatories.

other pairs of late-type spirals such as NGC 4027 – 4027A, and NGC 4618 – 4625 (Figure 2); (vi) the flatness of other magellanic spirals seen edge on, for example NGC 55 (Figure 3), NGC 1507, NGC 2188, NGC 4631 (deV. 1961; deV. and Freeman 1972).

The inclination of the equatorial plane of the LMC to the tangent plane (or of its spin axis to the line of sight) was initially determined to be about 27° from the apparent ellipticity of the outer isophotes in red light and from the shape of the outer loop (deV. 1954, 1957). It was soon confirmed by independent estimates from the ellipticity of HI contours (McGee and Milton 1966a), from the isopleths of the cluster distribution (McGee and Milton 1966b), and finally from the luminosity gradient of cepheids at constant period (deV. 1955a, 1980; Gascoigne and Shobbrook 1978; Martin, Warren and Feast 1979; Caldwell and Coulson 1985). The weighted mean of these various determinations is $27°.2 \pm 2°.4$. A slightly different value, $33° \pm 3°$, was derived by Feitzinger (1977) from an extensive review of the distributions of various classes of objects. A mean value of $30° \pm 2°$ may be adopted. [Values around 45° recently proposed by some authors from radio-continuum studies are certainly not applicable to the massive stellar disk, even if they were valid for the outlying gas corona.]

The position angle of the line of nodes of the disk with the tangent plane is well determined for objects close to the plane (young stars and HI), but Feitzinger *et al.* (1977) found a significant difference between the geometric major axis and the kinematic line of nodes (168° versus 188°) which they interpreted as evidence for a considerable transverse velocity of the LMC along the Magellanic Stream of 275 km/s relative to the LSR and 143 km/s relative to the galactic center. However, in a new discussion including additional material Duval (1983) showed that most of the discrepancy is in the kinematic line of nodes of the rather small sample of planetary nebulae (199°),

Figure 3. NGC 55 is a Magellanic, asymmetric barred spiral seen almost exactly edge on. The LMC would probably look very much like it if viewed at the same inclination and orientation giving an end-on view of the bar. Mt Stromlo Observatory (1954).

while HI, HII and stars give values in the range 168° − 187°, with an average of 178°, which is not significantly different from the geometric major axis. On the other hand, Freeman (1983) has reported strong evidence that the line of nodes defined by the velocities of *old* clusters differs by some 40° from the line of nodes defined by the young clusters and other disk objects. It may be that the difference reflects different rates of precession of systems of different flattening and ages as the LMC describes its orbit around the Galaxy. Note that the prominent axial bar is in position angle 120° and does not coincide with either of the principal axes.

Once the basic similarities between the Large and Small Clouds, and other late-type, asymmetric barred spirals are recognized, for example NGC 4618-4625, NGC 4027-4027A (Figure 2), it is not too difficult to define the major axis and inclination of the SMC, allowing for the distortion produced by the tidal prominence pointing toward the LMC. An early analysis of star count isopleths and optical isophotes (deV. 1955b, 1957a) gave for the major axis a position angle of 45° ± 3° and an inclination of 60° ± 3°. The axial bar is in position angle 25°, significantly different from the major axis of the disk. Again the radio data considered alone do not lead to a clear cut picture, because of confusion with the Magellanic Stream and the tidally displaced gas.

3. CLASSIFICATION AND MORPHOLOGY

Studies of the Clouds at Mount Stromlo and Sydney in the 1950's demonstrated that the Magellanic "irregulars" form a natural extension of the spiral sequence, both barred and non-barred, and that many of them display conspicuous rotational motions and spiral structure, albeit often asymmetric about the axial bar (deV. 1956a, 1959a). The Magellanic Clouds are the prototypes of stage Sm in the revised Hubble sequence: Sa - Sb - Sc - Sd - Sm - Im. Spiral structure is prominent, but somewhat irregular at Sd, vanishing or absent at Im, the only stage which fully deserves the "irregular" description.

The very-late or Magellanic stage of spiral structure is encountered in both families of spirals, SA (non-barred), such as NGC 45, 2574 and 4395, and SB (barred), such as NGC 1313, 2537, 4027, 4618 and, of course, the two Magellanic Clouds.

(a) Structure of the LMC. Direct photography (Figure 4 a,b), surface photometry and star counts (deV. 1955a, 1956b, 1964a) indicate the presence of three main regions (Figure 5): Region A, the conspicuous bar, about $3°5 \times 1°0 = 3 \times 0.8$ kpc with major axis in p.a. 120°. It is relatively poor in OB supergiant stars brighter than $M \approx -5$, in HII regions and in HI gas; it may be described as made up of an "aging" Population I component. Region B includes the main irregular and asymmetric spiral arm (B_1) on the north side of the bar and the short "embryonic" arms (B_2, B_3) emerging at the SW and SE ends of the bar (Figure 5). It is marked by the richest OB associations and concentrations of supergiant stars brighter than ≈ -5, the brightest and largest HII regions (excluding 30 Doradus), and some of the densest clouds of neutral hydrogen (again outside the 30 Dor complex) (McGee and Milton 1964, 1966a). These characteristics and the blue colors of these features, which make them especially prominent in ultraviolet photographs (Figure 1), are typical of a very young Type I population. [1]

Region C is the faint outer loop or whorl which is an extension of the dominant spiral arm (B) circling the main body and returning toward the bar near B_2 after a complete turn. This region is devoid of bright supergiants and associated HII regions and is very poor in neutral hydrogen. It was first detected by its unresolved luminosity on small-scale, long-exposure photographs (Figure 4b), but it is also well outlined by the distribution of carbon stars (Westerlund 1964) and by the distribution of outlying star clusters (Shapley and Lindsay 1963). The stellar population of this loop begins to be resolved only at m > 15 (M > −4). These characteristics and the redder

---

[1] The 30 Dor complex, at the NE end of the bar, is located where a third embryonic arm (B_4) appears in other SBm galaxies, e.g., NGC 4027 and NGC 4618. Suggestions that it is the "nucleus" of the LMC and the center of its spiral structure are in conflict with all studies of the distributions of light, stars, clusters, HI and with the velocity field, all centered near or slightly north of the bar center. The brightest HII region in a galaxy can appear anywhere, often far from the center, and is a short-lived phenomenon. To claim that 30 Dor is the center of the LMC is akin to claiming that NGC 604 is the center of M 33.

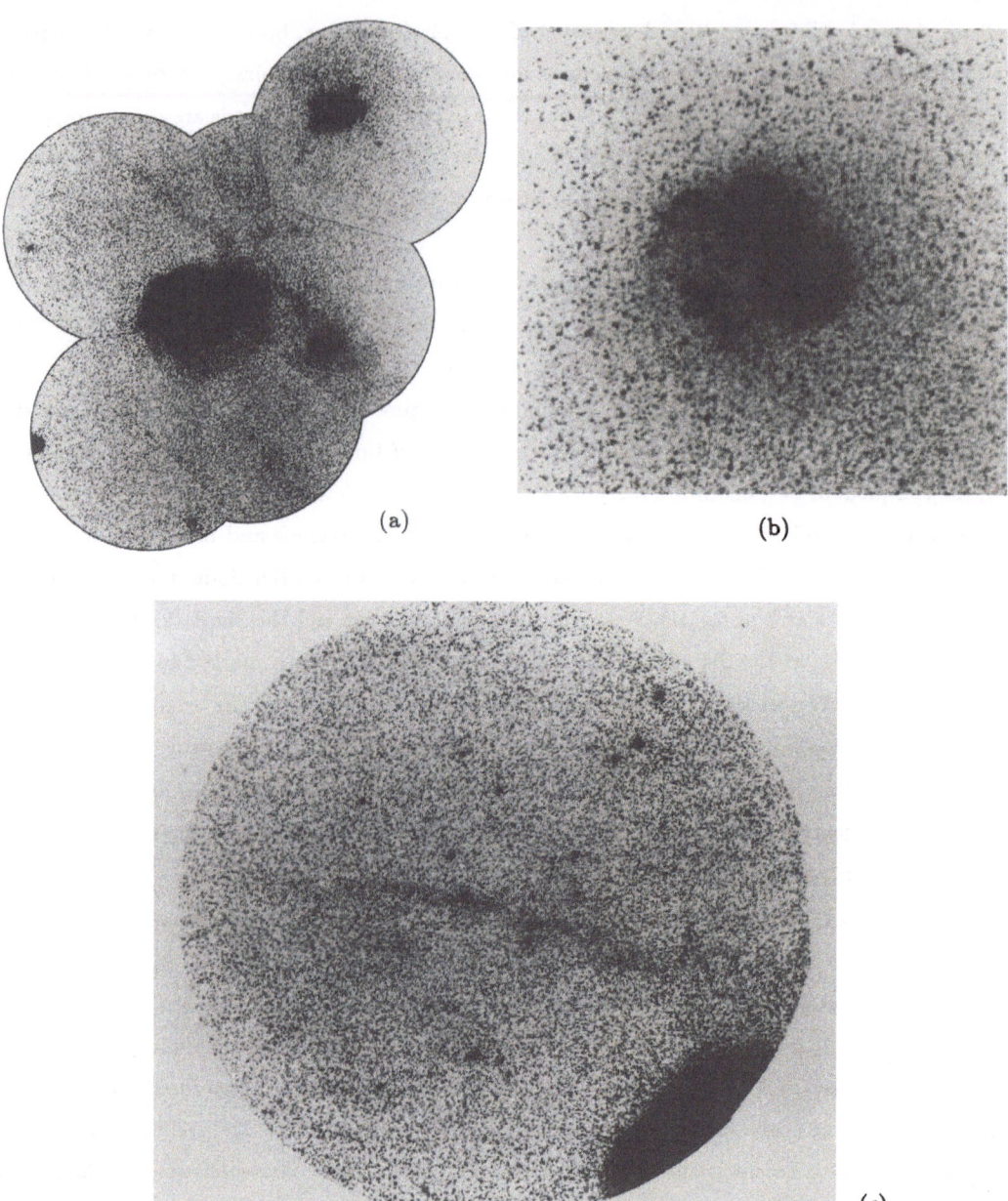

(a)

(b)

(c)

Figure 4. (a) Mosaic of long-exposure direct photographs of the LMC-SMC field showing faint outer arms and loop of LMC, prominence of SMC, and foreground galactic nebulosities. Mt Stromlo Observatory (1953). (b) Direct photograph of the LMC with a 35 mm camera showing spiral arms and loop. Mt Stromlo Observatory (1953). (c) Foreground galactic nebulosity ("cirrus") stretching north-west of LMC (overexposed at edge of field) photographed in blue light. Mt Stromlo Observatory (1953).

color of the region (Eggen and deV. 1956) are consistent with an old Type I population. This outer, asymmetric loop continuing the main arm is typical of the barred Magellanic systems and

224

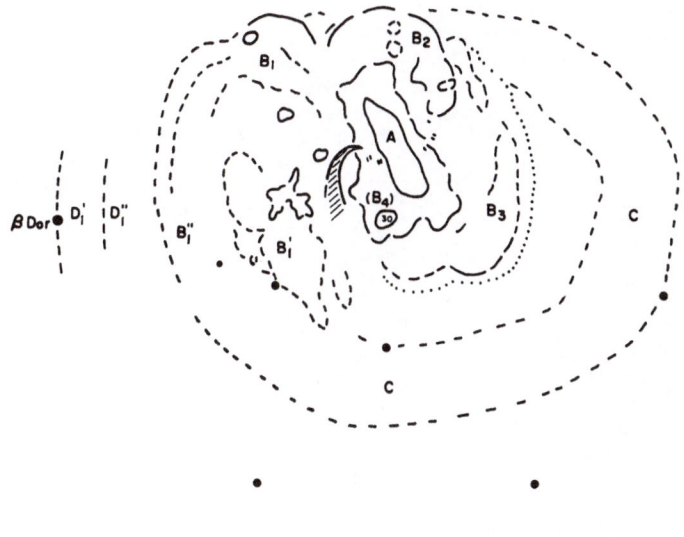

Figure 5. Schematic structure of the Large Magellanic Cloud and nomenclature of principal regions. Compare Figures 1, 2, 4.

is obvious in other galaxies, such as NGC 4027 and 4618. It was more difficult to detect in the LMC only because of its large angular diameter and confusion by the rich field of galactic stars in the foreground.

On the north side of the main body there are detached fragments of spiral arms as far north as β Doradus, over 7° from the center of the bar. On the south side the picture is confused by galactic nebulosities in the foreground (Figure 4a). On the west side a striking giant filament stretching over a 20° arc from 4^h45^m, −73° to 3^h25^m, −57° merges smoothly (in projection) with the LMC outer loop (Figure 4c). It was initially mistaken for a tidal arm (deV. 1954a, 1955a), but later photographs centered on the field and of better quality showed it to be a galactic nebulosity of unusual size and shape (deV. 1960a). This has been confirmed by inspection of the larger scale photographs taken in recent years with the 1.2-m UK Schmidt telescope. It coincides with an infrared "cirrus" in the IRAS survey and with a low-velocity HI filament at 21-cm (Cleary, Heiles and Haslam 1979, Heiles and Cleary 1979).

The near absence of neutral hydrogen in the eastern part of the outer loop of the LMC (at lower galactic latitudes) and the great pile-up of HI in the 30 Doradus region and across the east end of the bar, suggest that this peculiar distribution could be due to ram pressure pushing back

Figure 6. (a) Direct photograph of Small Magellanic Cloud in blue light (limiting magnitude = 16). Mt Stromlo Observatory (1953).

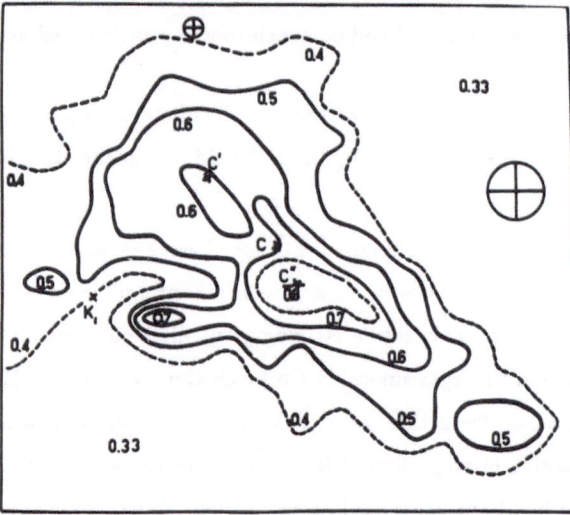

Figure 6. (b) Schematic structure of the Small Magellanic Cloud and equi-gradient contours of apparent photographic luminosity function between 14th and 16th magnitudes in SMC. Knot K_2 (Shapley's wing) is outside frame at left.

the gas as the leading edge of the LMC penetrates the outer corona of our Galaxy.[2] If so, this may provide another clue to the direction of the transverse motion of the LMC as it travels along the orbit marked by the Magellanic Stream, as well as an indication of the maximum extent of the gaseous corona of our Galaxy.

---

[2] I believe that this suggestion was first made by Frank Kerr, but the reference eludes me.

(b) Structure of the Small Cloud. The structure of the Small Cloud is more difficult to analyze because it is foreshortened in projection by the larger inclination angle (60° instead of 30°) and distorted by the asymmetric "wing" or prominence pointing toward the Large Cloud. Because the SMC is more distant than the LMC by 5-7 kpc and is separated from it by 21° ≈ 19 kpc (in projection on the sphere), this prominence is probably not in the equatorial plane of the system, but inclined to it by some 10°. Nevertheless, the basic structure of the SB(s)m spirals can still be recognized through analyses of direct photographs, surface photometry, star counts, cluster distributions, and radial velocities (deV. 1955b, 1957b,c; Maurice 1979; Dubois 1980; Azzopardi 1981). Thus, region A is the bar-like core of the system elongated in p.a. 45° through the optical center C ($00^h51^m, -73°1$) (Figure 6 a,b). The brightest part of the bar is south-west of it in a region (C'') where the gradient of the luminosity function between m = 14 (M = −5) and m = 16 (M = −3) and the concentration of HI are both high (deV. 1955b; Hindman 1964, 1967). Indeed, there is a close similarity between the HI distribution and the isopleths of star counts to m = 16 (M = −3).

Region B includes the highly-resolved north-east arm, corresponding to the main asymmetric arm (B_1) in the LMC. It is rich in blue supergiants (M < −5) and HII regions, and in it the gradient of the luminosity function is much lower than in region A (Figure 7). This is typical of a young population, whose presence is also indicated by a concentration of blue clusters in this region (deV. 1957b; Brück 1975).

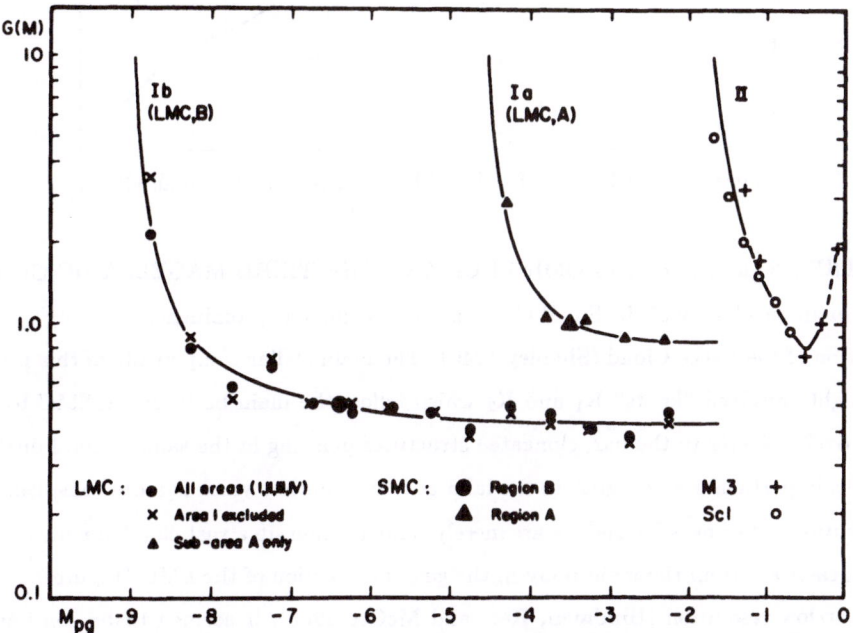

Figure 7. Gradients of stellar luminosity function in different parts of the Magellanic Clouds.

Region C, corresponding to the outer loop of the LMC, is more difficult to trace in the SMC because of the unfavorable inclination and possibly because of a warp of the main spiral arms. Its presence is nevertheless indicated by the elongated region populated by faint stars (m > 15) stretching over some 2° south-west of the bar. It is rich in cepheids and red clusters (Figure 8), but relatively poor in neutral hydrogen (Figure 9). It is not clear whether this formation is an extension of B_1 looping around the main body, or a separate spiral arm originating at the opposite end of the bar (that is, analogous to B_2).

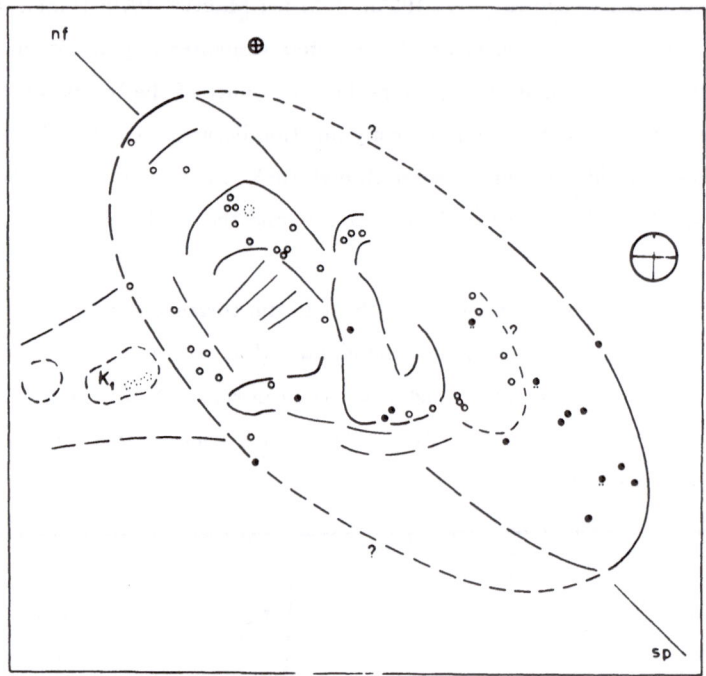

Figure 8. Distribution of red and blue clusters in the Small Cloud.

4. THE ASYMMETRIC PROMINENCE AND THE THIRD MAGELLANIC CLOUD

The main peculiarity of the Small Cloud is the asymmetric prominence or "wing" pointing in the direction of the Large Cloud (Shapley 1940). The main stellar components of this prominence are the highly resolved "knots" K_1 and K_2 which follow the main body of the SMC by 1°.8 and 5° respectively. Nearer to the bar, elongated structures pointing in the same general direction are discernible, in particular a "tongue" in the faint star distribution which is more conspicuous in the HI distribution. The knots K_1 and K_2 are merely condensations in a vast cloud of neutral hydrogen extending eastward from the main body in the general direction of the LMC (Figure 10). Initially observed at low resolution (Hindman, Kerr and McGee 1963), it seemed to be continuous, but turned out to be strongly patchy when examined at higher resolution and sensitivity (McGee and

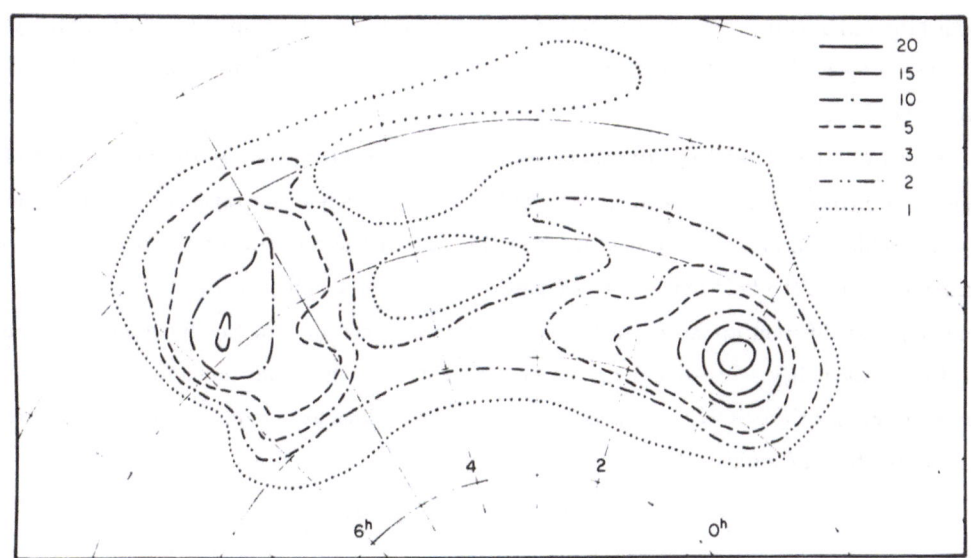

Figure 9. An early map of HI emission in and between the Magellanic Clouds (Hindman, Kerr and McGee 1950).

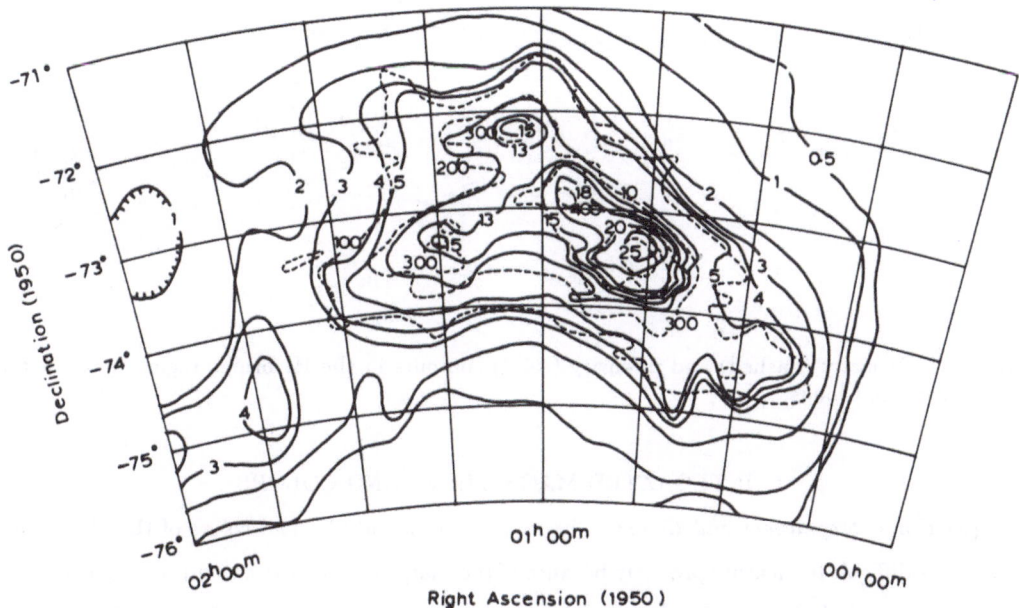

Figure 10. A high-resolution map of the distribution of neutral hydrogen density in the SMC and in the prominence (after Hindman 1967).

Milton 1966 a,b; Hindman 1967). The SMC wing has the characteristics of a young association of the NGC 2264 type and it is tempting to regard it as an incipient "Third Magellanic Cloud" in the process of formation in the densest part of the tidal prominence (deV. 1976). Beyond K_2 and toward the LMC scattered stars at distances consistent with membership in the LMC-SMC system have been reported (Kunkel 1979), but their integrated luminosity is too low to be detected on direct photographs.

Because the SMC and bridge region are overlapped in projection by the denser part of the Magellanic Stream (Figure 11), it is very difficult to disentangle the various components of the HI line profiles. Various ways of decomposing these profiles and of tracing the continuity of each component across the field have led to suggestions of expanding shells and/or separate concentrations of gas at different distances, including the extreme view that the Small Cloud is made up of two fragments widely separated along the line of sight. It is clear that much further work will be needed to clarify the relative locations in space of the various gas clouds and of the main stellar body of the Small Cloud.

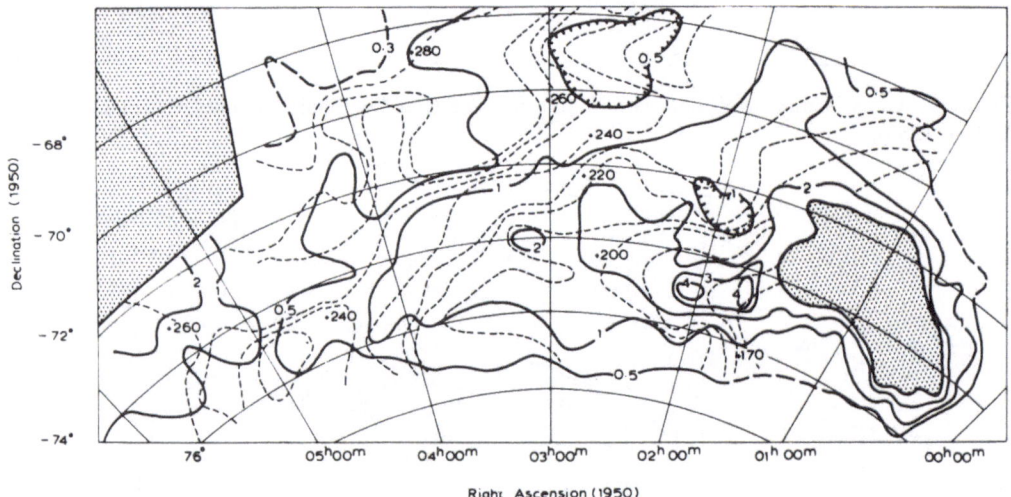

Figure 11. Velocity (dashed) and intensity (solid) contours in the HI bridge region between the two Magellanic Clouds.

5. INTEGRATED MAGNITUDES AND COLORS

(a) U,B,V Magnitudes and Colors. The integrated magnitudes and colors of the Magellanic Clouds are difficult to measure precisely because of their large angular extent and confusion by field stars, galactic nebulosities, dark nebulae and irregular foreground luminosity. Several photographic and photoelectric determinations define integrated luminosity curves in close agreement with the standard curve applicable to late-type galaxies (Figure 12), that is, essentially exponential surface brightness distributions (Figure 13), and to revised total magnitudes of $B_T = 0.91$ for the LMC and 2.70 for the SMC. These values are respectively 0.28 (LMC) and 0.09 (SMC) mag fainter than those generally adopted during the past three decades (deV. 1960) which depended on uncertain

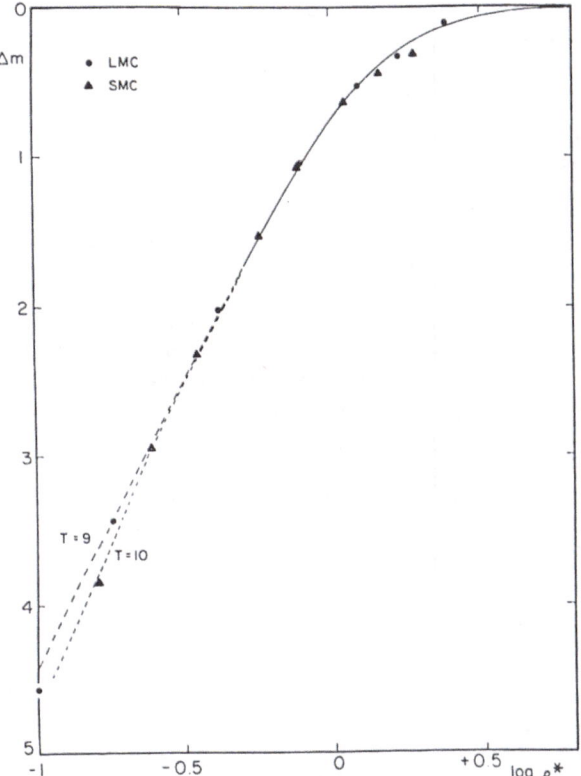

Figure 12. Normalized integrated luminosity curves of the Magellanic Clouds. Note agreement with standard curves for types 9 (Sm) and 10 (Im).

extrapolations. Similarly revised color indices are $(B - V)_T = 0.51$ (LMC) and 0.45 (SMC), and $(U - B)_T = 0.0$: (LMC) and -0.2 : (SMC). Note that the magnitude and colors of the SMC refer to the main body only, not including the knots K_1, K_2 for which no magnitude or color data are available as yet.

(b) Ultraviolet Colors. The integrated magnitudes and colors of the central regions of the Small Cloud have been derived by Nandy *et al.* (1978) from observations at four wavelengths between $\lambda 1550$ and $\lambda 2740$ Åwith the Sky Survey Telescope S2/68 in the TD-1 satellite. On isophote maps (Figure 14) derived from these observations the main features seen in the visible range are easily recognizable (the prominence on the north-west side is caused by the globular cluster 47 Tucanae). The UV magnitudes are defined as $m_\lambda = -2.5 \log I_\lambda - 21.1$, if the energy flux I_λ is in erg cm^{-2} s^{-1} Å^{-1}. ($V = 0$ corresponds to $I_V = 3.64 \times 10^{-9}$ units). By extrapolation of magnitude-aperture curves Nandy *et al.* calculated magnitudes in a standard reference area of 12 square degrees including $\sim 80\%$ of the B-luminosity of the Small Cloud (deV. 1960). The UV colors are consistent with those of a B9V star. However, because of the composite nature of the spectrum, no single star or color temperature can represent the spectrum over a large range of wavelengths.

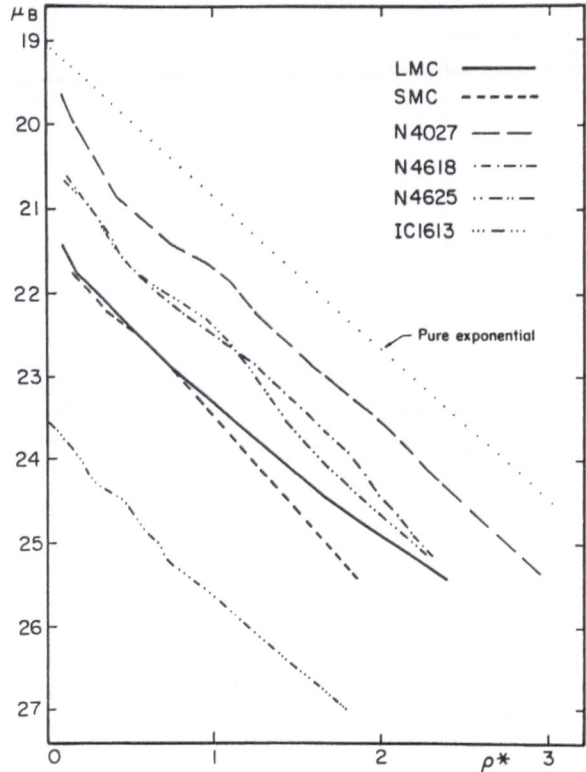

Figure 13. Mean surface brightness distributions in galaxies of the Magellanic type. Abscissae are in reduced units, normalized to the equivalent effective radius. Ordinates are in B mag per square arcsec. Note characteristic exponential distribution.

(c) *Color Distributions.* The color distributions in the Magellanic Clouds are rather peculiar. Compared to the average of other magellanic irregulars the SMC has an unusual color gradient, being bluer near the center (there is little or no systematic color trend in the average for magellanic irregulars; deV. *et al.* 1983). The LMC has no definite radial gradient in the mean, but has a strong transverse (north-south) gradient, being bluer north of the bar and redder south of the bar (Eggen and deV. 1956; deV. 1982, unpublished).

(d) *Absolute Magnitudes.* With the adopted distances the absolute magnitudes, corrected for extinction as in RC2, are about $M_T(B) = -18.0$ (LMC) and -16.4 (SMC) (mean : -17.7), in close agreement with the mean absolute magnitudes of their look-alike counterparts ("sosies") at greater distances. In particular NGC 4618, 4625 have $M_T(B) = -18.9$ and -15.5 (mean : -17.2), and NGC 4027, 4027A have $M_T = -18.9$ and -16.8 (mean : -17.8) on the short distance scale. This agreement disappears if the long distance scale of the RSA were adopted (LMC, SMC : -18.4, -17.0; NGC 4618, 4625 : $-19.4, -17.7$; NGC 4027, 4027A : $-21.1, -17.6$).

(e) *HI Line to B-band Luminosity Ratio.* With the revised magnitudes, the distance independent hydrogen index (as defined in RC2), measuring the ratio of 21-cm integrated flux to B-band luminosity, is HI = 2.06 for the LMC (versus 1.3 for the mean of its sosies) and 0.36

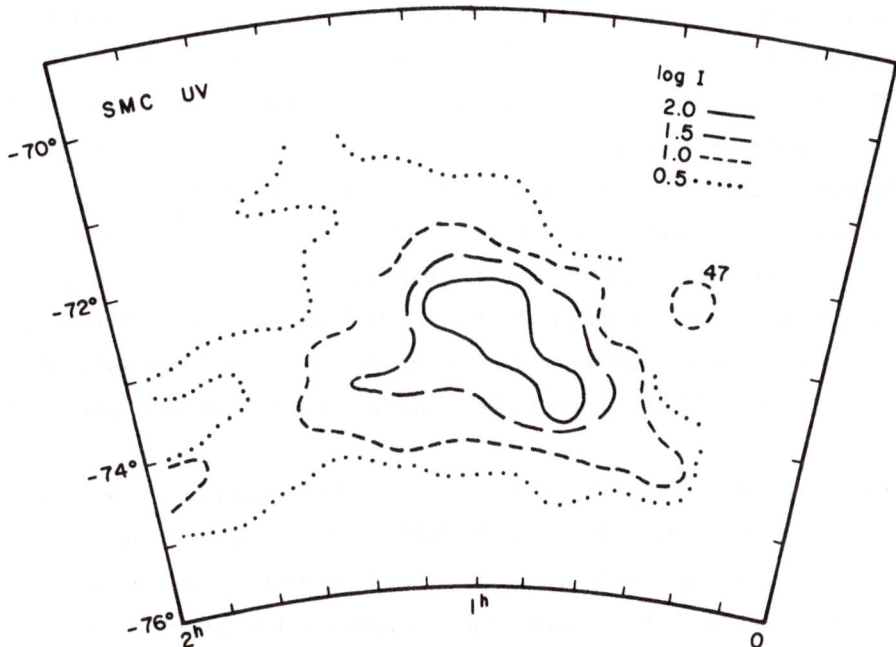

Figure 14. A low-resolution ultraviolet isophote map of the Small Cloud and prominence, derived from the data of Nandy *et al.* (1978).

for the SMC (versus 0.9 for its sosies). Converted to ratios of HI mass to B luminosity, via $\log(M_H/L_B) = 0.02 - 0.4HI$, this indicates that, per unit of luminosity, neutral hydrogen is deficient in the LMC by some 50%, while it is overabundant in the SMC by some 60%. This is consistent with the unusually red color of the LMC for its morphological type, and by the extreme blue color of the core of the SMC. In the latter case, however, separating the HI associated with the main body of the system from that in the prominence is precarious and subject to revision.

(f) Radio Continuum to B-band Luminosity Ratio. The radio index, RI, as defined in RC2, is a magnitude measure of the ratio of the integrated radio continuum flux S_R at the standard frequency (1415 MHz) to the optical B-band flux. The average value of this index for magellanic galaxies is $<RI>= +2.65$ (deV. 1977); with the revised magnitudes of the Clouds the corresponding values are +4.20 (LMC) and +2.29 (SMC) which indicate a near normal radio/light ratio for the SMC, but a deficiency of a factor ~ 4 for the LMC which may be related to its abnormally low gas ratio.

6. ROTATION AND MASS OF THE CLOUDS

(a) The Large Cloud. Rotation of the Large Magellanic Cloud was first suggested by Wilson (1918) as a possible interpretation of his measures of radial velocities of 17 emission nebulosities. An alternative interpretation in terms of a large transverse motion, advanced by Hertzsprung (1920,

1923), was generally accepted during the next three decades (Luyten 1928, Wilson 1944).

The rotational interpretation of the Lick velocities was advocated anew and independently (deV. 1954) after the detection at Mount Stromlo of extensive spiral structure. This interpretation was placed beyond doubt through our detailed analysis of the first observations made at Sydney of the 21-cm line emission from the Clouds (Kerr, Hindman and Robinson 1954, Kerr and deV. 1955, 1956) and was soon confirmed by a discussion of the optical velocities of 26 supergiant stars observed at Pretoria (Feast, Thackeray and Wesselink 1955).

Since that time an ever growing body of new optical and ratio observations has produced somewhat overwhelming information on velocity distributions in the Large and Small Clouds, most recently rediscussed by Feitzinger (1979), Feitzinger and Schmidt-Kaler (1980) and Duval (1983).

In the absence of a good model for the velocity field in asymmetric barred spirals of the Magellanic type, most analyses of the velocity field were based on the assumption of circular symmetry around a fictitious "radio" center (Kerr and de Vaucouleurs 1956) defined as the point on the major axis about which the rotation curve is approximately symmetric. This point, C_R, is displaced from the optical center, C, of the bar by about one degree (~ 0.8 kpc) toward the major arm. It is also the approximate center of symmetry of the optical rotation curve derived from HII regions, supergiant stars, planetary nebulae and star clusters (Feast 1961, 1964, 1968, Webster 1965, 1969). In most analyses it was also assumed that the major axis of the isophotes coincides within errors with the line of nodes of the plane of the disk with the tangent plane to the sky. However, recent analyses by Feitzinger et $al.$ (1980) and Duval (1983), suggest that the two lines may differ by 10 to 20 degrees. Freeman (1983) has reported that the lines of nodes of the systems of young and old clusters may differ by as much as 40 degrees. This suggests that analyses aiming at determining the mass of the LMC should be restricted to the young Population I component most closely associated with the current space orientation of the disk.

The isovelocity contours are illustrated in Figure 15. The maximum rotation velocity, corrected for inclination, is about 100 km/s, reached at a radial distance of 3 to 4 degrees (~ 3 kpc) from C_R. This essentially determines the mass calculated by conventional methods (no dark halo). Most estimates are in the range $9.5 < \log(M/M_\odot) < 10.4$. Many of the determinations calculated in the thin disk approximation and neglecting the mass of regions beyond 4° from the center and the velocity dispersion are only lower limits. With allowance for this incompleteness and reducing all determinations to the distance and inclination adopted here ($i = 30°, \Delta = 50$ kpc), a recent, unpublished review of the data suggests a total mass $\log(M/M_\odot) \sim 10.0 \pm 0.1$: (m.e.), corresponding to a mass/luminosity ratio, f_B =M/L= 4.5, a typical value for late-type galaxies. As usual, most of the uncertainty is in the choice of model, aggravated by the small inclination.

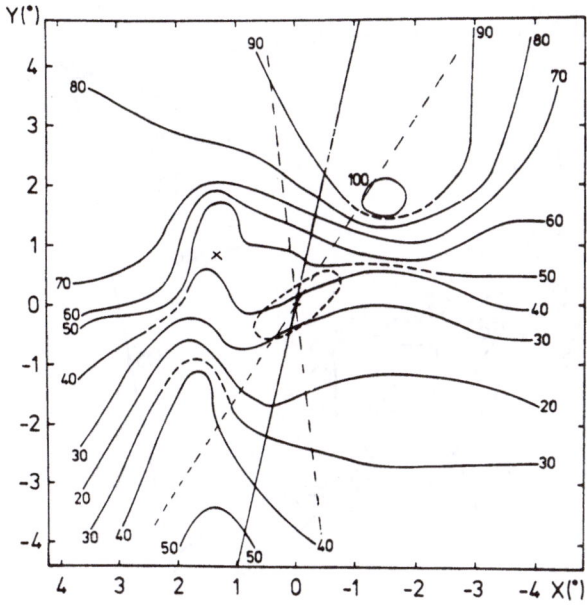

Figure 15. Smoothed isovelocity contours in LMC after Feitzinger *et al.* (1977). Note twist characteristic of barred spirals and distortion near 30 Doradus which is far from rotation center.

A further difficulty is the fact that the velocity field departs significantly from the circular symmetry assumed by the model. Although the nature of the departures due to the presence of an offset bar is now understood in general terms, detailed predictions of the velocity field are not yet available.

With an estimated total HI mass of 6×10^8 solar masses the HI to total mass ratio is $M_H/M_T = 0.06$, a value somewhat lower than typically found in magellanic systems, and about 27% of that of the SMC. This is consistent with the other indications of HI deficiency in the LMC.

(b) The Small Cloud. The rotation of the Small Cloud was first detected in the early Sydney observations (Kerr, Hindman and Robinson 1954) which indicated a strong velocity gradient along the major axis of the optical system (Kerr and deV. 1955, 1956) consistent with the interpretation of the Clouds as rotating disk systems. This was soon confirmed by optical studies of the radial velocities of individual members of the Small Cloud, including supergiant stars (Feast *et al.* 1961, Florsch 1972, Dubois 1975, Thackeray 1978, Ardeberg and Maurice 1979), HII regions (Feast 1970, Dubois 1975), and planetary nebulae (Feast 1968, Webster 1969). Since then the velocity field and HI line profiles have been studied in much greater detail with the 64-m Parkes telescope (Hindman 1967, McGee and Newton 1982 a,b). This new material was summarized and analyzed by Maurice (1979).

The analysis of the velocity field (Figure 16) is complicated by the presence of the gas associated with the prominence and it is difficult to disentangle in the multi-modal line profiles the components associated with the main disk system from those originating in the prominence. Many conflicting "shell" models have been proposed (Hindman 1967, Hindman and Balnaves 1967, Math-

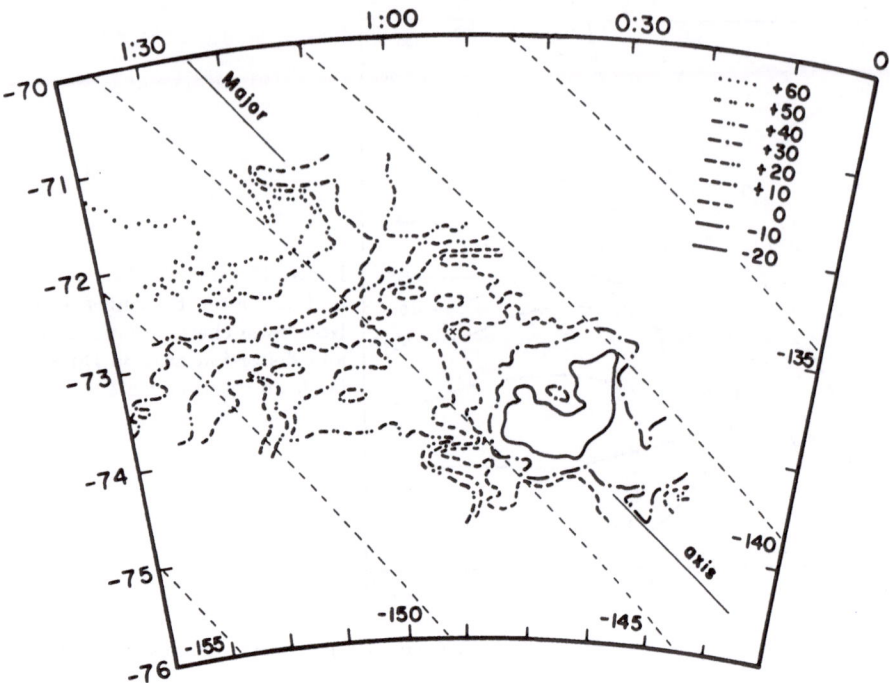

Figure 16. HI isovelocity contours in SMC, after Hindman (1967). C marks bar center.

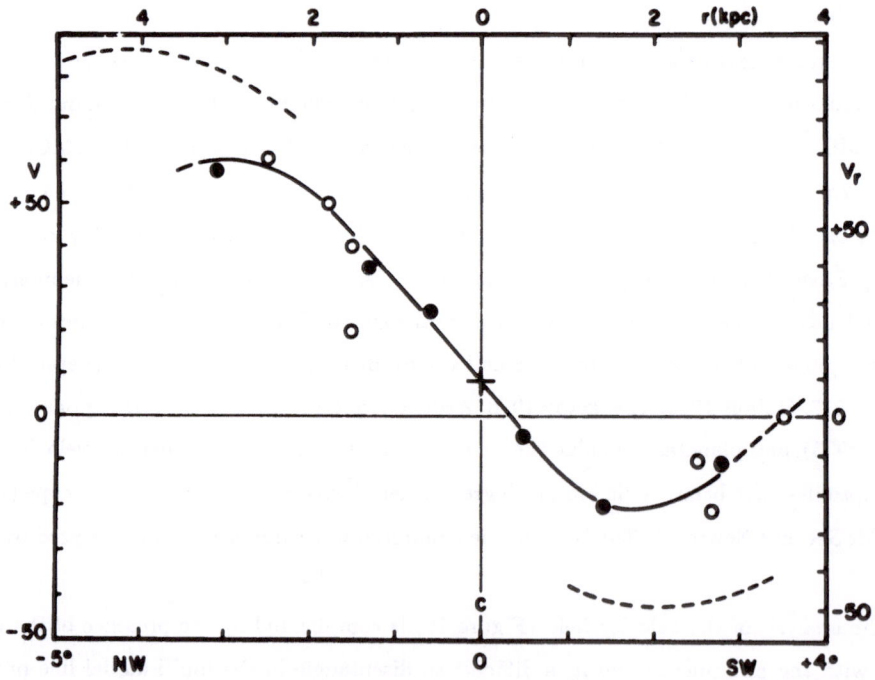

Figure 17. Rotation curve of the Small Cloud along the optical major axis, derived from data by Hindman (1967).

ewson 1977, Maurice 1979). Here we are mainly concerned with the rotation of the disk system whose major axis is in position angle 45°. Although a straightforward analysis of the velocity field in terms of a circularly symmetric rotating disk gives a position angle of 55° for the line of maximum velocity gradient, it is questionable that this represents the true position angle of the line of nodes of the disk system because of the contamination of the profiles by the prominence on the north-east side. A rotation curve derived from the mean velocities of HI line profiles within one degree from the major axis through C_o in p.a. 45° is shown in Figure 17. A slightly different choice of center and position angle does not alter seriously the amplitude of the rotation velocity (the uncertainty in the inclination angle is unimportant in this case) as long as the prominence or shell components are excluded. A maximum rotation velocity (corrected for inclination) of ~ 50 km/s is indicated at a distance of $\sim 2° \sim 2.0$ kpc from the center. This amplitude is consistent with the integrated line profile of the main body as defined by McGee and Newton (1981a) (Figure 18). By simple scaling of the models used for the Large Cloud this implies a total mass $M_T \sim 1.7 \times 10^9$ solar masses and a ratio of mass to luminosity $f_B = M/L \sim 3.2$, not significantly different from that of the Large Cloud.

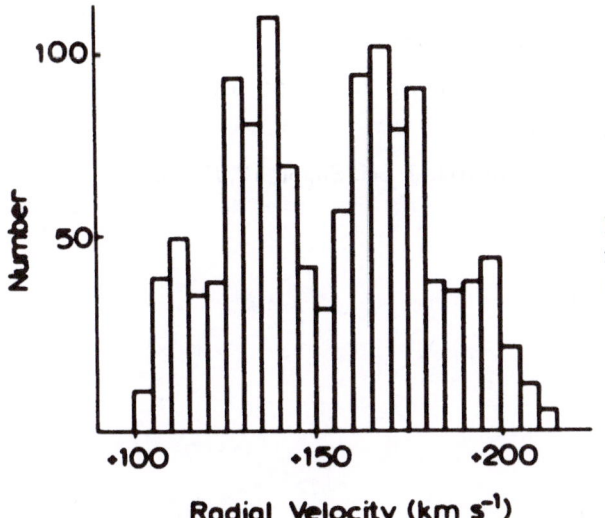

Figure 18. Pseudo-integrated HI line profile of SMC, after McGee and Newton (1981). Note multimodal distribution.

With an estimated total HI mass of 4×10^8 solar masses in the main system of the Small Cloud (Hindman, Kerr and McGee 1963, Hindman 1967), the ratio of HI to total mass is $M_H/M_T = 0.22$, or ~ 3.7 times that of the Large Cloud. This is, possibly, an overestimate because it includes some components of the line profiles that may be associated with the prominence rather than the main body. In any case it is clear that the Small Cloud is much richer in neutral hydrogen than the Large Cloud whether by comparison with its luminosity or its mass. This is also consistent with the color differences between the two Clouds.

(c) Magellanic System. To the masses of the two Magellanic Clouds proper should be added the masses of the stellar prominence (K_1, K_2), of the inter-cloud gas bridge and of the more distant Magellanic Stream which total less than 10^9 solar masses. Thus, the combined mass of the Magellanic System appears to be less than 15×10^9 solar masses. This is probably insufficient to account for the observed amplitude of the warp of the HI layer in the outer parts of the Galaxy. Whether this is evidence for a larger invisible mass associated with the Clouds or not is better left to others to argue.

In any case it is clear that our knowledge of the Magellanic Clouds made a giant step forward when 35 years ago Frank Kerr and his associates detected the 21-cm line emission from these nearest and most fascinating satellites of our Galaxy.

REFERENCES

Ardeberg, A. and Maurice, E. 1980, Astron. Ap. **91**, 53.

Azzopardi, M. 1981, thesis, Toulouse University.

Bruck, M. T. 1975, MNRAS **173**, 327.

— 1978, Astron. Ap. **68**, 181.

Caldwell, J.A.R. and Coulson, I. M. 1986, MNRAS **218**, 223.

Cleary, M.N., Heiles, C., and Haslam, C.G.T. 1979, Astr. Ap. Suppl., **36**, 95.

de Vaucouleurs, G. 1954a, Observatory, **74**, 23 and 158.

— 1954b, in *Problems of the Magellanic Clouds,* Austral. J. Sci. Suppl. **17**, No. 3.

— 1955a, A.J. **60**, 126.

— 1955b, A.J. **60**, 219.

— 1955c, Publ. Astron. Soc. Pacific **67**, 350.

— 1955d, Publ. Astron. Soc. Pacific **67**, 397.

— 1956a, Irish Astron. J. **4**, 13.

— 1956b, Austral. J. Phys. **9**, 90.

— 1957a, A.J. **62**, 69.

— 1957b, Publ. Astron. Soc. Pacific **69**, 252.

— 1959, Lowell Obs. Bull. **4**, 98.

— 1960a, Ap.J. **131**, 265.

— 1960b, Ap.J. **131**, 574.

— 1961, Ap.J. **133**, 405.

— 1964, in *The Galaxy and the Magellanic Clouds,* IAU Symposium No. 20, Canberra, 269.

— 1976,

— 1980, Publ. Astron. Soc. Pacific **92**, 576.

— and Freeman, K. C. 1972, in *Vistas in Astronomy*, ed. A. Beer, Oxford: Pergamon Press, Vol. 14, 163.

— and de Vaucouleurs, A. 1982, unpublished.

— de Vaucouleurs, A. and Buta, R. 1983, A.J., **88**, 939.

Dubois, P. 1975, Astron. Ap. **40**, 227.

— 1980, thesis, University Strasbourg I.

Duval, M.-F. 1983, dissertation, University Aix-Marseille.

Eggen, O. J. and de Vaucouleurs, G. 1956, Publ. Ast. Soc. Pacific, **68**, 421.

Feast, M. W. 1961, MNRAS **122**, 1.

— 1964, MNRAS **127**, 195.

— 1968, MNRAS **140**, 345.

— 1970, MNRAS **149**, 291.

Feast, M.W., Thackeray, A. D., and Wesselink, A.J. 1955, MNRAS **115**, 217.

— 1961, MNRAS **122**, 433.

Feitzinger, J. V., Isserstedt, I., and Schmidt-Kaler, Th. 1977, Astron. Ap. **57**, 265.

Feitzinger, J. V. and Weiss, G. 1979, Astron. Ap. Suppl. **37**, 575.

Florsch, A. 1972, Publ. Obs. Strasbourg, **2**, 129 pp.

Freeman, K. C., Illingworth, G., and Oemler, A., Jr. 1983, Ap.J. **272**, 488.

Gascoigne, S.C.B. and Kron, G. E. 1952, Publ. Astr. Soc. Pacific **64,**, 196.

Gascoigne, S.C.B. and Shobbrook, R. . 1978, Proc. Astron. Soc. Australia, **3**, 285.

Heidmann, J., Heidmann, N., and de Vaucouleurs, G. 1971, Mem. R.A.S. **75**, 85.

Heiles, C. and Cleary, M. N. 1979, Aust. J. Phys. Ap. Suppl. No. 47, 1.

Hertzsprung, E. 1920, MNRAS **80**, 782.

— 1923, MNRAS **83**, 348.

Hindman, J. V. 1964, in *The Galaxy and the Magellanic Clouds*, IAU Symp. 20, eds. F. J. Kerr and A. W. Rodgers, Canberra: Austral. Acad. Sci., 255.

Hindman, J. V. 1967, Austral. J. Phys. **20**, 147.

— and Balnaves, K. M. 1967, Austral. J. Phys. Suppl. No. 4, 38 pp.

—, Kerr, F. J. and McGee, R. X. 1963, Austral. J. Phys. **16**, 570.

Kerr, F. J., Hindman, J. V., and Robinson, B. J. 1954, Austral. J. Phys. **7**, 297.

Kerr, F. J. and de Vaucouleurs, G. 1955, Austral. J. Phys. **8**, 508.

— 1956, Austral. J. Phys. **9**, 90.

Luyten, W. J. 1928, Proc. Nat. Acad. Sci., Washington **14**, 24l.

Martin, W. L., Warren, P. R., and Feast, M. W. 1979, MNRAS **188**, 139.

Mathewson, D. S., Schwarz, M. P., Murray, J. D. 1977, Ap. J. (Letters), **217**, L5.

Maurice, E. 1979, thesis, University Aix-Marseille.

McGee, R. X. 1964, Austral. J. Phys. **17**, 515.

— and Milton, J. A. 1966a, Austral. J. Phys. **19**, 343.

— 1966b, Austral. J. Phys. Astrophys. Suppl. No. 2.

McGee, R. X. and Newton, L. M. 1982a, Proc. Astron. Soc. Austral. **4**, 189.

— 1982b, Proc. Astron. Soc. Austral. **4**, 308.

Nandy, K., Morgan, D. H., and Carnochan, D. J. 1978, MNRAS **186**, 421.

Shapley, H. 1940, Harvard Obs. Bull. No. 914.

— 1950, Publ. Obs. University Michigan **X**, 79.

— and Lindsay, E. M. 1963, Irish Astron. J. **6**, 74.

Shostak, G. S. and van der Kruit, P. C. 1982, Astron. Ap. **105**, 351.

Thackeray, A. D. 1978, MNRAS **184**, 699.

Webster, B. L. 1965, in *Symposium on the Magellanic Clouds*, Mount Stromlo, 29.

— 1969, MNRAS **143**, 79 and 97.

Westerlund, B. E. 1964, MNRAS **127**, 429.

Wilson, R. E. 1917, Publ. Lick Obs. **13**, 187.

— 1944, Publ. Astron. Soc. Pacific **56**, 102.

A CO SURVEY OF THE LARGE MAGELLANIC CLOUD

Patrick Thaddeus
Harvard-Smithsonian Center for Astrophysics
60 Garden St., Cambridge, MA 02138

Frank Kerr, in collaboration with J.V. Hindman and B.J. Robinson (1954), undertook the first 21-cm survey of the LMC, so I thought this meeting to commemorate Frank's retirement was appropriate for a brief presentation of the results of the first CO survey of molecular clouds in the LMC. A few CO observations toward some of the more prominent Population I objects in the LMC had been done prior to our work or were underway (Israel 1984; Israel *et al.* 1986; and references therein), but no systematic survey existed. It was clear to us from the previous work that CO is weak in the LMC relative to the Milky Way, owing presumably to the lower metallicity, and that a sensitive receiver and long integration times would be required to accomplish a useful inventory of the molecular clouds which the flamboyant star formation in the LMC suggested must surely exist.

Our survey was done with an improved copy of the 1.2 m millimeter-wave telescope with which we have been studying molecular clouds with CO in the northern sky since 1975 (Cohen *et al.* 1986). The southern telescope was installed in Chile on Cerro Tololo late in 1982, and the LMC survey was assigned a high priority: roughly 6 hours/day for the next two years were dedicated to observing its inner $6° \times 6°$ on a $7!5$ grid, with a typical integration of about 40 min per position. The line surveyed was the standard $1 \rightarrow 0$ CO transition at 115 GHz used for nearly all large-scale molecular cloud work in both the Milky Way and external galaxies. The beamwidth of the telescope at that frequency is $8!8$, so the sampling interval is somewhat less than one beamwidth (85%). Our spectrometer was a standard filter bank with 256 channels, each 0.5 MHz wide, for a resolution in radial velocity of 1.3 $km\ s^{-1}$ and a range of 333 $km\ s^{-1}$. Integrations were generally continued until the rms noise per channel was reduced to 0.06 K in radiation temperature (antenna temperature corrected for the beam efficiency of 0.82), but a few positions of particular interest were observed for as long as 10 hr.

Like most of the projects undertaken with the Chile Telescope the LMC survey was the result of a collaboration between our laboratory and scientists from the University of Chile. Richard Cohen was responsible for building the telescope and installing it in Chile; he provided a superb instrument—one that was extremely reliable and rugged with a sensitive liquid nitrogen cooled

Schottky heterodyne receiver well suited for operation at a distant site—and he also planned and supervised the observations. Joe Montani took most of the observations and Monica Rubio some, and she, with Guido Garay and Tom Dame, did the data analysis. Like other spectral line surveys, ours produced a data cube—intensity as a function of radial velocity and the two coordinates of position—which is not readily summarized in a single map; the most revealing summary map of the survey, shown in Figure 1, is obtained by integrating in radial velocity, since the integrated intensity, W_{CO}, is the best measure we have of the "total" molecular column density, N_{H_2} (with a contribution from He).

As Figure 1 shows, CO emission actually is extremely widespread in the LMC, but, as we suspected, it is also extremely weak, seldom exceeding 0.2 K in peak line intensity. Considerable care was required in defining the range of velocity integration to avoid being swamped by noise. Having determined by inspection that the CO line in terms of radial velocity is almost invariably contained in the 21-cm profile, we integrated over a window 30 km s^{-1} wide centered on the 21-cm peak, except in the vicinity of 30 Dor where it was widened to 40–50 km s^{-1} to take in a surprisingly large amount of high-velocity molecular gas ($\sim 4 \times 10^6 M_\odot$) apparently accelerated by the many young stars. Pains were taken to demonstrate that no detectable CO emission lay outside our range of velocity integration, but this obviously should be checked with future surveys at higher sensitivity and angular resolution. The main molecular clouds of the LMC are well defined at the full survey resolution, but to enhance the signal-to-noise and best exhibit the overall distribution of clouds, Figure 1 has been slightly smoothed to a resolution of 12'. Nearly 10% of the inner 38 deg^2 of the LMC we surveyed exhibits CO emission at this resolution. More probably exists over a larger area, since much of the emission we did detect is near the limit of our sensitivity; but beyond the boundary of the survey (dashed line) there is probably little, since nearly all the classical Population I objects of the LMC fall within.

With the aid of the unsmoothed data, the CO emission of the LMC has been partitioned into 40 molecular clouds, most fairly well defined, and a tentative cloud catalog produced with positions, peak CO line temperatures, radial velocities, velocity widths, cloud radii, and masses. In deriving masses from the CO luminosity, we followed the procedure generally adopted for our own and other galaxies and assumed that W_{CO} is proportional to N_{H_2} (e.g., Bloemen et al. 1986). We did not, however, adopt the Galactic value for the ratio $N_{H_2}/W_{CO} \equiv X_G$, because of the fourfold lower metallicity in the LMC and because that ratio yielded cloud masses systematically an order of magnitude smaller than virial masses, $5R\Delta v^2/8G \ln 2$. Calibrating X directly with the virial masses requires the measurement of cloud radii that are often near the limit of our angular resolution; we chose instead to calibrate against the empirical CO *luminosity-linewidth* relations (Dame *et al.* 1986), on the assumption that in the Galaxy and the LMC these are the result of

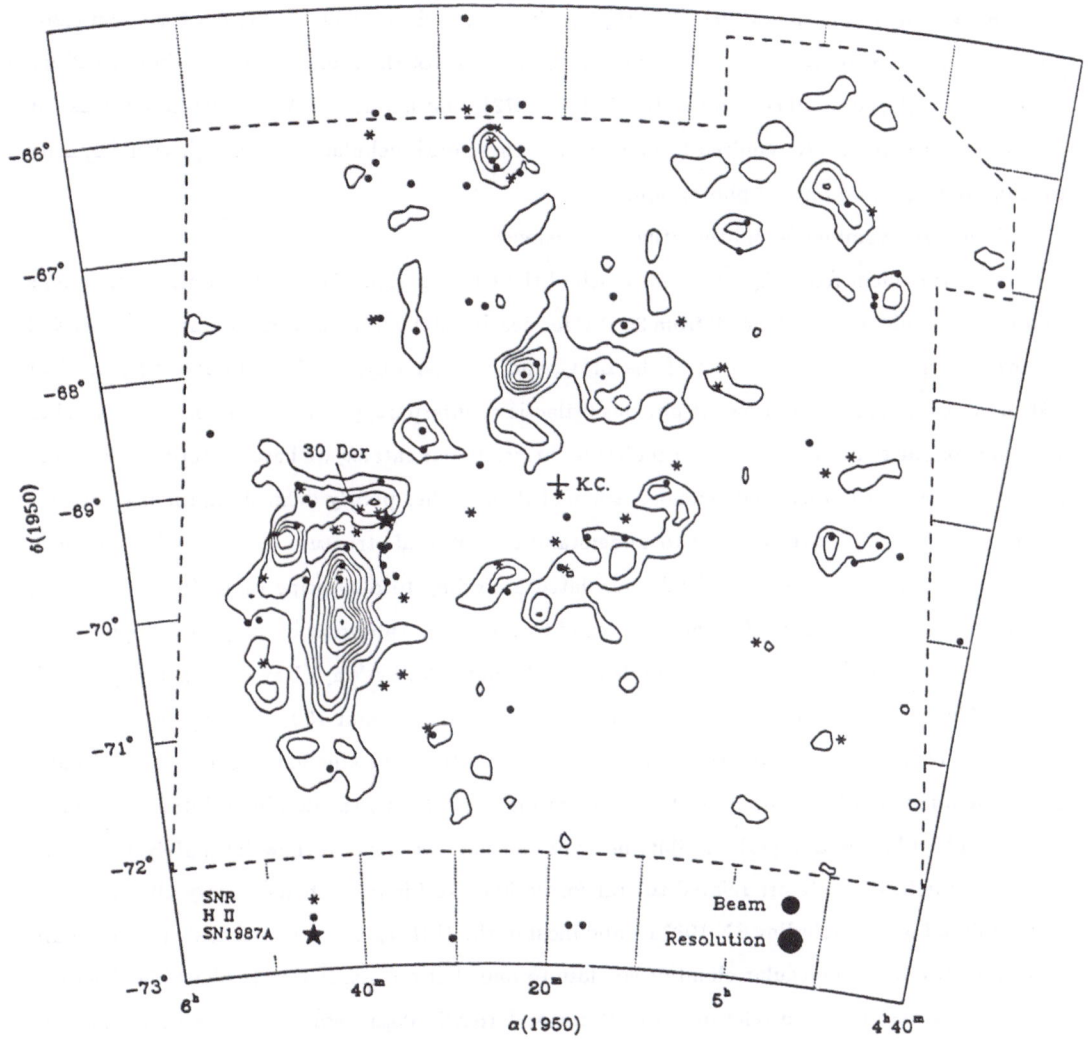

Figure 1: Map of velocity-integrated CO intensity in the LMC. The contour interval is 0.38 K km s^{-1}, which is about the 1.6 σ noise level. SNR's (Mathewson *et al.* 1983; 1984; 1985), supernova 1987a, and H II regions (6 cm continuum sources not identified as SNR's; McGee, Brooks, and Batchelor 1972) are shown for comparison. K. C. marks the kinematic center of the LMC (de Vaucouleurs and Freeman 1973).

an identical underlying *mass-linewidth* relation. This procedure yielded $X_{LMC} = 6X_G$ and cloud masses in satisfactory agreement with virial masses. Some of the molecular clouds in our catalog have been cataloged as dark nebulae by Hodge (1972), and a value of X at least as large as the one we derive is probably required to explain why these dark nebulae have enough visual opacity to stand out at all on optical photographs.

Some brief comments and tentative conclusions:

1. It is evident from Figure 1 that much of the CO emission of the LMC comes from a huge complex of clouds that stretches S from 30 Dor for nearly 2400 pc. This extraordinary object is well removed from the kinematic center of the LMC, and in other ways too is quite unlike the nuclear disk of our Galaxy; indeed, it has no obvious Galactic counterpart, perhaps being best compared to a *segment* of one of our crowded inner molecular arms. In our catalog we have tentatively dissected this complex into 10 individual "clouds," some of them rather poorly defined, and it is clear that higher resolution data are required to understand the internal structure of the 30 Dor complex. The masses of these component clouds calculated according to our recipe range from 3×10^6 to $18 \times 10^6 M_\odot$, for a total mass for the complex of $60 \times 60^6 M_\odot$—an order of magnitude higher than the masses of the largest molecular complexes in the outer Milky Way (Dame *et al.* 1986).

2. The total mass of the molecular clouds in the LMC is about $150 \times 10^6 M_\odot$, or about 30% of the mass of the H I. Since the H I extends much farther from the kinematic center, the molecular gas in the inner $R \approx 3$ kpc where CO is observed may actually be the dominant interstellar mass.

3. The LMC is a superb exhibit, perhaps the best we have, of how intimately molecular clouds on a galactic scale are related to star formation. As Figure 1 shows, nearly all the known SNR's of the LMC—including SN 1987a—and most of the H II regions detected as 6-cm continuum sources lie toward a molecular cloud or so close to one that an association is plausible. There is also an extremely good correlation with 21-cm and IRAS 100μm emission: in either case, CO delineates the *core* of the distribution; it almost always appears when the H I column density exceeds 2×10^{21}cm^{-2} and the 100μ flux exceeds 50 MJy sr^{-1}. A great deal more could be said on this general topic; an extensive intercomparison of CO and the other Population I in the LMC is underway.

4. There is a wealth of kinematic information in the CO data. From the radial velocities in our cloud catalog we have derived a rotation curve (Figure 2) which agrees extremely well with that found, e.g., by Feitzinger (1980) from a variety of other kinematic tracers, and we have derived as well quite accurate values for the systemic velocity of the LMC relative to the Galactic center $(69.1 \pm 0.8$ km s$^{-1})$ and for the position of the kinematic line of nodes of the LMC $(168° \pm 3°)$. Molecular clouds are of particular kinematic interest because they constitute a "cold" population; their line-of-sight velocity dispersion relative to the thin circularly rotating disk characterized by

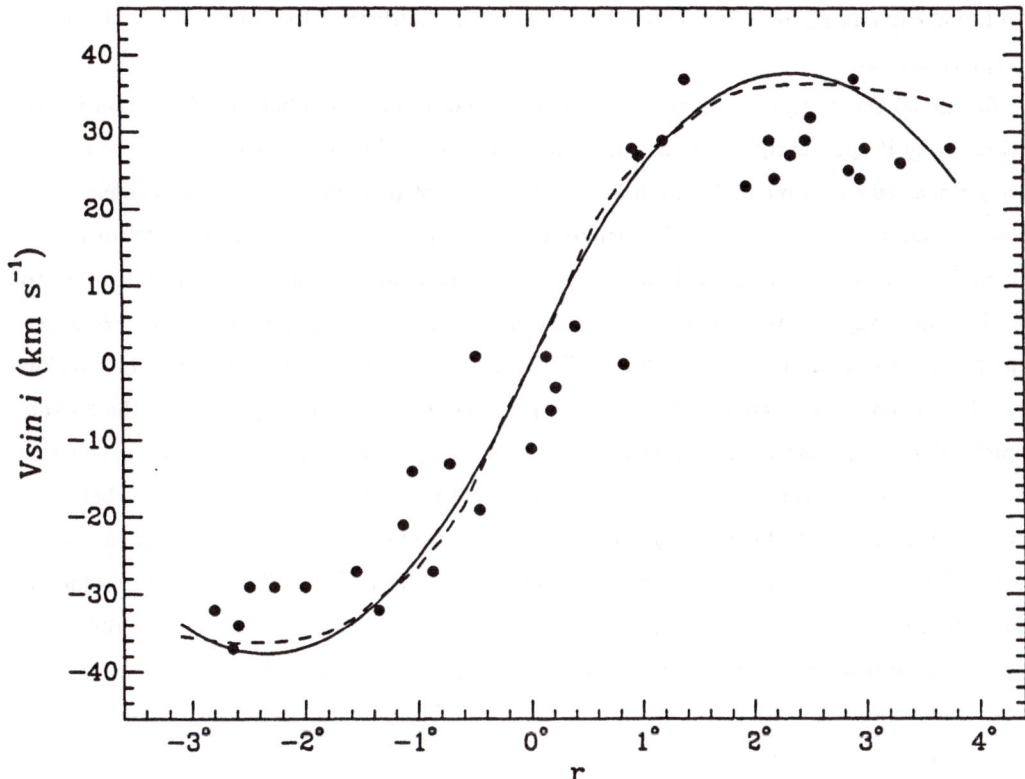

Figure 2: Radial velocities for LMC molecular clouds (LSR velocities corrected for a solar motion of 225 km s^{-1} around the Galactic center; Kerr and Lynden-Bell 1986) plotted versus projected angular distance along the kinematic line of nodes; the LMC systemic velocity has also been subtracted. The kinematic center was assumed to be at $\alpha = 5^{\mathrm{h}}\ 20^{\mathrm{m}}$, $\delta = -69°$ (de Vaucouleurs and Freeman 1973). The solid curve is the best-fit rotation curve (i.e., radial velocity versus angular distance along the kinematic line of nodes), given by the equation $V \sin i = 32.0\ r - 6.81\ r^2$. The dotted curve is the rotation curve of Feitzinger (1980; adapted from his Figure 15, using $i = 32°5$), determined from a variety of kinematic tracers. The scatter of points about the rotation curve (*solid line*) is not representative of the true dispersion of the fit: because of the differential rotation, only clouds lying precisely along the line of nodes are expected to fall on this line; the rest will have velocities somewhat less in absolute value.

the rotation curve in Figure 2 is only 6.2 km s^{-1}—much smaller than that of any other kinematic tracer, including H I.

5. The 30 Dor complex is, I suspect, a crucial clue to understanding star formation in the LMC and why it apparently occurs in large bursts. It is plausible that this object will be almost entirely dispersed on a time scale of a few tens of millions of years by the wholesale conflagration of star formation now observed in the vicinity of 30 Dor and around the entire northern rim of the complex. The end result is likely to be not merely the kind of stellar associations familiar to us in the Milky Way, but the vast collections of young stars over many hundreds of parsecs which Shapley called Constellations. Constellation III, located at the upper left in Figure 1 in a region where there is not a large amount of molecular gas, may be a good example of how the 30 Dor complex will evolve; it lies at the center of an extremely large hole in the H I distribution which is associated with a perturbation in the H I velocity field of about 15 km s^{-1}. This perturbation is most plausibly the result of expansion, and if so the energy of expansion is comparable to that of the extensive high velocity gas we observe near 30 Dor—of order 10^{52}ergs. This and other evidence suggests that star formation in the LMC is highly punctuated—characterized by the build up of exceptionally large molecular cloud concentrations, only one or a few at a time, which once formed are consumed and dissipated fairly quickly by a wave or burst of star formation. How and why these large gas concentrations form is essentially unknown.

REFERENCES

Bloemen, J. B. G. M., *et al.* 1986, *Astr. Ap.*, **154**, 25.
Cohen, R. S., Dame, T. M., and Thaddeus, P. 1986, *Ap. J. Suppl.*, **60**, 695.
Dame, T. M., Elmegreen, B. G., Cohen, R. S., and Thaddeus, P. 1986, *Ap. J.*, **305**, 892.
de Vaucouleurs, G., and Freeman, K. C., 1973, *Vistas Astron.*, **14**, 163.
Feitzinger, J. V. 1980, *Space Sci. Rev.*, **27**, 35.
Hodge, P. W. 1972, *Pub. Astron. Soc. Pacific*, **84**, 365.
Israel, F. P. 1984, in IAU Symposium 108: *Structure and Evolution of the Magellanic Clouds*, ed.
    S. van den Bergh and K. S. de Boer (Dordrecht: Reidel), p. 139.
Israel, F. P., and Burton, W. B. 1986, *Astr. Ap.*, **168**, 369.
Kerr, F. J., Hindman, J. V., and Robinson, B. J. 1954, *Australian J. Phys.*, **7**, 297.
Kerr, F. J., and Lynden-Bell, D. 1986, *M. N. R. A. S.*, **221**, 1023.
Mathewson, D. S. *et al.* 1983, *Ap. J. Suppl.*, **51**, 345.
Mathewson, D. S. *et al.* 1984, *Ap. J. Suppl.*, **55**, 189.
Mathewson, D. S. *et al.* 1985, *Ap. J. Suppl.*, **58**, 197.
McGee, R. X., Brooks, J. W., and Batchelor, R. A. 1972, *Austr. J. Phys.*, **25**, 581.

SOME SURPRISES IN THE DYNAMICS OF M33 AND M31

Vera C. Rubin
Department of Terrestrial Magnetism,
Carnegie Institution of Washington, Washington, D.C.

I. INTRODUCTION

Part of the fun and joy of doing extragalactic astronomy in Northwest Washington has been having Frank Kerr and his colleagues at the University of Maryland as neighbors. So I am offering this contribution as a thank you to Frank for many years of friendship. I am also stretching the subject matter of the symposium, stretching it enough so that M33 and M31 can be considered the far outer parts of our Galaxy.

One arcsec subtends 0.04 pc at the center of our Galaxy, 1 arcsec subtends 3.5 pc at M33. But 1 arcsec subtends 100 pc at the Virgo cluster. Thus our resolution from the ground, when we study M31 and M33, exceeds the resolution which will be available in studying Virgo cluster galaxies from the Hubble Space Telescope. For this reason, I have returned to studies of the inner parts of M33 and M31, to see what details we can learn of their morphology, chemistry, and dynamics, using the highest resolution available with modern detectors.

II. M33

I show in Fig. 1 an Hα image of the central region of M33, taken by Malcolm Smith with the Cerro Tololo 4-m telescope. The semistellar nucleus is unremarkable, and barely distinguishable from the superposed stars. The near-nuclear region abounds with Hα knots, spheres, and shells, but these do not appear to be organized into any large-scale over-all patterns at the light levels displayed in this image.

The chemistry of the nucleus is nearly unique among spiral galaxies, in showing Hα in deep sharp absorption and [NII]λ6548 and λ6583 in emission (Rubin and Ford 1987). However, this uniqueness is related to its proximity. For galaxies at larger distances, a nuclear spectrum contains light from the nucleus as well as light from the disk, so the disk Hα emission blends with the nuclear absorption to produce an Hα emission feature (Keel 1983, Filippenko and Sargent 1985). Immediately beyond the nucleus to the SW, prominent emission lines of Hα and [NII] comprise a characteristic disk spectrum. However, there is a curious SW/NE asymmetry. In contrast to the strong disk emission to the SW of the nucleus, NE of the nucleus the [NII] lines are weak, and Hα shows weakly in absorption. Hence, the spectrum NE resembles more a

Fig. 1. The nucleus of M33, from an Hα plate of M33 taken by Malcolm Smith at the Cerro Tololo 4-m telescope. The nucleus is the stellar-like object above the center; the blobby regions are Hα knots. The field shown covers about 2 arcmin; N is to the top, E to left.

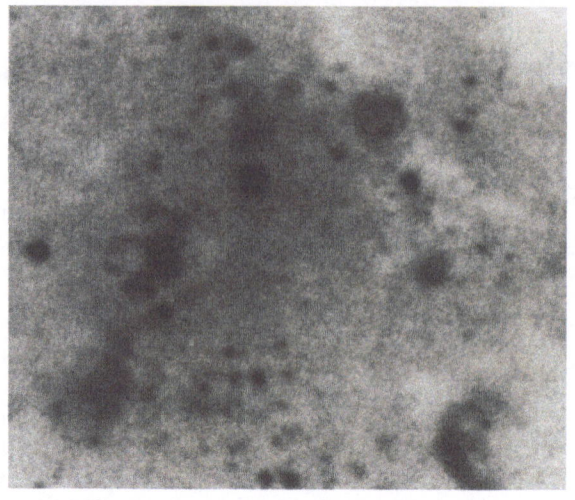

spectrum from the nucleus than a spectrum from the disk. Yet the decrease in surface brightness with distance from the nucleus is symmetric from the NE to the SW, making it unlikely that nuclear absorption is producing the curiously asymmetric line ratios. This observation gives added evidence to the supposition that nuclei of galaxies are chemically distinct entities, with properties signficantly different from their surrounding disks.

The asymmetry in line intensity ratios is coupled with a corresponding asymmetry in the velocities of the ionized gas. From a spectrum taken by Kent Ford and me with the Palomar double spectrograph plus CCD detector along the minor axis of M33, I have measured line-of-sight velocities. There is a steep velocity gradient across the nucleus to the NE, where the emission extends in to the edge of the nucleus. On the corresponding region to the SW, the nuclear gradient in the emission-line velocities is not present, but the velocities are virtually constant. From the velocities shown in Fig. 2(left), the center of dynamical symmetry is determined, and is located not at the nucleus, but it is displaced a few arcsec ($\sim$ 7 pc) toward the NE.

Apart from this asymmetry, the rotation curve shows the characteristic steep nuclear velocity gradient and the relatively flat velocities beyond. Yet here we are observing only the inner 2' of the total 70' visible diameter of M33. Why, then, are the velocities constant, when they represent such a small fraction, both in radius and in velocity, of the overall rotation? I show in Fig. 2(right) the entire rotation curve of M33 from the 21-cm observations of Newton (1980). For the 21-cm observations, the limited spatial resolution makes the inner velocities a guess, at best. The points plotted at small R in Fig. 2 (right) are the same points as shown in Fig. 2(left). From Newtonian dynamics, a flat rotation curve implies that density is falling as $1/r^2$, while a linearly rising rotation curve implies that the density is constant. Hence the small inner plateau suggests the presence of an inner mass component, $M(R<200 \text{ pc}) \sim 2 \times 10^7 \text{ M}_\odot$. It is the combination of

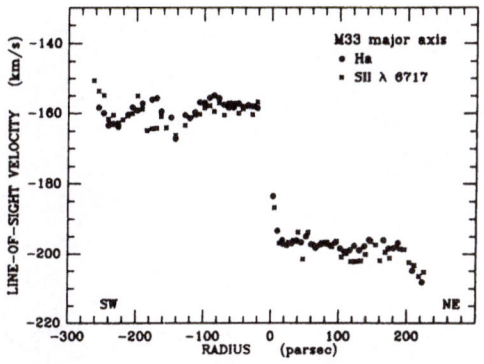 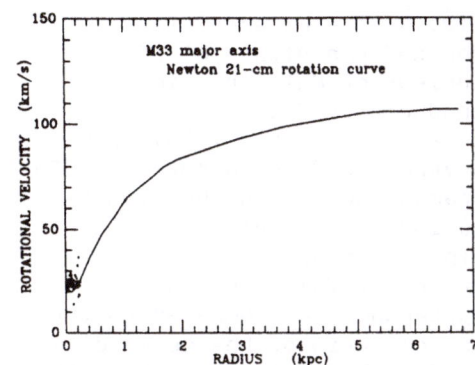

Fig. 2 (left) Observed velocities in the near-nuclear region of M33,
plotted as a function of distance from the semistellar nucleus. The
asymmetrical velocities near r=0 indicate that the center of dynamical
symmetry is displaced a few arcsec NE. (right) Rotation velocities in
M33 from the 21-cm observations of Newton (1980); the curve for r<0.1
kpc has been deleted. The splotch of points near the nucleus are all the
velocities from Fig. 2 (left).

the falling velocities produced by the gravitational field of this inner
mass, plus the rising velocities produced by the mass in the disk, which
combine to produce the approximately flat velocities. This central mass
is only 1/100 of the mass interior to 200 pc in our Galaxy, so is not a
remarkable entity (Oort, 1977)

It seems likely that M33, an Sc II-III galaxy with minimal bulge,
has a shallow central potential in which the relatively low mass nucleus
sloshes around. But the asymmetries we observe in M33, both in the
dynamics and in the chemistry, are representative of asymmetries which
are observed in numerous galaxies including our own, which turn up when
the dynamics and morphology are subjected to a detailed look. Another
such galaxy is M31, which I discuss below.

III. M31

During a period spanning five years, Ciardullo and his associates
regularly obtained CCD images of the near-nuclear regions of M31 in
order to search for nova. One filter used covered the [NII] and Ha
emission. The frames were taken with the Kitt Peak 36-inch telescope.
Ultimately, hundreds of frames were available, from which Ciardullo has
produced a composite image of the ionized gas near the nucleus of M31.
This image is reproduced in Fig. 3. (An earlier less detailed image has
been published by Jacoby et al. 1985). The bright spiral covers a region
of about 4' along the major axis, or about one-thirtieth of the optical
disk of M31.

Morphologically, the ionized gas resembles a turbulent spiral, with
finely patterned filaments and neatly parallel features. Overall, the
aspect ratio of the gas appears more face-on than does the outer disk of

Fig. 3. The Hα and [NII] ionized gas within 8' of the nucleus of M31. This image was produced by Ciardullo (Ciardullo et al. 1988) from several hundred narrow-band CCD frames taken with the Kitt Peak 36-inch telescope. Note the fine filamentary spiral, and its relatively face-on appearance, as compared to the aspect ratio of the outer disk of M31. N is to top, E to left.

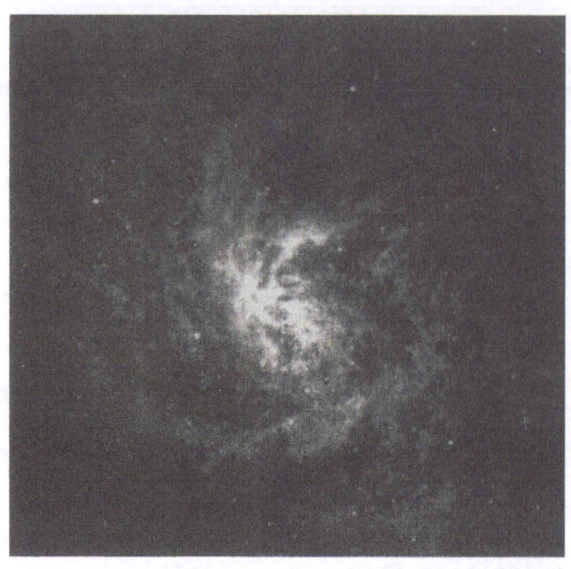

stars and gas, for which i=77°. If we approximate the distribution of the gas as a disk, then the NE part of the spiral is inclined about 45 degrees to the line-of-sight; the SW region is significantly more face on. In particular, the SW arm, which extends over 500 pc in length, appears almost circular on the plane of the sky, suggesting an inclination of no more than 20°. This asymmetry, NE to SW, is one of the many striking features of Fig. 3.

The velocity field of the excited gas near the nucleus of M31 was studied by Rubin and Ford (1971), without the benefit of an image of the nuclear gas. Interestingly, one major feature of the velocity field is a severe asymmetry. On the NE, a pattern typical of a rotating disk is observed; the viewing angle suggested by the velocity pattern is near 45°. To the SW, the velocity pattern is almost flat, as would be expected in viewing a nearly face-on disk. Hence both the morphology and the dynamics suggest that the small central gas disk in M31 in not in the principal plane of the galaxy, and that this inner gas disk experiences a pronounced twist from the NE to the SW.

The gas layer is very thin, probably less than 80pc thick. This is indicated by the narrow emission lines. We show in Fig. 4 a spectrum in the Hα region, taken along the major axis of M31 with the Palomar 200-inch double spectrograph. The narrow Hα emission sits within a broad Hα absorption line arising from the combined light of the stellar population. An indication of the differences between the stellar and the gas velocities can be obtained by noting that the emission lines weave across the broad absorption. The intensity ratio of the two SII lines decreases rapidly from the nucleus, indicating that the gas density falls off by at least two orders of magnitude in the central arcmin of the galaxy.

Fig. 4. The Hα region of
the spectrum near the
nucleus pf M31, taken with
the Palomar 200 and double
spectrograph. The narrow Hα
emission sits within the
broad Hα absorption arising
from the bulge stars. The
emission lines at longer
and shorter wavelength than
Hα arise from [NII].

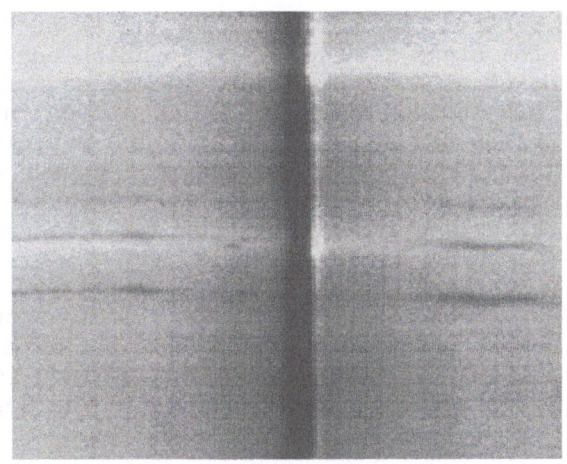

The mass of this ionized gas near the nucleus is small, of order
1500 M☉ (Jacoby et al. 1985). Brinks (1984) observed no H I within 500
pc of M31's nucleus; Stark (1985) found little or no CO. Combined with
estimates of the near nuclear dust (Gallagher and Hunter 1981), an upper
limit of M<10^6 M☉ seems likely for the total interstellar mass in the
region. This small mass would collect in less than 10^7 years from
planetaries alone. Hence it is likely that a galactic wind (Mathews and
Baker 1971) rids M31 of most of its nuclear gas. This expanation is
consistent with expansion motions observed in its nuclear gas (Rubin and
Ford 1971; Ciardullo et al. 1988).

For a galactic wind to operate, the heating from supernovae must
overcome the forbidden line cooling of the gas. Since the efficiency of
collisional cooling goes as the square of the density, in the innermost
regions of M31 the gas should be dense enough to resist the wind. Here,
the gravitational collapse of the gas into a disk is expected. Away from
the nucleus, as the gas density decreases, the rate of cooling becomes
less and less efficient. Since the bulge is about 20 times brighter than
the disk in this region, the gravitational potential of the bulge must
far exceed that of the disk. Hence it is not surprising that the small
gas disk aligns itself in a plane not that of the outer disk.

Overall, it is the asymmetries near the nucleus of M31 which seem
most remarkable. Yet asymmetrical features, which are seen on all scales
and in all components of M31, have been insufficiently emphasized
previously. Dressler and Richstone (1987) have shown that M31's
kinematical center is displaced 0.5" (2 pc) SW of the optical nucleus.
It has long been known that eccentricities and position angles of M31's
elliptical isophotes change rapidly within the first few minutes from
the nucleus (Kent 1983). Our observations described above demonstrate a
morphological and kinematical asymmetry in the bulge's ionized gas, and
suggest that a warp has twisted the gas south of the nucleus into a more
face-on orientation. On scales of a kiloparsec, McElroy's (1983)
observations of the stellar velocity field shows a gross difference

between the velocities NE and SW of the nucleus. North of the nucleus, the stars show a rising rotation curve; south of the nucleus the curve is much flatter. These velocities remarkably mimic those in the ionized gas, but on a larger scale. And on the largest scales of several kiloparsecs, the HI velocities (Brinks 1984) again show a significant displacement of the kinematic center of the galaxy. While a triaxial bulge structure in M31 might produce some of these effects, asymmetries clearly exist also in the disk, where the orbits are believed to be circular.

Studies of the nuclear regions of the nearest spirals offer an opportunity to learn the details of morphology, chemistry, and kinematics, that cannot be obtained for more distant objects. We are continuing the observations in the hope of solving some of the puzzles we have uncovered.

References

Brinks, E. 1984, Ph. D. diff., Rijksuniversiteit te Leiden.
Ciardullo, R., Rubin, V. C., Jacoby, G. H., Ford, H., and Ford, W. K. Jr., 1988, A. J., submitted.
Dressler, A. and Richstone, D. O. 1987, Ap. J., in press.
Filippenko, A. and Sargent, W.L.W. 1985, Ap. J. Suppl., **57**, 503.
Gallagher, J. S. and Hunter, D. A. 1981, A. J., **86**, 1312.
Jacoby, G. H., Ford, H., and Ciardullo, R. 1985, Ap. J., **290**, 136.
Keel, W. C. 1983 Ap. J., **268**, 632.
Kent, S. A. 1983, Ap. J., **266**, 562.
Oort, J. 1977, An. Rev. Astron. Ap., **15**, 295.
Mathews, W. G. and Baker, J. C. 1971, Ap. J., **170**, 241.
McElroy, D. B. 1983, Ap. J., **270**, 485.
Newton, K. 1980, Mon. Not. Roy. Ast. Soc., **190**, 689.
Rubin, V. C. and Ford, W. K. Jr. 1971, Ap. J., **170**, 25.
Rubin, V. C. and Ford, W. K. Jr. 1986, Ap. J. Letters, **305**, L35.
Stark, A. A. 1985, in I.A.U. Symp. 106, The Milky Way Galaxy, ed H. van Woerden, R. J. Allen and W. B. Burton (Dordrecht:Reidel), p. 445.

MOLECULAR ARMS IN THE OUTER DISK OF M51: STRUCTURE AND ORIGINS

F. Verter
NAS/NRC Research Associate
NASA Goddard Space Flight Center
Greenbelt, MD 20771

M. L. Kutner
Physics Department
Rensselaer Polytechnic Institute
Troy, NY 12180-3590

I. OBSERVATIONS

This is the first study to focus on molecular gas in the <u>outer</u> disk of an <u>external</u> spiral.   We have used the NRAO 12m telescope to make a CO(2-1) map of 23 points covering the outermost southwest arm/interarm region in M51. Our 30" beam is adequate to resolve arm and interarm regions in this portion of the M51 disk. Our map points are primarily spaced every beamwidth on a square grid, except at the inner edge of the outermost arm, where observations were taken at half-beamwidth spacing to confirm the peaking of the molecular emission on the dust lane along the inner edge of the arm. The observations range in galactocentric radius from 1.95 to 4.25 arcmin, or 5.4 to 11.9 kpc at a distance of 9.6 Mpc. In figure 1 a contour map of our data is superimposed on a photo of M51.

Throughout the inner disk of M51 ($R \leq 2.6$ arcmin), observations with resolutions of 45" to 11" have found the molecular arm/interarm contrast to be about 1.5 (Rydbeck et al. 1985, Lord et al. 1987). Resolution of 11" should be adequate to resolve interarm regions, such as they are, in the inner galaxy.   In the outer disk of M51, we find that the molecular arm/interarm contrast increases with galactocentric radius, with a mean integrated emission ratio of 4 in our map region. Figure 2 plots integrated CO emission as a function of M51 galactocentric radius for the detections of Rydbeck et al. (1985; open squares) and both the Verter and Kutner detections (closed circles) and upper limits (closed triangles). Rydbeck et al. (1985) observed CO(1-0) emission in the northwest quadrant at 33" resolution, whereas Verter and Kutner observed CO(2-1) emission in the southwest quadrant at 30" resolution.   The overall shape of CO emission vs. radius appears roughly exponential, as first noted by

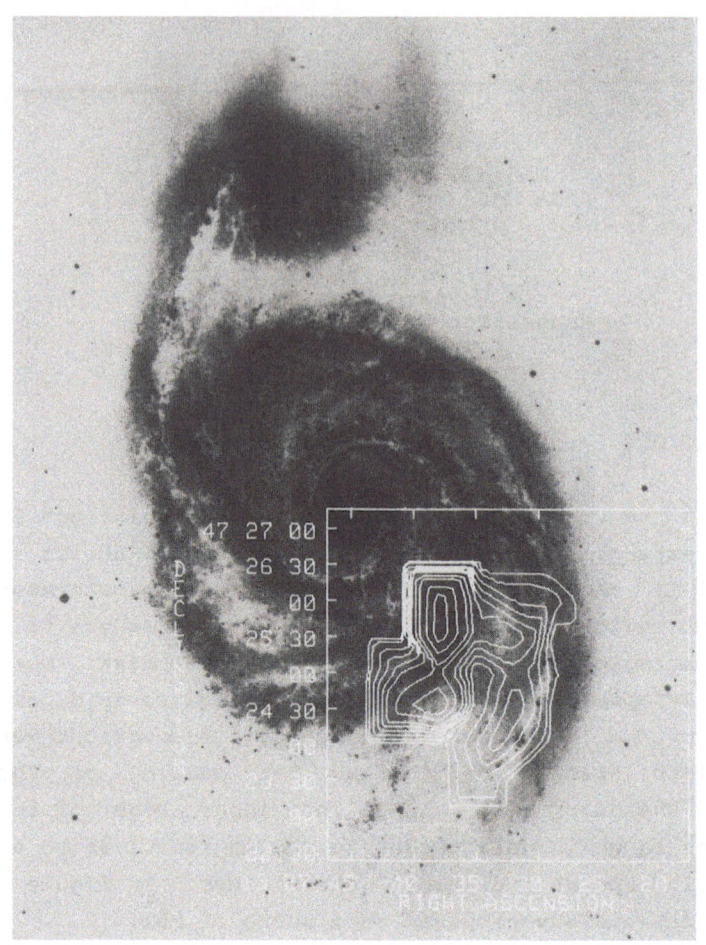

Figure 1

The integrated CO emission contours of Verter and Kutner are superimposed on a photo of M51. The contour levels are 0.1, 0.2, 0.4, 0.6, 0.8, 1.0, 1.4, 1.8, 2.6, and 3.4 K km/s. The average rms noise in our map spectra is 0.020 K.

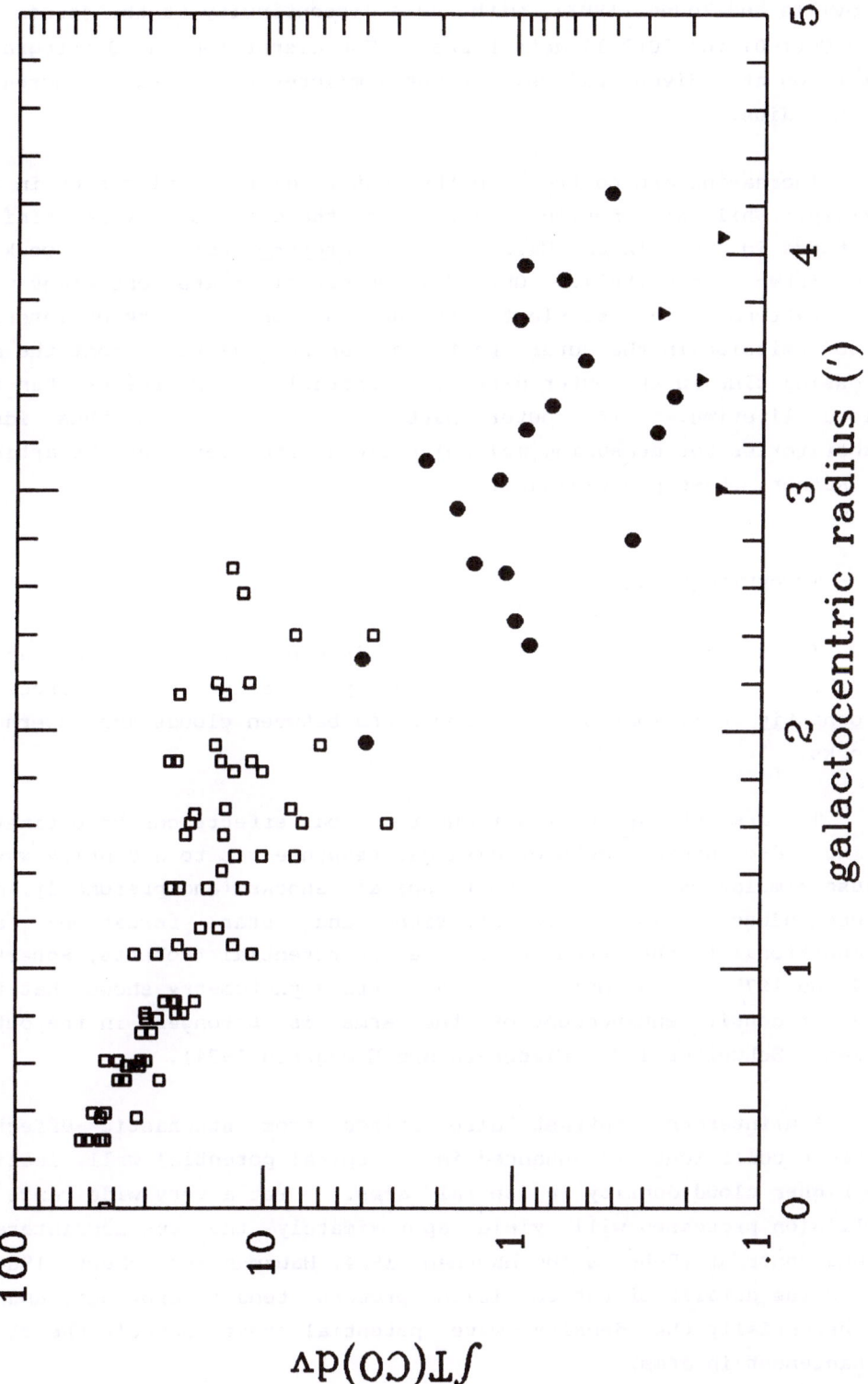

Scoville and Young (1983), with some discontinuity at the joining of the CO(1-0) and CO(2-1) data bases. The dispersion in CO integrated emission at a given radius, or the arm/interarm contrast, increases with radius.

Increasing arm contrast in the outer disk is also seen in the Galaxy. While star-forming regions in the outer Galaxy are similar to those in the inner Galaxy, the _interarm_ emission is weak to nonexistent (Mead 1986). Our M51 observations are consistent with this pattern. The simplest explanation for the lack of interarm cloud emission in the outer parts of spiral disks is that the arm crossing time in the outer disk is comparable to or greater than the cloud lifetime. The outer parts of galaxies are thus ideal laboratories for measuring molecular cloud lifetimes (see the article by Kutner, these proceedings).

II. THEORETICAL FRAMEWORK

The spiral structure of the ISM is controlled by a combination of the _global_ organizing force of density waves and the net effect of _stochastic_ collisions between clouds and between clouds and supernova shocks.

The arm/interarm contrast due to global effects can be estimated using hydrodynamic models of disk gas response gas to a density wave. These simulations indicate that spiral shocks (and presumably, any shock-induced cloud agglomeration and star formation) are proportional to the strength of the arm potential (Roberts, Roberts, and Shu 1975). In M51, near-IR surface photometry shows that the stellar density enhancement of the arms is strongest in the outer galaxy (Schweizer 1976, Elmegreen and Elmegreen 1984).

Arm/interarm contrast also arises from stochastic effects, because collisions are enhanced in the spiral potential well, leading to higher cloud density in spiral arms. Yet a very wide range of collision processes will yield approximately the same arm/interarm cloud contrast (Roberts and Hausman 1984, Hausman and Roberts 1984). Hence the details of the collision process tend to drop out, and it is essentially the density wave potential that controls the cloud enhancement in arms.

REFERENCES

Elmegreen, D. M., and Elmegreen, B. G. 1984, <u>Ap. J. Supp.</u> 54, 127.

Hausman, m. A., and Roberts, W. W. Jr. 1984, <u>Ap. J.</u> 282, 106.

Lord, S. D., Strom, S. E., and Young, J. S. 1987, FCRAO preprint to appear in Proc. 2nd International IRAS Conference "Star Formation in Galaxies".

Mead, K. N. 1986, Ph.D. thesis, Rensselaer Polytechnic Institute

Roberts, W. W. Jr., and Hausman, M. A. 1984, <u>Ap. J.</u> 277, 744.

Roberts, W. W. Jr., Roberts, M. S., and Shu, F. H. 1975, <u>Ap. J.</u> 196, 381.

Rydbeck, G., Hjalmarson, A., and Rydbeck, O. E. H. 1985, <u>A. Ap.</u> 144, 282.

Schweizer, F. 1976, <u>Ap. J. Supp.</u> 31, 313.

Scoville, N. Z., and Young, J. S. 1983, <u>Ap. J.</u> 265, 165.

SEARCHING FOR GALAXIES IN THE ZONE OF AVOIDANCE

Patricia A. Henning
Astronomy Program
University of Maryland
College Park, MD 20742

Abstract

Studies of external galaxies are hindered by the fact that 20-25% of the sky lies in the "zone of avoidance" where galaxies are hidden by the Galactic disk. Dr. Frank J. Kerr and I have conducted a pilot survey at 21 cm to determine the ease with which galaxies may be discovered behind the Milky Way.

Introduction

A large portion of the sky lies in the so-called "zone of avoidance" where galaxies are obscured by dust and confused by the high star density in the Galactic plane. Because of this, we can study only 3/4 of the visible universe. A census of galaxies at 21 cm behind the Milky Way could uncover new local-group galaxies and would aid studies of large-scale structure of extragalactic systems. For instance, the local supercluster and at least two of the big "sheets" of galaxies disappear into the obscured region where they can no longer be followed. We have conducted a pilot search to determine the feasibility of a full scale survey in the zone of avoidance.

Observing Strategy

The observations were carried out on the NRAO 91-m telescope at Green Bank, West Virginia. The half-power beamwidth of the telescope is 10.8 arcminutes when observing at 21 cm. We searched a velocity range of 300 - 7200 km/s, with a velocity resolution of 22 km/s. This precluded our finding any local group members, but had the advantage of avoiding the Galactic signal which is, of course, extremely strong in the zone of avoidance. The Galactic signal looks like a strong, narrow spike when observed with such a wide-band system, and causes a large-amplitude ringing in the autocorrelator. We wished to avoid this complication in our early work, but will extend our velocity range to low negative velocities in the future.

We observed points about 1°.0 apart along lines of constant declination through the Galactic plane clustered around -10°, 10°, 30°, 47°, and 60°. Because the 91-m is a transit instrument, we tracked each point for the maximum time of 4^m sec δ . As a control experiment, we used the same blind search method in the "clear region", away from the zone of avoidance.

Results

At this time, 1920 points have been observed in the hidden region, with 16 new galaxies detected. Most of these show the usual steep-sided, two-horned or flat-topped profile typical of spirals, while the others are probably irregular or dwarf galaxies. In the clear region, we have observed 860 points and detected 11 galaxies (Kerr and Henning 1987). This is a detection rate of about 1%, which is what is expected from the frequency of known galaxies elsewhere in the sky. Also, we have several other possible detections slated for re-observation. It should be noted that the positions of these galaxies are not precisely known; all that can be said is that each galaxy is somewhere inside a particular beam area, so the positions are uncertain by 5 to 10 arcminutes. Six typical HI profiles are shown in figure 1.

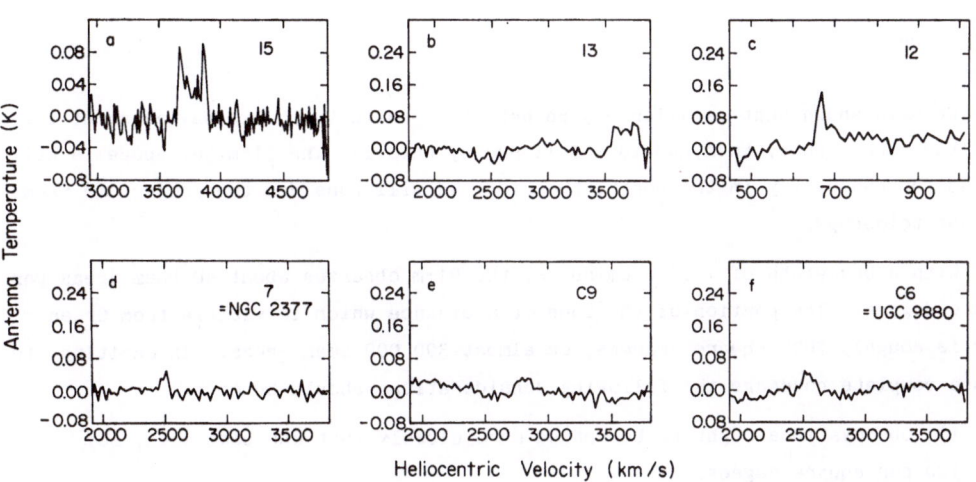

Figure 1:   A selection of typical HI profiles. Figures 1a-1d are galaxies in the zone of avoidance; figures 1e and 1f are galaxies in the clear region.

A search of the Palomar Observatory Sky Survey prints provided a rather surprising result. Two of our hidden galaxies are visible on the prints, and one of these is even an NGC object. We happened to come across by chance one of the few optically observed galaxies near the Galactic equator.

Seven of the 11 clear region galaxies that we detected are listed in catalogs. The identifications depend mainly on a close coincidence in velocity and a few of the identified galaxies lie well away from the center of the beam. No identifications could be found for the remaining 4 galaxies, but there are one or more small, uncatalogued galaxies visible on the POSS prints in each beam area. We will not know which galaxies correspond to our detections until we make higher resolution 21 cm observations, presumably with the VLA, in order to determine the positions and characteristics more precisely.

Another method of detecting galaxies beyond the Galactic disk is by use of the far infrared. We examined the IRAS Point Source Catalog and found at least one source in each of our beam areas. However, these correlations are most likely meaningless since the density of IRAS sources is extremely high at low galactic latitudes. Furthermore, in most cases it is impossible to differentiate between galaxies and stars by their colors because most of the fluxes are limits, due to confusion in the Galactic disk.

Far more useful is the IRAS Coadded data. We had maps made of the regions surrounding most of our beam areas, and we find an object with the spectrum of a galaxy in each case. This has provided further support of the reality of our weaker detections.

The Future

We have shown that galaxies can be detected behind the Milky Way at about the expected rate. What, then, would a full survey entail? The 91-m telescope is well suited for the survey in the north, therefore we will consider the project in terms of that telescope.

With a beamwidth of 10.8 arcminutes, the 91-m observes about 40 beam areas per square degree. The portion of the zone of avoidance which is visible from Green Bank is roughly 7000 square degrees, or almost 300,000 beam areas. In addition, to ensure complete coverage the following considerations should be made:

i) Observations ought to be spaced more closely that one beam area, say 100 per square degree.

ii) For studying large-scale structure, the velocity range should be extended to 10 - 12,000 km/s or more. This would require taking multiple observations at each point, and integrating.

iii) For a full survey, southern observations are needed, presumably to be made with the Parkes 64-m dish.

iv) Because the interference from the sun severely reduces telescope sensitivity, a definitive survey should be done only at night.

Clearly, a full survey of the zone of avoidance would be a truly immense project. The time required to complete the survey could be reduced by the development of a multifeed, multi-backend system, making the survey a more tenable project. Should such a system be made operational, the survey would still require many years and many astronomers.

Another way to find galaxies behind the Milky Way is to begin with the IRAS Point Source Catalog, choose objects with the spectra of galaxies, and observe them at 21 cm. This would be a faster method, but not as systematic and complete. The IRAS galaxy candidates seem to be exclusively spirals, whereas our 21 cm blind search method is not as selective. Also, the IRAS survey cannot be used within a few degrees of the galactic equator because of the confusion in the Milky Way.

The two methods should be used in tandem, with careful consideration given to the strengths and weaknesses of each.

Kerr, F.J., and Henning, P.A., 1987, **Ap. J.** (Letters), in press.

VII. THE HISTORICAL FRANK KERR

SOME REMINISCENCES OF FRANK KERR AND THE WORK ON THE MILKY WAY AND THE MAGELLANIC CLOUDS

G. de Vaucouleurs, University of Texas

Frank Kerr and I first met in 1952, when he was on the staff of the CSIRO Radiophysics Division, then housed on the grounds of the University of Sydney, and I was working at Mount Stromlo Observatory as the first Research Fellow of the newly created Astronomy Department of the Australian National University in Canberra.

Frank Kerr and the Magellanic Clouds

Frank Kerr, Jim Hindman, and Brian Robinson had just detected the HI line emission from the Magellanic Clouds (AJ **58**, 218, 1953; Austral. J. Phys. **7**, 297, 1954) with the 36-foot reflector (which they were still calling an "aerial") at Potts Hill, N.S.W. It had a beamwidth of 1.5 degrees at half-maximum, good enough to give about 100 resolution elements (the word "pixel" had not been invented yet) in the area of the Large Cloud. The receiver had a bandwidth of 40 kHz (kilo-cycles per second, as we said then) giving a respectable velocity resolution of 8 km/s.

It was during one of his first visits to Mount Stromlo that Frank Kerr showed us his map of velocity residuals across the LMC which, as I saw immediately, revealed the typical pattern of a rotating disk galaxy. This was obvious to me because I had just re-analyzed the old Lick velocities of a score of HII regions and re-interpreted them in terms of a general rotation consistent with the spiral structure which I had detected on direct photographs and by means of star counts (Obs. **74**, 23, 158, 1954; AJ **60**, 126, 1955). Remember that in those days the party line was that the Clouds were "irregular" galaxies and that the velocity gradient across the LMC was the reflection of a large transverse motion.

After a couple of years of mutual visits to Sydney and Canberra we produced two papers which placed beyond reasonable doubt the rotation of the two Clouds and gave first estimates of their masses from rotation and velocity dispersion, 3×10^9 and $1.3 \times 10^9 M_\odot$ (Austral. J. Phys. **8**, 508, 1955; **9**, 90, 1956). These early estimates were too low and were soon revised to 25×10^9 for the LMC with the help of a few additional optical velocities secured at Mt Stromlo in 1955 (Ap. J. **131**, 574, 1960). This revision, in turn, probably overshot the mark since later re-discussions, including independent confirmation of the rotation of the LMC from stellar velocities by M. Feast, A. D. Thackeray, and A. J. Wesselink at Pretoria, led to revised estimates of 15×10^9 and $3 \times 10^9 M_\odot$ (Vistas in Astron. 1972, **14**, 232).

Frank Kerr and the Warp of the Galactic Plane

In a letter dated May 29, 1956, Frank Kerr wrote of a complimentary postcard from Harlow Shapley to whom we had sent reprints of our joint papers on the Magellanic Clouds:

"I have just read with much interest your and de Vaucouleurs' big paper on Mag Cloud motions. Very fine, even if provisional in spots. I shall get a reference to it in my Inner Metagalaxy monograph" (he did).

In the same letter Frank also reported the preliminary results of his big sweep of the galactic plane in the HI line leading to his discovery of the warp of the plane, probably by interaction with the Magellanic Clouds:

"The surface obtained in this way is not a plane, although it is fairly close to one. The main deviation from a plane appears to be a dip downwards in the region closest to the Magellanic Clouds, and a (smaller) rise upwards on the side furthest from the Clouds. ... The effect is qualitatively like that of a "Magellanic tide," but is quantitatively several hundred times too great for a point mass at the centre of the LMC."

This was for our small mass estimate; the discrepancy was much less with the larger values.

Here are some photographs of Frank Kerr and Jim Hindman on the roof of the Radiophysics Laboratory in March 1957, displaying their brand new maps of the spiral structure of the Galaxy and of the warped equatorial plane outside the solar radius. [The photographs are somewhat over-exposed because of the well-known extra-strength of the "Australian Sun."]

A few months later Frank Kerr stopped in Flagstaff to visit my wife and I at Lowell Observatory, and here he is with Antoinette on the rim of the Grand Canyon on a sunny winter day. Since we cannot be present today, this little memento will convey to Frank, with the memory of better days, our greetings and best wishes for many years of a happy, healthy and still productive retirement.

Frank Kerr and Radio Waves:
From Wartime Radar to Interstellar Atoms

Woodruff T. Sullivan, III
Department of Astronomy
University of Washington

My task tonight is a pleasant one, for Frank is my mentor and I am delighted to join in honoring him. My goal is to review the accomplishments and fascinating story of his career, as well as to convey in the short time available some of the spirit and essence of Frank's scientific work, in the process giving you a feeling for the milieu in which his attitudes toward research were born. I will be focussing on the pre-1960 period because (1) these years set the stage for all that followed, (2) I suspect that most of you have little knowledge of Frank's career then, and (3) I have developed a peculiar attitude that the older something in radio astronomy is, the more I'm interested in it. I also realize that there is a great mixture of Frank's friends and colleagues here and so I will endeavor to explain scientific points when necessary and to limit the astronomical and radio jargon. So here follows a subjective look at a small portion of Frank's career–keep in mind that he has written well over a hundred scientific papers and just to read the titles of these would fill up the rest of the time.

First of all, a little about his personal background. In the 1850's his ancestors came to Australia from England and Ulster, Ireland, whence the name Kerr. (Note that they did not come any earlier–for an Australian this is always important to establish!) In 1913 his father was a Rhodes Scholar at Oxford and then got caught up in World War I, and in 1916 his fiancee set sail from Melbourne to get married, leading in 1918 to Frank's birth in England. A year later the family went back to Victoria, where Frank spent much of his childhood. There he was taught among other things that he was the *right* kind of Irishman, and never to be caught eating fish in public on Friday. His father and two uncles were medical doctors and so it would have been natural for Frank to have continued along that line, but he got the physics bug in high school and received his bachelor's degree in 1938 at Melbourne University in what was then still called the Department of Natural Philosophy (Fig. 1). Just before World War II broke out he stayed on for a masters degree and then was all set up for a scholarship for a Ph.D. at the Cavendish Laboratory in England–I remind you that the first Ph.D. in Australia was not given until 1948 and so you had to go overseas to get an advanced degree–but the war came on and changed all those plans. His first publications were based on his thesis work on the refractive index of air and water vapor

Fig. 1. Frank J. Kerr as an undergraduate at Melbourne University (ca 1935). (Photo courtesy of Maureen Kerr).

at the ultra short wavelength of five meters, a wavelength fully ten times shorter than had been measured before. Two papers with identical titles came out of his thesis, but the one in *Nature* used the term "indexes" in the title and the other in the *Proceedings of the Physical Society* used "indices" (Kerr 1941, 1943). Now we all know how Frank cares for the English language–it must have driven him absolutely nuts!

World War II came on in 1939 and the secret of radar, which had been developed in England, was passed on to Commonwealth countries such as Australia, and so the Radiophysics Laboratory was founded at that time as a part of CSIR, the Council for Scientific and Industrial Research. This was a very unusual CSIR division–it was called "Radiophysics" while most of them then had names like Animal Nutrition or Economic Entomology, practical things that Australia needed. What did the Lab do? At first it adapted British radars, but then as the war went on and it gained more confidence and expertise, whole new designs were produced. Frank's work involved developing switches and wave guides and transmission lines, field-testing of models, and at the end of the war conducting a mammoth project not unlike some of his later surveys. It was a statistical study at radar stations all round Australia of what was called super-refraction: radio waves anomalously bent in the atmosphere when the weather conditions were right, allowing one to detect targets much farther away than usual (Kerr 1948).

By war's end the Radiophysics Lab had a staff of about three hundred and at its core arguably the greatest concentration of technical talent on the continent. Nobody wanted to break it up, but what were all these people going to do? It was very different from the situation in the U.S., where places like the Radiation Lab at MIT *were* broken up, and in England where the Telecommunications Research Establishment was greatly diminished in size–the best people wanted to get back to the universities where the prestige was. That's where they had come from and that's where they wanted to get back to–they didn't want to be civil servants anymore. In Australia, however, there was little research going on in the universities and the best researchers wanted to stay in the Radiophysics Lab where the action was. So they cast around for how they could turn their talent to peacetime pursuits and many areas were pursued. A few examples: vacuum physics, ionosphere work (such as Frank first did), rain and cloud physics (obviously of practical import in dry Australia), navigation radars, weather radars, and mathematical physics (one of the first early digital computers [CSIRAC] was developed there). Among all this was a strange thing called "extraterrestrial noise," which by 1951 grew to include about one-third of the staff and which, along with the cloud physics, certainly became the most scientifically productive area. To give you an idea of how this extraterrestrial noise was looked upon at the end of the war, let me quote from a prospectus written at the time:

> Little is known of this noise and a comparatively simple series of observations on radar and short wavelengths might lead to the discovery of new phenomena or to the introduction of new techniques. For example, it is practicable to measure the sensitivity of a radar receiver by the change in output observed when the aerial is pointed in turn at the sky and at a body at ambient temperature.

You see that it was then being sold as something that would have some practical outcome, as CSIR was accustomed to dealing with.

The Radiophysics Lab was run by Taffy Bowen, an aggressive go-getter personality, astute in the ways of politics and scientific management. He had been one of the original British radar pioneers in the 1930's and the Lab thrived under his leadership over the next twenty-five years. His righthand assistant, in charge of all work on solar and cosmic noise, was Joe Pawsey. Pawsey got his degree at Cambridge under Ratcliffe in the 1930's and was an amazing man but who unfortunately died in 1963 before I ever got the chance to meet him. He was Frank's scientific mentor (Fig. 2). Not only Frank, but one after another of his colleagues from those years have testified to the vital guidance they received from Pawsey. He didn't write many papers himself, but almost every radio astronomy paper from that era gives more than the usual token acknowledgement to Pawsey's influence on the research. Bowen and Pawsey developed the Radiophysics Lab over the succeeding decade to a degree of international prominence previously unknown for *any* scientific group in Australia.

Fig. 2. Joseph L. Pawsey (left) and Kerr in the CSIR Radiophysics Laboratory (ca 1948). (Photo courtesy of Sally Atkinson and the CSIRO Radiophysics Division; also Figs. 4, 5, and 13).

Let me now diverge and take a very brief look at early radio astronomy more generally. There's no time for details, but the story begins in the 1930's when Karl Jansky at Bell Laboratories in New Jersey accidentally discovered radio noise from the Milky Way. His discovery constitutes what we now see as the beginning of radio astronomy, but at the time it languished and was not followed up except by one remarkable person, Grote Reber. In 1937 he built a 30 ft diameter dish in his backyard on his own time and with his own money–you can imagine what his neighbors said! Reber, however, was the only one that paid any serious attention, and so this "cosmic noise" just lay there until World War II. In 1942 a secret report was circulated after James Stanley Hey found that the air-warning radars round the English coast, which had seemed to be all jammed simultaneously, were in fact suffering not from new German capabilities, but from the Sun! This report, however, had only limited circulation during the war and it seems that what really piqued the Australians' interest at war's end was rather what was called the "Norfolk Island effect," an independent discovery in 1945 of the Sun's radiation by a radar installation on an isolated island between Australia and New Zealand. When this report came into Sydney, Pawsey got quite excited and jumped onto the radio sun. Many of his first observations were made from a cliff named Dover Heights, within the city limits of Sydney, where simple antennas were pointed toward the Sun and made some of the very basic radio discoveries concerning the Sun, for instance, the existence of a one-million-degree corona.

During the postwar period there were three places that dominated radio astronomy: the Radiophysics Lab, the Cavendish Laboratory group under Martin Ryle's leadership in Cambridge, England, and Bernard Lovell's Jodrell Bank station near Manchester. But it is important to note that the Radiophysics Lab's effort was much larger–three or four times larger than either of the others. I shall mention just some of the names of the key people: Pawsey, Payne-Scott, Christiansen, Wild, Smerd (all working on the Sun), Mills and Bolton making fundamental discoveries about radio sources, Piddington working on the Galaxy and on theory in general, and then there was Frank Kerr. Now you think I'm going to say "working on the 21 centimeter line," but no, his first experiment to do with astronomy was bouncing 15 meter-long radio waves off the Moon!

The first lunar radar had been done in 1946 by the U.S. Army Signal Corps. But in 1947-48, together with Alec Shain and at a field station outside of Sydney called Hornsby, Frank carried out the first detailed analysis of lunar echoes. Actually, his main motivation was not astronomy, but ionospheric research–he and Shain wanted to see how the upper atmosphere of the earth was affecting the radio waves as they went out and came back in. They found that the echo strengths varied from second to second (Fig. 3), as well as on longer time scales (Kerr, Shain and Higgins 1949; Kerr and Shain 1951). These changes arose not only in the ionosphere, but also from the moon itself rocking back and forth in motions that astronomers call librations. And this is when Frank first had to start digging into astronomy books, to find out about these complex motions of the moon. Having done this, his appetite was whetted for other possibilities, such as trying radar off the outer atmosphere of the Sun or off the planets. He did a detailed study of this and the resultant paper (Kerr 1952) is a classic in the field. But it just didn't look like it was going

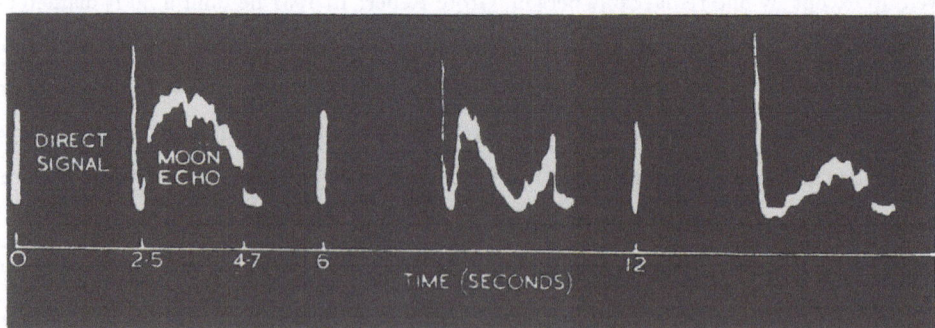

Fig. 3. Three successive lunar echoes obtained in 1948 with a transmitted frequency of 20 MHz and pulse length of 2.2 sec; the lunar echo begins 2.5 sec after the transmitted pulse. Note the varying intensities for each echo, caused by lunar librations. (Kerr and Shain 1951).

to be doable without a very large antenna and so that was the end of astronomical radar at the Radiophysics Lab. In fact the Sun was not detected until 1959, with the planets following a few years later–it's just much farther to the planets than to the Moon, and radar signals fall off as $(1/r^4)$.

Having finished this lunar radar, Frank decided that he needed to learn some astronomy, so he got a Fulbright grant and spent a year at Harvard University in 1950-51. By an amazing coincidence also there (besides people like Harlow Shapley, Donald Menzel and Fred Whipple) was Henk van de Hulst from Leiden Observatory, who was a Visiting Professor for a semester. Furthermore, in the Physics Department was Ed Purcell and his graduate student "Doc" Ewen (Fig. 4), who were then searching for the 21 centimeter hydrogen line. Frank took a course in radio astronomy taught by van de Hulst and the collected notes from that course were published in a mimeograph form that became the bible of its time–the first textbook in radio astronomy and now a very nice benchmark for anyone studying the development of the field.

From that course Frank learned that in 1944 van de Hulst had predicted that there should be a

Fig. 4. The "21 centimeter club" gathered in Sydney at the 1952 URSI General Assembly: (left to right) Kerr, Paul Wild, Jim Hindman, "Doc" Ewen, "Lex" Muller, and "Chris" Christiansen.

spectral transition in a hydrogen atom whenever the spins of its proton and electron flipped relative to each other. Whenever this happened, an atom would emit at the particular radio wavelength of 21.1 centimeters. Van de Hulst worked out that enough hydrogen atoms were expected between the stars that one should be able to measure excess radio intensity if a receiver were tuned precisely to 21.1 cm. Now he did this while the Nazis were occupying Holland and published the paper shortly after the war in *Nederlandsch Tijdschrift voor Natuurkunde*, in Dutch. You can imagine that it therefore didn't get much circulation, but the news did get around enough so that Ewen and Purcell had heard of the prediction and were searching for this line. Now, of course, the Dutch too planned to search for the line but Holland was in very bad straits after the war–it was a very poor country at that time. Nor did they have any radar engineering talent, because of course they had had no radar laboratories during the war. Things were going extremely slowly even though Jan Oort, the distinguished Professor at Leiden, world expert on the structure of the Milky Way since the 1920's, was making major organizational efforts. But as late as 1950 they still had no radio observations to talk about.

Now the essential problem with all of galactic structure is that we are in the middle of a "fog" of interstellar dust particles. Oort realized that if one could find a radio spectral line then one would be able to see through all of these dust particles blocking out the stars that are behind. He was also very enthusiastic because it turns out that if a hydrogen cloud is moving toward or away from us, the wavelength of the line is shifted slightly, something that could easily be measured with radio techniques. This would allow getting the rotation of the entire Milky Way and mapping out where the interstellar clouds were–for the first time perhaps the entire Galaxy could be "seen."

As it turned out, the Dutch, despite all their efforts, were not the first to find the hydrogen line. It was instead Ewen, with his horn antenna stuck out of a window at Harvard University. After a year of hard work with his receiver, he was able to establish that when the Milky Way came by daily, he indeed got a signal at 21 centimeter wavelength–it had to be hydrogen. To be exact, this discovery came on the 25th of March 1951. Seven weeks later the Dutch also found the line and it is often said that this was a big race that the Dutch unfortunately lost even though they had started a long time before. And it's also often said that the race was lost because the Dutch had a catastrophic fire that burned all their equipment in early 1950. But in fact if you look at it I think the fire actually helped the Dutch. It came about a year before they detected the line and it was partly what led to the dismissal of their first electrical engineer and the hiring of Lex Muller (Fig. 4), who was much more talented. It is also fascinating to look at the fact that it was van de Hulst at Harvard who acted as a catalyst for the discovery by both sides, because he was feeding hot electronic techniques from Harvard over to the Dutch at the same time that he was telling the Americans what to expect in the Galaxy. For instance, the key thing he told the

Dutch was about the technique of frequency switching (they hadn't been doing that) while he told the guys at Harvard, "Don't frequency-switch over only 10 hertz or so because you'll subtract the line from itself–it's going to be a broad line in the interstellar medium, not a narrow one as in a laboratory cell" (these were physicists, not astronomers). So you can see that if it was a "race," the competitors certainly cooperated with each other more than is usual in such a contest.

Now back to Frank. He got all caught up in this excitement and after the discovery wrote to Pawsey saying, "We've also got to do this." Pawsey immediately assigned Chris Christiansen and Jim Hindman (Fig. 4) to the task and with their technical expertise it didn't take them more than six weeks to slap together a receiver and find the line. So it was that the Australians too joined in on a triplet of notes in *Nature* (Ewen and Purcell 1951, Muller and Oort 1951, Christiansen and Hindman 1951). I'd like to quote from a letter that Frank wrote to Pawsey in that (northern) spring of 1951, containing prophetic statements. At this time he was searching around for what to do when he went back to Australia and he said:

> I have in fact got very interested in the subject [of the hydrogen line], though close contact with van de Hulst and Ewen, and would welcome the opportunity of entering this field. The main reason I had not mentioned it before as a possible project for me was not through any lack of interest, but because I had thought you might well regard it as in the field of those already engaged in galactic work. However, as you imply in your letter, the study of this line would seem to offer a full-time job for somebody for quite a long time, provided the intensity is great enough to do things with. One thinks in terms of a thorough survey of the sky, at both low and high resolution, measuring for each position the intensity, line-breadth, and line shift.

How's that for a precis of much of Frank's subsequent career?! At the end of the letter he further says, "I am now taking the opportunity of pumping van de Hulst about knowledge of interstellar things."

So Frank went back fully primed with all this knowledge from Harvard and indeed Pawsey set him up as the 21 centimeter man and they built a large antenna, much more maneuverable and larger than the one Christiansen and Hindman had used for their brief survey of the southern Milky Way. In Figure 5 we see this 36 ft dish at the Potts Hill field station, and that's not Frank out at the focus, but Brian Robinson, who collaborated with him on the first work. This thing was built in less than a year in the shops of the Radiophysics Lab. Meanwhile Frank and Hindman and Robinson developed a *four*-channel receiver (all previous 21 cm work had been laboriously carried out with only a single channel), but they had all kinds of trouble making it work, and had finally to use only one channel of 40 kHz width for the survey.

Fig. 5. The 36 ft diameter dish used by Kerr and collaborators for most of their 21 cm hydrogen line work in the 1950's. Working at the focus is Brian Robinson (ca 1953).

Rather than start with the Milky Way, Frank decided to do something more distinctly southern and thus chose to look for hydrogen in the Clouds of Magellan. Each of the Magellanic Clouds is a glowing patch of sky about the size of your fist at arm's length, each a small galaxy orbiting around our own Galaxy. Frank reasoned that they probably should contain hydrogen, since our own Galaxy does, and indeed in March 1953 his team detected the Magellanic Clouds and he excitedly wrote a letter to Shapley announcing it. This was the first time that hydrogen had been detected in an extragalactic object. They surveyed both Clouds and radio astronomers will be interested to note that they had a system temperature of 1500 degrees and a beam of one and a half degrees. The surprising thing from their survey was that the Small Cloud had two-thirds as much hydrogen as the Large Cloud, even though it was much fainter and had much less dust. Both Clouds also had large hydrogen sizes when compared to the optical. Figure 6 shows you the contour maps that they made (Kerr, Hindman and Robinson 1954).

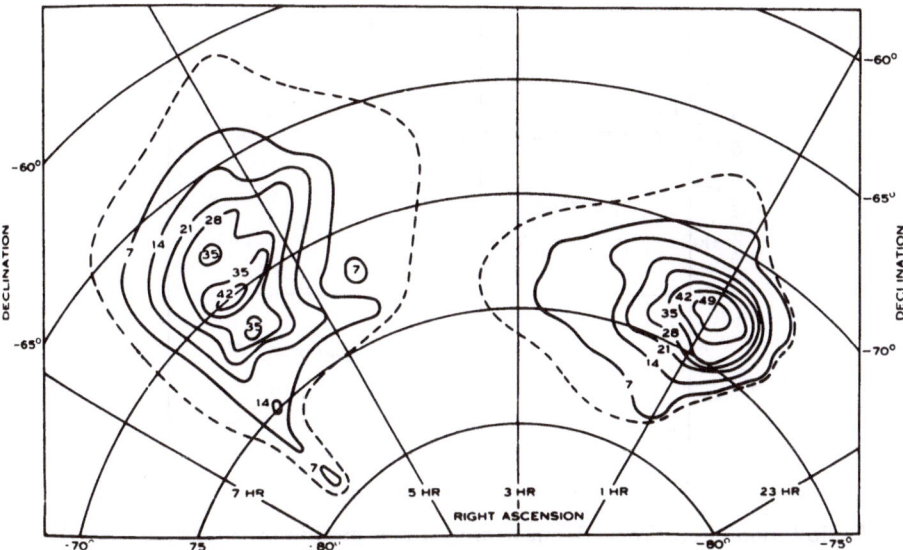

Fig. 6. Contours of 21 cm hydrogen intensity, integrated over velocity for the Magellanic Clouds (unit $= 10^{-16}Wm^{-2}ster^{-1}$). The dashed lines indicate the outer limits of the regions where hydrogen was detected. (Kerr, Hindman and Robinson 1954).

The other thing they could do was look at the motions of the hydrogen and that was particularly exciting because now one could see whether the Clouds were rotating or what they were doing. Frank teamed up with Gerard de Vaucouleurs, a young optical astronomer at Mt. Stromlo and already an expert on the Magellanic Clouds. Now you have to realize that up until this time a grand total of only *seventeen* radial velocities had been measured for the Large Cloud, and even those dated back to 1917! To this Frank's group was now adding two hundred more, and with a much better sky distribution. From the earlier velocities nobody knew what was going on, whether the Large Cloud was rotating or in random motion or whether it was just translational motion. But now with the hydrogen-line data they were able to derive a decent rotation and mass for each Cloud (the faster they spun the more massive they were), and they estimated them to be $1-4 \times 10^9$ solar masses (Fig. 7). (Note that at this time there were in the whole sky only about half a dozen other galaxies that had masses, another gauge of the state of knowledge on galaxies.) In the end they favored Shapley's scheme wherein a galaxy spins faster and faster as it evolves, and therefore

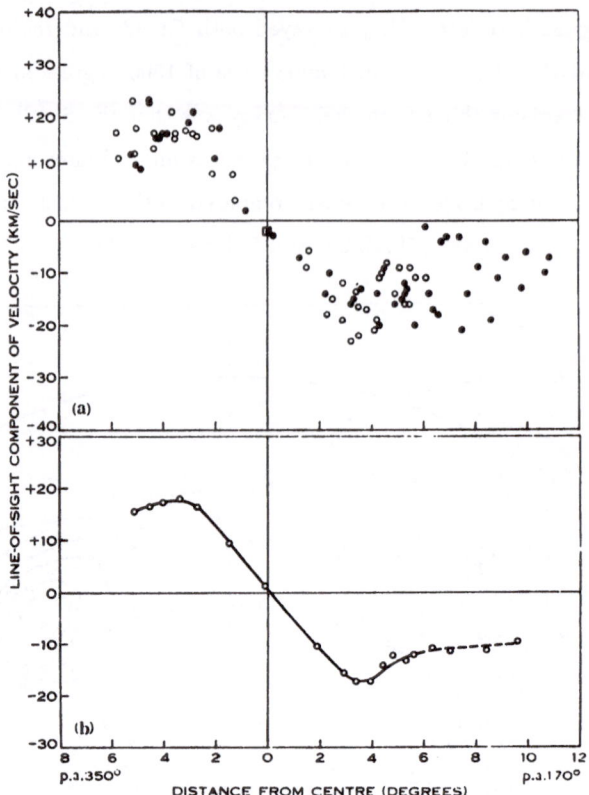

Fig. 7. Rotation curve from 21 cm hydrogen line profiles of the Large Magellanic Cloud. Filled circles represent measures along the major axis, open circles those off axis. The bottom curve is an overall average. (Kerr and de Vaucouleurs 1955).

each of the Magellanic Clouds was a very young system that would later become a faster-spinning spiral and eventually an elliptical galaxy spinning fastest of all (Kerr and de Vaucouleurs 1955, 1956).

While all this was going on, the Dutch were forging ahead mapping the part of the sky that they could see, the so-called northern side of the Milky Way, and Oort was getting impatient with Frank because Oort needed the other half of the galaxy in order to put the whole picture together. And eventually in 1954 Frank did survey the southern Milky Way, now that the four-channel receiver was working. But it still suffered from instabilities and these data were never published. In 1955, however, the survey was repeated and the first good results emerged. The most interesting thing was that the galactic plane was warped–it was in fact not a plane but had a warp in it. Figure 8 shows how the warp extends to almost a thousand parsecs above and below the inner galactic plane, going way out beyond the solar circle. And it turned out that where it went below the plane was toward the Magellanic Clouds–Frank couldn't get away from the Magellanic Clouds even when he tried! So Frank immediately suspected that the Clouds' tidal forces might

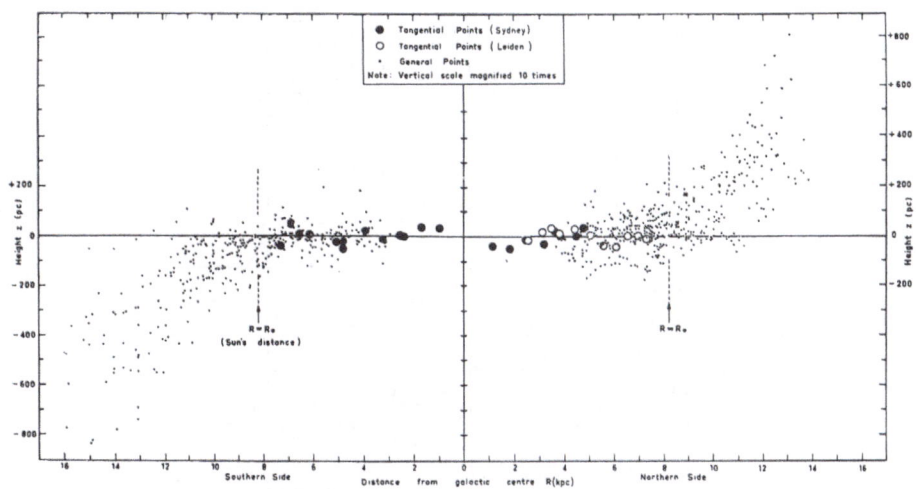

Fig. 8. The warp of the galactic plane (refered to the new 1958 IAU system of coordinates) as measured with 21 cm hydrogen line data from Sydney and Leiden. "Northern" and "southern" sides refer to the galactic hemispheres on either side of the galactic center. Note that the vertical scale is magnified 10 times over the horizontal. (Gum, Kerr and Westerhout 1960).

be causing this disturbance, but when he plugged in the numbers it just didn't seem to work–and in fact we are today still not certain what's causing this warp, as has been mentioned at this conference a couple of times. This result was first presented (Kerr 1957a) at a symposium in 1956 honoring the fact that Bart Bok, an expatriate Dutchman whose specialty was galactic structure, was coming to be the Director at Mt. Stromlo. As Frank has said earlier this evening, Bok had a major influence on his career–he was a champion of combining radio and optical observations in all fields, especially for spiral structure, and this was the beginning of his influence on Frank. Bok liked this result so much that he asked Frank if he could present it on Frank's behalf to the American Astronomical Society in a few months (Kerr 1957b), and it happened that at that very same meeting Bernie Burke, based on other data, independently came up with the same discovery (Burke 1957).

Another thing done at that time was to determine the exact location of the inner plane of the Galaxy – if you're going to claim that something is warped, then you have to know what is normal. Frank together with Colin Gum found that the inner part of the Galaxy was an amazingly good plane – they found that the hydrogen layer had a degree of thin-ness and flatness about like *Time* magazine, with the *mean* position of the hydrogen even better than that – more like a sheet of paper. And the sun, by accident (we *think* it is by accident!), turned out to be, as far as could be determined, exactly in this plane. Now this all tied in with a committee of the International Astronomical Union that was then trying to determine whether galactic coordinates should be revised. They were grappling with how to change the 30-year-old coordinate system because the

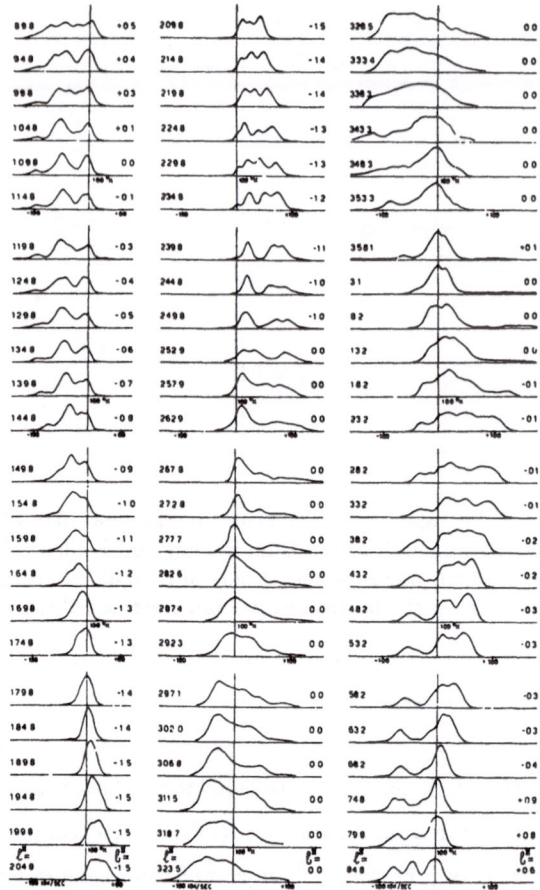

Fig. 9. Hydrogen-line profiles (Plots of intensity versus velocity) measured with 1.5-2.5 degree beams at the indicated longitudes (new system) all round the galactic plane. These profiles were taken by Kerr's group in Sydney and Oort's in Leiden. (Kerr and Westerhout 1965).

galactic center was at a longitude of 332°, which was a nuisance, and the old galactic plane was also off. In the end they did change the coordinate system in 1958, flipping the plane about a degree and changing the zero of longitude, and it was largely based on this new-fangled 21 centimeter line that this fundamental shift in astronomical coordinates was made (Gum, Kerr and Westerhout 1960).

But of course the main motivation for the hydrogen survey was spiral structure and here we begin a long series of studies that Frank got involved with, an ever- changing parade of diagrams of the Milky Way's structure produced by changing data and/or changing ideas of how to interpret the data. The basic data from both Sydney and Leiden were plots at each galactic longitude of the hydrogen-line intensity as a function of its velocity (Fig. 9). With a model of the Galaxy's rotation (itself largely determined from the 21 cm data) the bumps and peaks on these plots could be translated into hydrogen clouds with specific locations in space. Frank spent six months in

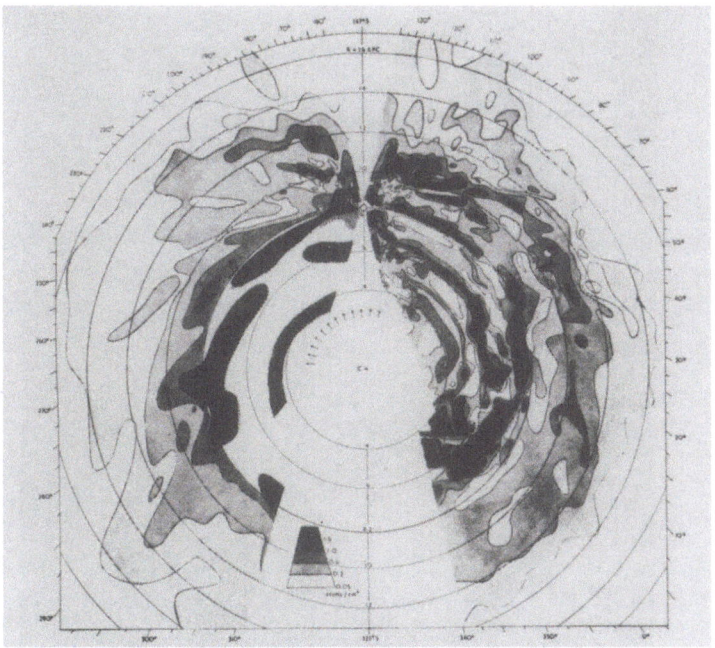

Fig. 10. Regions of high neutral hydrogen density as projected on to the galactic plane (old coordinate system). The righthand side of the diagram is derived from Leiden data and the left side from Sydney observations. The sun S (top center) is located 8.2 kpc from the galactic center C. (Oort, Kerr and Westerhout 1958).

Leiden in 1957 and this led to the northern and southern sides being combined in a classic paper by Oort, Kerr and Westerhout (1958), one that produced the diagram of Figure 10–mother's milk for us apprentices who got involved in this in the 60's. For the decade of "That Was The Week That Was" this *was* the Galaxy that was. These kinds of spiral arms were seen in other galaxies and so were fully expected in our own Galaxy–the arms of Orion, Perseus, Cygnus, Sagittarius and Carina. You notice that the Australian side of the Milky Way looks qualitatively different from the Dutch side. That came about because the two camps had different antennas and receivers and different sets of procedures to interpret the data. For instance the Dutch liked to correct for the beamwidth of their antenna and for the finite bandwidth of their receiver much more so than did the Australians. The philosophy of the various observers also influenced which features should be connected with which, and even the method of graphical presentation changed one's perception of the Galaxy–compare Figure 11 with Figure 10, both from the same paper. As the 60's passed and the 70's commenced these kinds of problems caused people to temper their belief in the details and to realize that the assumptions being made were sometimes shaky and yet greatly influenced the location and orientation of the various arms. In particular it became clear that a bump in a profile did not necessarily mean a cloud in space. Despite this later caution, this work by Frank and

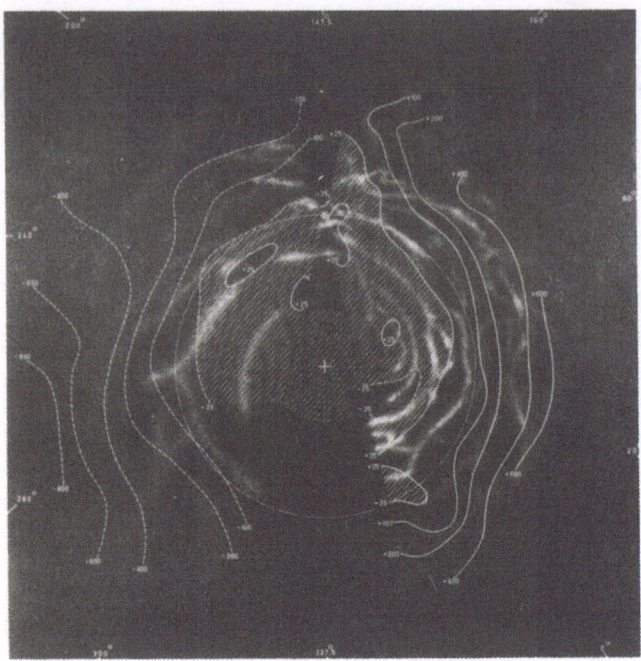

Fig. 11. Another means of presentation of the hydrogen features of the Galaxy, based on the same density contour map as for Figure 10. Also shown are contour lines corresponding to the deviation of the hydrogen (in parsecs) above and below the inner plane. (Oort, Kérr and Westerhout 1958)

others was seminal and indeed is still fundamental for anyone trying to understand the structure of our Galaxy.

Another neat thing that Frank turned up was that the Galaxy seemed to be rotating differently in the southern part of the sky than in the northern part (Kerr 1962a; Fig. 12). Now that didn't make any sense at all–it's all the same galaxy, so why should it be different in Sydney than in Leiden? He was convinced that it was not an artifact of the different survey techniques and proposed that we are moving away from the galactic center at a relatively low speed–something we didn't suspect before–and that this motion created the apparent difference in the Galaxy's rotation. In fact Frank proposed that perhaps all of the gas in the Galaxy is expanding, because certainly the central region of the Galaxy was known at that time to be expanding.

A major chunk of Frank's career was involved with the "Giant Radio Telescope," the GRT as it was called. As radio astronomy matured in the 50's its character began to change from each person doing everything-digging holes, pouring concrete, building aerials and receivers, and doing all the astrophysics–to specialists building large antennas and complex receivers and other specialists using them. In Australia this trend was exemplified by the GRT, which was a hoped-for dish that Bowen pushed for starting in the early 50's. Frank was instrumental in putting together

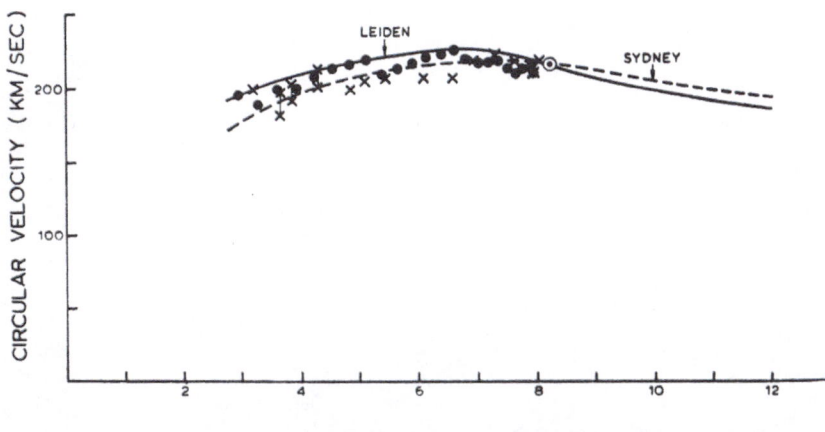

DISTANCE FROM GALACTIC CENTRE (KPC)

Fig. 12. Rotation curves of the Galaxy, derived from 21 cm hydrogen tangential- point analysis, based on Leiden (dots) and Sydney (crosses) data. The sun is represented by ⊙. (Kerr 1962).

a little green book, a propaganda booklet, that was given wide circulation, and was sent to all Members of Parliament. It extolled everything that had been done at the Radiophysics Lab and argued that they should get a lot of money to build a giant dish. And in fact they did get the money, from both American foundations and the Australian government, and in the late 50's Frank was involved in the site survey. They picked the sleepy outback town of Parkes, around 230 miles from Sydney, a very auspicious site for an Australian astronomical establishment as the land was owned by a person named Australia "Austy" James Helm (because he had been born on the day the Australian Federation came into being) and the Mayor was named Cecil Moon! Mr. Moon held all town business in the local pub, and Frank tells me this is where he learned that one can drink an awful lot of Swan lager in the heat of the outback.

In the years before the telescope came into operation in 1961, Frank was head of a committee that oversaw development of the interface between telescope and user. One of the things he was most concerned about was that users be able to accurately read numbers from the controls, so he pushed to have everything digital–today that seems routine, but at that time it was the first telescope that had any large degree of digitization (Fig. 13). Immediately upon its commissioning the dish made all kinds of contributions: for instance, an accurate position for 3C 273 (from a lunar occultation) that then led to its identification as a quasar, and the discovery of linear polarization in Centaurus A (Minnett 1962; Kerr 1962b).

Frank continued with his hydrogen-line surveys in the Milky Way, especially toward the galactic center, and now he had 48 channels and a 14 arcminute beam to work with. But during this period Frank began to get wanderlust, as did several other of the senior researchers at Radiophysics

Fig. 13. Frank Kerr at the control desk of the newly-commissioned 210 ft diameter Parkes radio telescope (ca 1961).

who found themselves, so to speak, outgrowing the Lab as they sought bigger and bigger projects and more and more independence. In Frank's case he had been awarded a Doctor of Science degree by Melbourne University in 1962 and now he specifically sought an astronomical setting in academia. So it was that in 1966 he came here to College Park and to Gart Westerhout to update the combined northern and southern Milky Way surveys, as he had done in Leiden a decade before. He also came to teach and to advise, at which he turned out to be masterful, and indeed he was soon surrounded by a coterie of graduate students. Now it turns out that it is at this juncture that Frank's and my world lines intersected for the first time: my account now becomes less detailed and more subjective.

At Maryland Frank supervised and collaborated in a tremendous variety of projects. Still central to his work was large-scale galactic structure and he sent students like Ron Harten, Dave Ball, Ed Grayzeck, Phil Bowers and Pete Jackson to exotic places like Australia, Argentina and West Virginia to gather the data, now with as many as 1024 receiver channels to play with. He was also involved in several special projects, of which I might point out his work with Jill Knapp on the gas-to-dust ratio toward globular clusters, with Sandy Sandqvist on lunar occultations of the galactic center, and with Pete Jackson on detecting ionized hydrogen recombination lines in the general interstellar medium. By the way, that signal is 5000 times weaker than the signals from the Magellanic Clouds that he had worked on 15 years before, yet both experiments were, in their times, right at the hairy edge of what one could do.

But in his desire for academia, I suspect Frank didn't realize the maelstrom he was entering in the flower child, anti-Vietnam-war days of the late 60's. For instance there was one grad student who walked around in the halls barefoot at that time, but her Scottish brogue (much stronger then than today) and science were intriguing, so that's how she survived. There was another student who was always listening to baseball games while reducing data, and was even known to skip a lecture for an important game (after all, Frank, it *was* the first time the Red Sox had gone to the World Series in twenty years!), but Frank liked cricket and they shared a love for the English language, and so he survived. There was a third guy who insisted on wearing lederhosen, and on one occasion a kilt–I never was able to figure out how he survived!

As we trace Franks's career we can see several themes. First, he has always been attracted to and excelled in the basic, fundamental kinds of observational projects, the ones that are not glamorous but in the long run contribute so vitally to the development of our science. In this sense he is a member of the Dutch school both in this methodology and in the direct influence from van de Hulst, Oort, Bok, and Westerhout. Together these men over a decade or so gave us *a new Galaxy*, another in the line of new views starting with William Herschel almost two centuries ago, continuing with Kapteyn, and culminating with Shapley. Frank also, unlike many of his colleagues from the early years of radio astronomy, has always had the Dutch approach of integrating as completely as possible radio and optical work. This is why, unlike others who left the Radiophysics Lab, he headed for an *astronomy* department, and one led by an ex-Dutchman at that. Another forte of Frank's is represented by the many excellent review articles and talks that he has produced over the years, once again an unpopular task, but a vital one to the development of our field. Two of particular importance appeared in the *Stars and Stellar Systems* compendium (Kerr and Westerhout 1965, Kerr 1968). He is also first-class as an organizer of cooperative research programs, as a scientific and academic administrator, as an editor (including two IAU Symposia [Kerr and Rodgers 1964, Kerr and Simonson 1974]), and as bibliographer–his work in this last regard has been of immeasurable aid to my own historical research (McKechnie *et al.* 1957, 1963). And to so many of us he has been, and continues to be, a wise counsellor, teacher and friend.

In conclusion, I note that Australia's history has been linked with astronomy from the start–Captain Cook's first voyage, which led to the discovery of Australia's eastern coast, was as much to observe the 1769 transit of Venus across the solar disk in Tahiti as it was to explore for what was called *Terra Australis Incognita*. But 175 years passed before Australians became part of the first rank of world astronomical research. And when this happened it was through a small group of men, one of whom was Frank Kerr, and in a most unlikely manner, for they did their astronomy not with glass lenses but with metal rods and giant dishes.

REFERENCES

Kerr, F. J. 1941, *Refractive Indexes of Gases at High Radio Frequencies*, Nature **148**, 751, 752.

Kerr, F. J. 1943, *Refractive Indices of Gases at High Radio Frequencies*, Proc. Phys. Soc. **55**, 92-98.

Kerr, F. J. 1948, *Radio Superrefraction in the Coastal Regions of Australia*, Australian J. Sci. Res. A1, 443-463.

Kerr, F. J., Shain, C. A., and Higgins, C. S. 1949, *Moon Echoes and Penetration of the Ionosphere*, Nature **163**, 310-313.

Ewen, H. I. and Purcell, E. M.1951, *Radiation from Galactic Hydrogen at 1420 Mc/s*, Nature **168**,

Muller, C. A. and Oort, J. H. 1951, *The Interstellar Hydrogen Line at 1420 Mc/s and an Estimate of Galactic Rotation*, Nature **168**, 357-358.

(Christiansen, W. N. and Hindman, J. V. 1951), (Announcement of Detection of the 21 cm Hydrogen Line), Nature **168**, 358.

Kerr, F. J. and Shain, C. A. 1951, *Moon Echoes and Transmission Through the Ionosphere*, Proc. IRE **39**, 230-242.

Kerr, F. J. 1952, *On the Possibility of Obtaining Radar Echoes from the Sun and Planets*, Proc. IRE **40**, 660-666.

Kerr, F. J., Hindman, J. V., and Robinson, B. J. 1954, *Observations of the 21-cm Line from the Magellanic Clouds*, Australian J. Phys. **7**, 297-314.

Kerr, F. J. and de Vaucouleurs, G. 1955, *Rotation and Other Motions of the Magellanic Clouds from Radio OBservations*, Australian J. Phys. **8**, 508-522.

Kerr, F. J. and de Vaucouleurs, G. 1956, *The Masses of the Magellanic Clouds from Radio Observations*, Australian J. Phys. **9**, 90-111.

Kerr, F. J. 1957a, *Radio Evidence on the Large-scale Structure of Galaxies*, pp. 78-82 in *Symposium on Radio Astronomy* (held September 1956), ed. CSIRO Radiophysics Lab. (Melbourne: CSIRO).

Kerr, F. J. 1957b, *A Magellanic Effect on the Galaxy* [abstract], Astron. J. **62**, 93.

Burke, B. F. 1957, *Systematic Distortion of the Outer Regions of the Galaxy* [abstract], Astron. J. **62**, 90.

McKechnie, M., Kerr, F. J., and Shain, C. A. 1957, *Radio Astronomy Bibliography, 1954-1956*, Radiophysics Lab Report RPP 610 (292 pp.).

Oort, J. H., Kerr, F. J., and Westerhout, G. 1958, *The Galactic System as a Spiral Nebula,* Mon. Not. Roy. Astron. Soc. **118**, 379-389.

Gum, C. S., Kerr, F. J., and Westerhout, G. 1960, *A 21-cm Determination of the Principal Plane of the Galaxy,* Mon. Not. Roy. Astron. Soc. **121**, 132-149.

Kerr, F. J. 1962a, *Galactic Velocity Models and the Interpretation of 21-cm Surveys,* Mon. Not. Roy. Astron. Soc. **123**, 327-345.

Kerr, F. J. 1962b, *210-foot Radio Telescope's First Results,* Sky and Tel. **24**, 254-260.

Minnett, H. C. 1962, *The Australian 210-foot Radio Telescope,* Sky and Tel. **24**, 184-189.

McKechnie, M.; Kerr, F. J., and Hill, E. R. 1963, *Radio Astronomy Bibliography, 1957-1960,* Radiophysics Lab Report RPL 168 (309 pp.).

Kerr, F. J. and Rodgers, A. W. (eds.) 1964, *The Galaxy and the Magellanic Clouds,* Proc. of IAU Symp. No. 20 (March 1963) (Canberra: Australian Academy of Science).

Kerr, F. J. and Westerhout, G. 1965, *Distribution of Hydrogen in the Galaxy,* pp. 167-202 in Galactic Structure (eds. A. Blaauw and M. Schmidt), Vol. 5 of *Stars and Stellar Systems* (Chicago: Univ. of Chicago Press).

Kerr, F. J 1968., *Radio-line Emission and Absorption by the Interstellar Gas,* pp. 575-622 in Nebulae and Interstellar Matter (eds. B. M. Middlehurst and L. H. Aller), Vol. 7 of *Stars and Stellar Systems* (Chicago: Univ. of Chicago Press).

Kerr, F. J. and Simonson, S. C., III (eds.) 1974, *Galactic Radio Astronomy,* Proc. of IAU Symp. No. 60 (September 1973) (Dordrecht: Reidel).

Index of Authors

List of Participants

M. F. A'Hearn . . . . University of Maryland
R. J. Allen . . . . University of Illinois
H. Alvarez . . . . University of Chile
S. A. Balbus . . . . University of Virginia
D. L. Ball . . . . Computer Sciences Corporation
T. M. Bania . . . . Boston University
F. N. Bash . . . . University of Texas
S. A. Baum . . . . University of Maryland/NRAO
R. A. Bell . . . . University of Maryland
L. Blitz . . . . University of Maryland
P. F. Bowers . . . . Naval Research Laboratory
P. B. Boyce . . . . American Astronomical Society
J. Brand . . . . Max-Planck-Institut für Radioastronomie
V. Bremenkamp . . . . Associated Universities Incorporated
M. M. Briley . . . . University of Maryland
R. L. Brown . . . . National Radio Astronomy Observatory
W. B. Burton . . . . University of Leiden
J.A.R. Caldwell . . . . Mount Wilson and Las Campanas Observatories
J. B. Carlson . . . . Center for Archaeoastronomy
U. Carsenty . . . . University of Maryland
T. Codippily . . . . World Bank
E. Dahlstrom . . . . unaffiliated
G. Deming . . . . University of Maryland
F. X. Désert . . . . NASA/Goddard Space Flight Center
G. E. Dieter . . . . University of Maryland
J. R. Dorfman . . . . University of Maryland
L. L. Dressel . . . . Rice University
B. G. Elmegreen . . . . IBM Thomas J. Watson Research Center
D. M. Elmegreen . . . . IBM Thomas J. Watson Research Center
M. Fich . . . . University of Waterloo
C. E. Fichtel . . . . NASA/Goddard Space Flight Center
K. C. Freeman . . . . Mount Stromlo and Siding Spring Observatories
T. E. Gergely . . . . National Science Foundation
P. Giovanoni . . . . University of Maryland
B.-Y. Gir . . . . University of Maryland
S. J. Goldstein . . . . University of Virginia
H. Grayber . . . . George Mason University
E. J. Grayzeck . . . . University of Nevada, Las Vegas/GSFC
H. Gursky . . . . Naval Research Laboratory
J. P. Harrington . . . . University of Maryland
M. G. Hauser . . . . NASA/Goddard Space Flight Center
R. J. Havlen . . . . National Radio Astronomy Observatory
T. M. Heckman . . . . University of Maryland
C. E. Heiles . . . . University of California, Berkeley
A. P. Henderson . . . . Manhattan College
P. A. Henning . . . . University of Maryland

M. R. Sloan . . . . University of Maryland
E.v.P. Smith . . . . Virginia Commonwealth University
T. J. Sodroski . . . . University of Maryland
A. A. Stark . . . . AT&T Bell Laboratories
F. C. Stark . . . . University of Maryland
K. A. Strand . . . . Washington, D. C.
W. T. Sullivan, III . . . . University of Washington
S. Terebey . . . . California Institute of Technology
P. Thaddeus . . . . Harvard-Smithsonian Center for Astrophysics
A. Toomre . . . . Massachusetts Institute of Technology
J. D. Trasco . . . . University of Maryland
N. D. Tyson . . . . University of Maryland
N. R. Vandenberg . . . . Interferometrics, Inc.
P. A. Vanden Bout . . . . National Radio Astronomy Observatory
G. L. Verschuur . . . . Greenbelt, Maryland
F. Verter . . . . NASA/Goddard Space Flight Center
J. M. Vrtilek . . . . NASA/Goddard Space Flight Center
K. W. Weiler . . . . Naval Research Laboratory
D. G. Wentzel . . . . University of Maryland/National Science Foundation
G. Westerhout . . . . U. S. Naval Observatory
R. Wielen . . . . Astronomisches Rechen-Institut, Heidelberg
B. A. Williams . . . . University of Delaware
A. S. Wilson . . . . University of Maryland
G. Wu . . . . University of Maryland
C. Yuan . . . . City College of New York

Lecture Notes in Mathematics

Lecture Notes in Physics

Astronomy and Astrophysics Library

K. Rohlfs

Tools of Radio Astronomy

1986. 127 figures. XII, 319 pages.
ISBN 3-540-16188-0

H. Scheffler, H. Elsässer

Physics of the Galaxy and Interstellar Matter

Translated from the German by A. H. Armstrong
1987. 207 figures. XI, 492 pages.
ISBN 3-540-17314-5

P. Lena

Observational Astrophysics

Translated from the French by A. R. King
1988. 200 figures. Approx. 350 pages.
ISBN 3-540-18433-3

M. Harwit

Astrophysical Concepts

2nd edition. 1988. 175 figures. XV, 625 pages.
ISBN 3-540-96683-8

G. L. Verschuur, K. I. Kellermann (Eds.)

Galactic and Extragalactic Radio Astronomy

2nd edition. 1988. 207 figures. XXII, 698 pages.
ISBN 3-540-96575-0

S. Laustsen, C. Madsen, R. M. West

Exploring the Southern Sky

A Pictorial Atlas from the European Southern Observatory (ESO)

1987. 240 photographs, partly in colour, 31 diagrams and a Fould-out-Plate. VI, 274 pages. ISBN 3-540-17735-5

Contents: Foreword. Introduction. Acknowledgements. A Guide to This Book.
Part 1: The Universe and Its Galaxies
A Look into Fornax. Galaxy Types in an Ordered Sequence. The Local Group. The Sculptor Group. Multiple Galaxies. Clusters of Galaxies. Peculiar Galaxies.
Part 2: The Milky Way Galaxy
Panorama of the Milky Way. The Milky Way from Orion to Puppis. The Milky Way from Vela to Carina. The Milky Way from Crux to Norma. The Milky Way from Scorpius to Scutum. Milky Way Objects at High Galactic Latitude.
Part 3: Minor Bodies in the Solar System
Meteorites and Minor Planets. Comets.
Part 4: The Southern Sky and ESO
A European Organization for Astronomy. The La Silla Observatory. The Headquarters in Garching. The Next Generation Telescopes. Glossary. Plate Data. Index of Objects.

Exploring the Southern Sky is not only a superb pictorial atlas and guide to the astronomical opportunities of the Southern Hemisphere but also a treasured addition to every stargazer's library. It is the timely new way to expand your astronomical horizons and add to your picture of the Universe.

German edition available by Birkhäuser-Verlag, Basle.

Springer-Verlag
Berlin Heidelberg New York
London Paris Tokyo

Springer

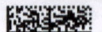